Yamaha RD and DT 125 LC Owners Workshop Manual

by Jeremy Churchill

with an additional Chapter on the RD/DT 125 LC II and III models

by Pete Shoemark

Models covered

RD125 LC. 123cc. Introduced June 1982
RD125 LC II. 123cc. Introduced February 1985
RD125 LC III. 123cc. Introduced November 1986
DT125 LC. 123cc. Introduced April 1982
DT125 LC II. 123cc. Introduced February 1985
DT125 LC III. 123cc. Introduced March 1986

ISBN 978 1 85010 417 9

(887-5P5)

Haynes Group Limited
Haynes North America, Inc

www.haynes.com

Acknowledgements

Our thanks are due to Jim Patch of Yeovil Motorcycle Services Ltd, Yeovil, Somerset and Fran Ridewood and Co. of Wells, Somerset, who supplied the machines featured throughout this manual; to Mitsui Machinery Sales (UK) Ltd who supplied the necessary service information and gave permission to reproduce many of the line drawings used, and to the staff of Mitsui's Technical Support Group who provided solutions to a number of problems.

We would like to thank the Avon Rubber Company, who kindly supplied information and technical assistance on tyre fitting, NGK Spark Plugs (UK) Ltd for information on spark plug maintenance and electrode conditions, and Renold Ltd for advice on chain care and renewal.

About this manual

The purpose of this manual is to present the owner with a concise and graphic guide which will enable him to tackle any operation from basic routine maintenance to a major overhaul. It has been assumed that any work would be undertaken without the luxury of a well-equipped workshop and a range of manufacturer's service tools.

To this end, the machine featured in the manual was stripped and rebuilt in our own workshop, by a team comprising a mechanic, a photographer and the author. The resulting photographic sequence depicts events as they took place, the hands shown being those of the author and the mechanic.

The use of specialised, and expensive, service tools was avoided unless their use was considered to be essential due to risk of breakage or injury. There is usually some way of improvising a method of removing a stubborn component, providing that a suitable degree of care is exercised.

The author learnt his motorcycle mechanics over a number of years, faced with the same difficulties and using similar facilities to those encountered by most owners. It is hoped that this practical experience can be passed on through the pages of this manual.

Where possible, a well-used example of the machine is chosen for the workshop project, as this highlights any areas which might be particularly prone to giving rise to problems. In this way, any such difficulties are encountered and resolved before the text is written, and the techniques used to deal with them can be incorporated in the relevant section. Armed with a working knowledge of the machine, the author undertakes a considerable amount of research in order that the maximum amount of data can be included in the manual.

A comprehensive section, preceding the main part of the manual, describes procedures for carrying out the routine maintenance of the machine at intervals of time and mileage. This section is included particularly for those owners who wish to ensure the efficient day-to-day running of their motorcycle, but who choose not to undertake overhaul or renovation work.

Each Chapter is divided into numbered sections. Within these sections are numbered paragraphs. Cross reference throughout the manual is quite straightforward and logical. When reference is made 'See Section 6.10' it means Section 6, paragraph 10 in the same Chapter. If another Chapter were intended, the reference would read, for example, 'See Chapter 2, Section 6.10'. All the photographs are captioned with a section/paragraph number to which they refer and are relevant to the Chapter text adjacent.

Figures (usually line illustrations) appear in a logical but numerical order, within a given Chapter. Fig. 1.1 therefore refers to the first figure in Chapter 1.

Left-hand and right-hand descriptions of the machines and their components refer to the left and right of a given machine when the rider is seated normally.

Motorcycle manufacturers continually make changes to specifications and recommendations, and these, when notified, are incorporated into our manuals at the earliest opportunity.

Whilst every care is taken to ensure that the information in this manual is correct no liability can be accepted by the author or publishers for loss, damage or injury caused by any errors in or omissions from the information given.

Contents

Left-hand view of the 1982 RD125 LC

RD125 LC engine/gearbox unit

Left-hand view of the 1982 DT125 LC

DT125 LC engine/gearbox unit

Introduction to the Yamaha RD125 LC and DT125 LC models

Yamaha first came to the public's attention in the early 1960s with their entry into road racing, competing largely with the other Japanese factories in the smaller capacity classes. The experience gained, initially with four-cylinder machines and then with twins, produced the world-beating TD series of racing machines, these becoming TZs with the adoption of water-cooling in the early 1970s. A series of machines was produced in 125, 200, 250 and 350 cc capacities as a direct offshoot of the racing programme, the lessons learned on the racetrack being applied very quickly to boost the performance and reliability of the road-going machines; these developed into the very successful and popular RD series of air-cooled parallel twins. In 1980 the development continued with the adoption of water cooling on the larger machines, these being known as the RD250 LC and RD350 LC.

From the mid 1970s onwards, Yamaha became increasingly interested in the off-road forms of motorcycle sport. Again, a racing programme was started which provided the experience to develop a very successful range of machines for sale to the general public. In motocross, this led to the YZ series, in enduro, the T series, and in trials, the TY series. While Yamaha machines have never achieved the complete domination of off-road sport that the TZ series have managed in the past few years of road-racing they have been very successful nevertheless, and have been used by competition riders at all levels from novice upwards. Part of this second line of development is the DT series of trail bikes. These machines, with their air-cooled, single cylinder engines, have been produced since the early 1970s to cater for the rider who ventures only occasionally off the road.

It seemed entirely logical from Yamaha's point of view to run the two lines of development together, the results of which were shown when the RD125 LC and DT125 LC were announced in the winter of 1981/2. Although it came as a surprise to see a single cylinder RD model Yamaha, doubts as to such an engine's power output were soon dispelled when it was revealed that the new engine, in unrestricted RD model form, produced approximately 25% more power than the air-cooled parallel twin it replaced. From the off-road enthusiasts' point of view, fears of the extra weight brought about by the use of water cooling were allayed when it was revealed also that the new engine weighed over 1 kg (2 lb) less than the old air-cooled single cylinder unit.

Both machines are finished in accordance with the latest styling trends, the RD model being fitted with the type of handlebar fairing and belly fairing that proved so popular as after-sales accessories for the RD250/350 LC machines, and the DT model being very similar in appearance to the current YZ series of motocross machines. Maximum use is made of plastic components to keep the overall weight down and while the cycle parts are apparently the same on both machines, the RD model employs a smaller version of the suspension, brakes and wheels fitted to the RD250/350 LC machines to give it superb steering and road-holding qualities. The DT model, on the other hand, employs long-travel suspension components derived from the motocross and enduro machines to give it an off-road potential far superior to any previous bike.

In both cases the machines were produced and sold in a form which took full advantage of the engine's potential power output. To comply with fresh legislation which gradually came into force from 1982 to 1983 in the UK, this power output was restricted to a maximum of 9 kw (12.2 bhp) on certain models.

In both cases the restricted and unrestricted versions are externally identical and can be distinguished only by their frame numbers. To assist owners in identification the frame number with which each version of each model began its production run is given below:

RD125 LC	Unrestricted version	10W - 000101
	Restricted version	12A - 000101
DT125 LC	Unrestricted version	10V - 000101
	Restricted version	12W - 000101

Dimensions and weight

	RD125 LC	DT125 LC
Overall length	1990 mm (78.35 in)	2135 mm (84.05 in)
Overall width	735 mm (28.94 in)	820 mm (32.28 in)
Overall height	1190 mm (46.85 in)	1195 mm (47.05 in)
Seat height	775 mm (30.51 in)	840 mm (33.07 in)
Wheelbase	1295 mm (50.98 in)	1345 mm (52.95 in)
Ground clearance	185 mm (7.28 in)	270 mm (10.63 in)
Curb weight	112 kg (246.96 lb)	108 kg (238.14 lb)
Dry weight	98 kg (216.09 lb)	97 kg (213.89 lb)

Ordering spare parts

Before attempting any overhaul or maintenance work it is important to ensure that any parts likely to be required are to hand. Many of the more common parts such as gaskets and seals will be available off the shelf from the local Yamaha dealer, but often it will prove necessary to order more specialised parts well in advance. It is worthwhile running through the operation to be undertaken, referring to the appropriate Chapter and Section of this book, so that a note can be made of the items most likely to be required. In some instances it will of course be necessary to dismantle the assembly in question so that the various components can be examined and measured for wear and in these instances, it must be remembered that the machine may have to be left dismantled while the replacement parts are obtained.

It is advisable to purchase almost all new parts from an official Yamaha dealer. Almost any motorcycle dealer should be able to obtain the parts in time, but this may take longer than it would through the official factory spares arrangement. It is quite in order to purchase expendable items such as spark plugs, bulbs, tyres, oil and grease from the nearest convenient source.

Owners should be very wary of some of the pattern parts that might be offered at a lower price than the Yamaha originals. Whilst in most cases these will be of an adequate standard, some of the more important parts have been known to fail suddenly and cause extensive damage in the process. A particular danger in recent years is the growing number of counterfeit parts from Taiwan. These include items such as oil filters and brake pads and are often sold in packaging which is almost indistinguishable from the manufacturer's own. Again, these are often quite serviceable parts, but can sometimes be dangerously inadequate in materials or construction. Apart from rendering the manufacturer's warranty invalid, use of sub-standard parts may put the life of the rider (or the machine) at risk. In short, where there are any doubts on safety grounds purchase parts **only** from a reputable Yamaha dealer. The extra cost involved pays for a high standard of quality and the parts will be guaranteed to work effectively.

Most machines are subject to continuous detail modifications throughout their production run, and in addition to annual model changes. In most cases these changes will be known to the dealer but not to the general public, so it is essential to quote the engine and frame numbers in full when ordering parts. The engine number is embossed in a rectangular section of the crankcase just below the left-hand carburettor, and the frame number is stamped on the right-hand side of the steering head.

Owners of the LC models may wish to modify their machines for road or competition use, but before doing so remember that the machine's warranty may be affected. Always choose accessories from a reputable manufacturer. If a modified exhaust system is being considered make sure that it is of good design and construction and that it has been shown to improve performance. A reputable manufacturer will provide information on any carburettor jet changes required to suit his particular system.

Location of engine number

Location of frame number

Safety first!

Professional motor mechanics are trained in safe working procedures. However enthusiastic you may be about getting on with the job in hand, do take the time to ensure that your safety is not put at risk. A moment's lack of attention can result in an accident, as can failure to observe certain elementary precautions.

There will always be new ways of having accidents, and the following points do not pretend to be a comprehensive list of all dangers; they are intended rather to make you aware of the risks and to encourage a safety-conscious approach to all work you carry out on your vehicle.

Essential DOs and DON'Ts

DON'T start the engine without first ascertaining that the transmission is in neutral.

DON'T suddenly remove the filler cap from a hot cooling system – cover it with a cloth and release the pressure gradually first, or you may get scalded by escaping coolant.

DON'T attempt to drain oil until you are sure it has cooled sufficiently to avoid scalding you.

DON'T grasp any part of the engine, exhaust or silencer without first ascertaining that it is sufficiently cool to avoid burning you.

DON'T allow brake fluid or antifreeze to contact the machine's paintwork or plastic components.

DON'T syphon toxic liquids such as fuel, brake fluid or antifreeze by mouth, or allow them to remain on your skin.

DON'T inhale dust – it may be injurious to health (see *Asbestos* heading).

DON'T allow any spilt oil or grease to remain on the floor – wipe it up straight away, before someone slips on it.

DON'T use ill-fitting spanners or other tools which may slip and cause injury.

DON'T attempt to lift a heavy component which may be beyond your capability – get assistance.

DON'T rush to finish a job, or take unverified short cuts.

DON'T allow children or animals in or around an unattended vehicle.

DON'T inflate a tyre to a pressure above the recommended maximum. Apart from overstressing the carcase and wheel rim, in extreme cases the tyre may blow off forcibly.

DO ensure that the machine is supported securely at all times. This is especially important when the machine is blocked up to aid wheel or fork removal.

DO take care when attempting to slacken a stubborn nut or bolt. It is generally better to pull on a spanner, rather than push, so that if slippage occurs you fall away from the machine rather than on to it.

DO wear eye protection when using power tools such as drill, sander, bench grinder etc.

DO use a barrier cream on your hands prior to undertaking dirty jobs – it will protect your skin from infection as well as making the dirt easier to remove afterwards; but make sure your hands aren't left slippery. Note that long-term contact with used engine oil can be a health hazard.

DO keep loose clothing (cuffs, tie etc) and long hair well out of the way of moving mechanical parts.

DO remove rings, wristwatch etc, before working on the vehicle – especially the electrical system.

DO keep your work area tidy – it is only too easy to fall over articles left lying around.

DO exercise caution when compressing springs for removal or installation. Ensure that the tension is applied and released in a controlled manner, using suitable tools which preclude the possibility of the spring escaping violently.

DO ensure that any lifting tackle used has a safe working load rating adequate for the job.

DO get someone to check periodically that all is well, when working alone on the vehicle.

DO carry out work in a logical sequence and check that everything is correctly assembled and tightened afterwards.

DO remember that your vehicle's safety affects that of yourself and others. If in doubt on any point, get specialist advice.

IF, in spite of following these precautions, you are unfortunate enough to injure yourself, seek medical attention as soon as possible.

Asbestos

Certain friction, insulating, sealing, and other products – such as brake linings, clutch linings, gaskets, etc – contain asbestos. *Extreme care must be taken to avoid inhalation of dust from such products since it is hazardous to health.* If in doubt, assume that they *do* contain asbestos.

Fire

Remember at all times that petrol (gasoline) is highly flammable. Never smoke, or have any kind of naked flame around, when working on the vehicle. But the risk does not end there – a spark caused by an electrical short-circuit, by two metal surfaces contacting each other, by careless use of tools, or even by static electricity built up in your body under certain conditions, can ignite petrol vapour, which in a confined space is highly explosive.

Always disconnect the battery earth (ground) terminal before working on any part of the fuel or electrical system, and never risk spilling fuel on to a hot engine or exhaust.

It is recommended that a fire extinguisher of a type suitable for fuel and electrical fires is kept handy in the garage or workplace at all times. Never try to extinguish a fuel or electrical fire with water.

Note: *Any reference to a 'torch' appearing in this manual should always be taken to mean a hand-held battery-operated electric lamp or flashlight. It does* **not** *mean a welding/gas torch or blowlamp.*

Fumes

Certain fumes are highly toxic and can quickly cause unconsciousness and even death if inhaled to any extent. Petrol (gasoline) vapour comes into this category, as do the vapours from certain solvents such as trichloroethylene. Any draining or pouring of such volatile fluids should be done in a well ventilated area.

When using cleaning fluids and solvents, read the instructions carefully. Never use materials from unmarked containers – they may give off poisonous vapours.

Never run the engine of a motor vehicle in an enclosed space such as a garage. Exhaust fumes contain carbon monoxide which is extremely poisonous; if you need to run the engine, always do so in the open air or at least have the rear of the vehicle outside the workplace.

The battery

Never cause a spark, or allow a naked light, near the vehicle's battery. It will normally be giving off a certain amount of hydrogen gas, which is highly explosive.

Always disconnect the battery earth (ground) terminal before working on the fuel or electrical systems.

If possible, loosen the filler plugs or cover when charging the battery from an external source. Do not charge at an excessive rate or the battery may burst.

Take care when topping up and when carrying the battery. The acid electrolyte, even when diluted, is very corrosive and should not be allowed to contact the eyes or skin.

If you ever need to prepare electrolyte yourself, always add the acid slowly to the water, and never the other way round. Protect against splashes by wearing rubber gloves and goggles.

Mains electricity and electrical equipment

When using an electric power tool, inspection light etc, always ensure that the appliance is correctly connected to its plug and that, where necessary, it is properly earthed (grounded). Do not use such appliances in damp conditions and, again, beware of creating a spark or applying excessive heat in the vicinity of fuel or fuel vapour. Also ensure that the appliances meet the relevant national safety standards.

Ignition HT voltage

A severe electric shock can result from touching certain parts of the ignition system, such as the HT leads, when the engine is running or being cranked, particularly if components are damp or the insulation is defective. Where an electronic ignition system is fitted, the HT voltage is much higher and could prove fatal.

Tools and working facilities

The first priority when undertaking maintenance or repair work of any sort on a motorcycle is to have a clean, dry, well-lit working area. Work carried out in peace and quiet in the well-ordered atmosphere of a good workshop will give more satisfaction and much better results than can usually be achieved in poor working conditions. A good workshop must have a clean flat workbench or a solidly constructed table of convenient working height. The workbench or table should be equipped with a vice which has a jaw opening of at least 4 in (100 mm). A set of jaw covers should be made from soft metal such as aluminium alloy or copper, or from wood. These covers will minimise the marking or damaging of soft or delicate components which may be clamped in the vice. Some clean, dry, storage space will be required for tools, lubricants and dismantled components. It will be necessary during a major overhaul to lay out engine/gearbox components for examination and to keep them where they will remain undisturbed for as long as is necessary. To this end it is recommended that a supply of metal or plastic containers of suitable size is collected. A supply of clean, lint-free, rags for cleaning purposes and some newspapers, other rags, or paper towels for mopping up spillages should also be kept. If working on a hard concrete floor note that both the floor and one's knees can be protected from oil spillages and wear by cutting open a large cardboard box and spreading it flat on the floor under the machine or workbench. This also helps to provide some warmth in winter and to prevent the loss of nuts, washers, and other tiny components which have a tendency to disappear when dropped on anything other than a perfectly clean, flat, surface.

Unfortunately, such working conditions are not always available to the home mechanic. When working in poor conditions it is essential to take extra time and care to ensure that the components being worked on are kept scrupulously clean and to ensure that no components or tools are lost or damaged.

A selection of good tools is a fundamental requirement for anyone contemplating the maintenance and repair of a motor vehicle. For the owner who does not possess any, their purchase will prove a considerable expense, offsetting some of the savings made by doing-it-yourself. However, provided that the tools purchased meet the relevant national safety standards and are of good quality, they will last for many years and prove an extremely worthwhile investment.

To help the average owner to decide which tools are needed to carry out the various tasks detailed in this manual, we have compiled three lists of tools under the following headings: *Maintenance and minor repair*, *Repair and overhaul*, and *Specialized*. The newcomer to practical mechanics should start off with the simpler jobs around the vehicle. Then, as his confidence and experience grow, he can undertake more difficult tasks, buying extra tools as and when they are needed.

In this way, a *Maintenance and minor repair* tool kit can be built-up into a *Repair and overhaul* tool kit over a considerable period of time without any major cash outlays. The experienced home mechanic will have a tool kit good enough for most repair and overhaul procedures and will add tools from the specialized category when he feels the expense is justified by the amount of use these tools will be put to.

It is obviously not possible to cover the subject of tools fully here. For those who wish to learn more about tools and their use there is a book entitled *Motorcycle Workshop Practice Manual* (Bk no. 1454) available from the publishers of this manual.

As a general rule, it is better to buy the more expensive, good quality tools. Given reasonable use, such tools will last for a very long time, whereas the cheaper, poor quality, item will wear out faster and need to be renewed more often, thus nullifying the original saving. There is also the risk of a poor quality tool breaking while in use, causing personal injury or expensive damage to the component being worked on.

For practically all tools, a tool factor is the best source since he will have a very comprehensive range compared with the average garage or accessory shop. Having said that, accessory shops often offer excellent quality tools at discount prices, so it pays to shop around. There are plenty of tools around at reasonable prices, but always aim to purchase items which meet the relevant national safety standards. If in doubt, seek the advice of the shop proprietor or manager before making a purchase.

The basis of any toolkit is a set of spanners. While open-ended spanners with their slim jaws, are useful for working on awkwardly-positioned nuts, ring spanners have advantages in that they grip the nut far more positively. There is less risk of the spanner slipping off the nut and damaging it, for this reason alone ring spanners are to be preferred. Ideally, the home mechanic should acquire a set of each, but if expense rules this out a set of combination spanners (open-ended at one end and with a ring of the same size at the other) will provide a good compromise. Another item which is so useful it should be considered an essential requirement for any home mechanic is a set of socket spanners. These are available in a variety of drive sizes. It is recommended that the $\frac{1}{2}$-inch drive type is purchased to begin with as although bulkier and more expensive than the $\frac{3}{8}$-inch type, the larger size is far more common and will accept a greater variety of torque wrenches, extension pieces and socket sizes. The socket set should comprise sockets of sizes between 8 and 24 mm, a reversible ratchet drive, an extension bar of about 10 inches in length, a spark plug socket with a rubber insert, and a universal joint. Other attachments can be added to the set at a later date.

Maintenance and minor repair tool kit

Set of spanners 8 – 24 mm
Set of sockets and attachments
Spark plug spanner with rubber insert – 10, 12, or 14 mm
as appropriate
Adjustable spanner
C-spanner/pin spanner
Torque wrench (same size drive as sockets)
Set of screwdrivers (flat blade)
Set of screwdrivers (cross-head)
Set of Allen keys 4 – 10 mm
Impact screwdriver and bits
Ball pein hammer – 2 lb
Hacksaw (junior)
Self-locking pliers – Mole grips or vice grips
Pliers – combination
Pliers – needle nose
Wire brush (small)
Soft-bristled brush
Tyre pump
Tyre pressure gauge
Tyre tread depth gauge
Oil can
Fine emery cloth
Funnel (medium size)
Drip tray
Grease gun
Set of feeler gauges
Brake bleeding kit
Strobe timing light
Continuity tester (dry battery and bulb)
Soldering iron and solder
Wire stripper or craft knife
PVC insulating tape
Assortment of split pins, nuts, bolts, and washers

Repair and overhaul toolkit

The tools in this list are virtually essential for anyone undertaking major repairs to a motorcycle and are additional to the tools listed above. Concerning Torx driver bits, Torx screws are encountered on some of the more modern machines where their use is restricted to fastening certain components inside the engine/gearbox unit. It is therefore recommended that if Torx bits cannot be borrowed from a local dealer, they are purchased individually as the need arises. They are not in regular use in the motor trade and will therefore only be available in specialist tool shops.

Plastic or rubber soft-faced mallet
Torx driver bits
Pliers – electrician's side cutters
Circlip pliers – internal (straight or right-angled tips are available)
Circlip pliers – external
Cold chisel
Centre punch
Pin punch
Scriber
Scraper (made from soft metal such as aluminium or copper)
Soft metal drift
Steel rule/straight edge
Assortment of files
Electric drill and bits
Wire brush (large)
Soft wire brush (similar to those used for cleaning suede shoes)
Sheet of plate glass
Hacksaw (large)
Valve grinding tool
Valve grinding compound (coarse and fine)
Stud extractor set (E-Z out)

Specialized tools

This is not a list of the tools made by the machine's manufacturer to carry out a specific task on a limited range of models. Occasional references are made to such tools in the text of this manual and, in general, an alternative method of carrying out the task without the manufacturer's tool is given where possible. The tools mentioned in this list are those which are not used regularly and are expensive to buy in view of their infrequent use. Where this is the case it may be possible to hire or borrow the tools against a deposit from a local dealer or tool hire shop. An alternative is for a group of friends or a motorcycle club to join in the purchase.

Valve spring compressor
Piston ring compressor
Universal bearing puller
Cylinder bore honing attachment (for electric drill)
Micrometer set
Vernier calipers
Dial gauge set
Cylinder compression gauge
Vacuum gauge set
Multimeter
Dwell meter/tachometer

Care and maintenance of tools

Whatever the quality of the tools purchased, they will last much longer if cared for. This means in practice ensuring that a tool is used for its intended purpose; for example screwdrivers should not be used as a substitute for a centre punch, or as chisels. Always remove dirt or grease and any metal particles but remember that a light film of oil will prevent rusting if the tools are infrequently used. The common tools can be kept together in a large box or tray but the more delicate, and more expensive, items should be stored separately where they cannot be damaged. When a tool is damaged or worn out, be sure to renew it immediately. It is false economy to continue to use a worn spanner or screwdriver which may slip and cause expensive damage to the component being worked on.

Fastening systems

Fasteners, basically, are nuts, bolts and screws used to hold two or more parts together. There are a few things to keep in mind when working with fasteners. Almost all of them use a locking device of some type; either a lock washer, lock nut, locking tab or thread adhesive. All threaded fasteners should be clean, straight, have undamaged threads and undamaged corners on the hexagon head where the spanner fits. Develop the habit of replacing all damaged nuts and bolts with new ones.

Rusted nuts and bolts should be treated with a rust penetrating fluid to ease removal and prevent breakage. After applying the rust penetrant, let it 'work' for a few minutes before trying to loosen the nut or bolt. Badly rusted fasteners may have to be chiseled off or removed with a special nut breaker, available at tool shops.

Flat washers and lock washers, when removed from an assembly should always be replaced exactly as removed. Replace any damaged washers with new ones. Always use a flat washer between a lock washer and any soft metal surface (such as aluminium), thin sheet metal or plastic. Special lock nuts can only be used once or twice before they lose their locking ability and must be renewed.

If a bolt or stud breaks off in an assembly, it can be drilled out and removed with a special tool called an E-Z out. Most dealer service departments and motorcycle repair shops can perform this task, as well as others (such as the repair of threaded holes that have been stripped out).

Spanner size comparison

Jaw gap (in)	Spanner size	Jaw gap (in)	Spanner size
0.250	$\frac{1}{4}$ in AF	0.945	24 mm
0.276	7 mm	1.000	1 in AF
0.313	$\frac{5}{16}$ in AF	1.010	$\frac{9}{16}$ in Whitworth; $\frac{5}{8}$ in BSF
0.315	8 mm	1.024	26 mm
0.344	$\frac{11}{32}$ in AF; $\frac{1}{8}$ in Whitworth	1.063	$1\frac{1}{16}$ in AF; 27 mm
0.354	9 mm	1.100	$\frac{5}{8}$ in Whitworth; $\frac{11}{16}$ in BSF
0.375	$\frac{3}{8}$ in AF	1.125	$1\frac{1}{8}$ in AF
0.394	10 mm	1.181	30 mm
0.433	11 mm	1.200	$\frac{11}{16}$ in Whitworth; $\frac{3}{4}$ in BSF
0.438	$\frac{7}{16}$ in AF	1.250	$1\frac{1}{4}$ in AF
0.445	$\frac{3}{16}$ in Whitworth; $\frac{1}{4}$ in BSF	1.260	32 mm
0.472	12 mm	1.300	$\frac{3}{4}$ in Whitworth; $\frac{7}{8}$ in BSF
0.500	$\frac{1}{2}$ in AF	1.313	$1\frac{5}{16}$ in AF
0.512	13 mm	1.390	$\frac{13}{16}$ in Whitworth; $\frac{15}{16}$ in BSF
0.525	$\frac{1}{4}$ in Whitworth; $\frac{5}{16}$ in BSF	1.417	36 mm
0.551	14 mm	1.438	$1\frac{7}{16}$ in AF
0.563	$\frac{9}{16}$ in AF	1.480	$\frac{7}{8}$ in Whitworth; 1 in BSF
0.591	15 mm	1.500	$1\frac{1}{2}$ in AF
0.600	$\frac{5}{16}$ in Whitworth; $\frac{3}{8}$ in BSF	1.575	40 mm; $\frac{15}{16}$ in Whitworth
0.625	$\frac{5}{8}$ in AF	1.614	41 mm
0.630	16 mm	1.625	$1\frac{5}{8}$ in AF
0.669	17 mm	1.670	1 in Whitworth; $1\frac{1}{8}$ in BSF
0.686	$\frac{11}{16}$ in AF	1.688	$1\frac{11}{16}$ in AF
0.709	18 mm	1.811	46 mm
0.710	$\frac{3}{8}$ in Whitworth; $\frac{7}{16}$ in BSF	1.813	$1\frac{13}{16}$ in AF
0.748	19 mm	1.860	$1\frac{1}{8}$ in Whitworth; $1\frac{1}{4}$ in BSF
0.750	$\frac{3}{4}$ in AF	1.875	$1\frac{7}{8}$ in AF
0.813	$\frac{13}{16}$ in AF	1.969	50 mm
0.820	$\frac{7}{16}$ in Whitworth; $\frac{1}{2}$ in BSF	2.000	2 in AF
0.866	22 mm	2.050	$1\frac{1}{4}$ in Whitworth; $1\frac{3}{8}$ in BSF
0.875	$\frac{7}{8}$ in AF	2.165	55 mm
0.920	$\frac{1}{2}$ in Whitworth; $\frac{9}{16}$ in BSF	2.362	60 mm
0.938	$\frac{15}{16}$ in AF		

Standard torque settings

Specific torque settings will be found at the end of the specifications section of each chapter. Where no figure is given, bolts should be secured according to the table below.

Fastener type (thread diameter)	kgf m	lbf ft
5mm bolt or nut	0.45 – 0.6	3.5 – 4.5
6 mm bolt or nut	0.8 – 1.2	6 – 9
8 mm bolt or nut	1.8 – 2.5	13 – 18
10 mm bolt or nut	3.0 – 4.0	22 – 29
12 mm bolt or nut	5.0 – 6.0	36 – 43
5 mm screw	0.35 – 0.5	2.5 – 3.6
6 mm screw	0.7 – 1.1	5 – 8
6 mm flange bolt	1.0 – 1.4	7 – 10
8 mm flange bolt	2.4 – 3.0	17 – 22
10 mm flange bolt	3.0 – 4.0	22 – 29

Choosing and fitting accessories

The range of accessories available to the modern motorcyclist is almost as varied and bewildering as the range of motorcycles. This Section is intended to help the owner in choosing the correct equipment for his needs and to avoid some of the mistakes made by many riders when adding accessories to their machines. It will be evident that the Section can only cover the subject in the most general terms and so it is recommended that the owner, having decided that he wants to fit, for example, a luggage rack or carrier, seeks the advice of several local dealers and the owners of similar machines. This will give a good idea of what makes of carrier are easily available, and at what price. Talking to other owners will give some insight into the drawbacks or good points of any one make. A walk round the motorcycles in car parks or outside a dealer will often reveal the same sort of information.

The first priority when choosing accessories is to assess exactly what one needs. It is, for example, pointless to buy a large heavy-duty carrier which is designed to take the weight of fully laden panniers and topbox when all you need is a place to strap on a set of waterproofs and a lunchbox when going to work. Many accessory manufacturers have ranges of equipment to cater for the individual needs of different riders and this point should be borne in mind when looking through a dealer's catalogues. Having decided exactly what is required and the use to which the accessories are going to be put, the owner will need a few hints on what to look for when making the final choice. To this end the Section is now sub-divided to cover the more popular accessories fitted. Note that it is in no way a customizing guide, but merely seeks to outline the practical considerations to be taken into account when adding aftermarket equipment to a motorcycle.

Fairings and windscreens

A fairing is possibly the single, most expensive, aftermarket item to be fitted to any motorcycle and, therefore, requires the most thought before purchase. Fairings can be divided into two main groups: front fork mounted handlebar fairings and windscreens, and frame mounted fairings.

The first group, the front fork mounted fairings, are becoming far more popular than was once the case, as they offer several advantages over the second group. Front fork mounted fairings generally are much easier and quicker to fit, involve less modification to the motorcycle, do not as a rule restrict the steering lock, permit a wider selection of handlebar styles to be used, and offer adequate protection for much less

money than the frame mounted type. They are also lighter, can be swapped easily between different motorcycles, and are available in a much greater variety of styles. Their main disadvantages are that they do not offer as much weather protection as the frame mounted types, rarely offer any storage space, and, if poorly fitted or naturally incompatible, can have an adverse effect on the stability of the motorocycle.

The second group, the frame mounted fairings, are secured so rigidly to the main frame of the motorcycle that they can offer a substantial amount of protection to motorcycle and rider in the event of a crash. They offer almost complete protection from the weather and, if double-skinned in construction, can provide a great deal of useful storage space. The feeling of peace, quiet and complete relaxation encountered when riding behind a good full fairing has to be experienced to be believed. For this reason full fairings are considered essential by most touring motorcyclists and by many people who ride all year round. The main disadvantages of this type are that fitting can take a long time, often involving removal or modification of standard motorcycle components, they restrict the steering lock and they can add up to about 40 lb to the weight of the machine. They do not usually affect the stability of the machine to any great extent once the front tyre pressure and suspension have been adjusted to compensate for the extra weight, but can be affected by sidewinds.

The first thing to look for when purchasing a fairing is the quality of the fittings. A good fairing will have strong, substantial brackets constructed from heavy-gauge tubing; the brackets must be shaped to fit the frame or forks evenly so that the minimum of stress is imposed on the assembly when it is bolted down. The brackets should be properly painted or finished – a nylon coating being the favourite of the better manufacturers – the nuts and bolts provided should be of the same thread and size standard as is used on the motorcycle and be properly plated. Look also for shakeproof locking nuts or locking washers to ensure that everything remains securely tightened down. The fairing shell is generally made from one of two materials: fibreglass or ABS plastic. Both have their advantages and disadvantages, but the main consideration for the owner is that fibreglass is much easier to repair in the event of damage occurring to the fairing. Whichever material is used, check that it is properly finished inside as well as out, that the edges are protected by beading and that the fairing shell is insulated from vibration by the use of rubber grommets at all mounting points. Also be careful to check that the windscreen is retained by plastic bolts which will snap on impact so that the windscreen will break away and not cause personal injury in the event of an accident.

Having purchased your fairing or windscreen, read the manufacturer's fitting instructions very carefully and check that you have all the necessary brackets and fittings. Ensure that the mounting brackets are located correctly and bolted down securely. Note that some manufacturers use hose clamps to retain the mounting brackets; these should be discarded as they are convenient to use but not strong enough for the task. Stronger clamps should be substituted; car exhaust pipe clamps of suitable size would be a good alternative. Ensure that the front forks can turn through the full steering lock available without fouling the fairing. With many types of frame-mounted fairing the handlebars will have to be altered or a different type fitted and the steering lock will be restricted by stops provided with the fittings. Also check that the fairing does not foul the front wheel or mudguard, in any steering position, under full fork compression. Re-route any cables, brake pipes or electrical wiring which may snag on the fairing and take great care to protect all electrical connections, using insulating tape. If the manufacturer's instructions are followed carefully at every stage no serious problems should be encountered. Remember that hydraulic pipes that have been disconnected must be carefully re-tightened and the hydraulic system purged of air bubbles by bleeding.

Two things will become immediately apparent when taking a motorcycle on the road for the first time with a fairing – the first is the tendency to underestimate the road speed because of the lack of wind pressure on the body. This must be very carefully watched until one has grown accustomed to riding behind the fairing. The second thing is the alarming increase in engine noise which is an unfortunate but inevitable by-product of fitting any type of fairing or windscreen, and is caused by normal engine noise being reflected, and in some cases amplified, by the flat surface of the fairing.

Luggage racks or carriers

Carriers are possibly the commonest item to be fitted to modern motorcycles. They vary enormously in size, carrying capacity, and durability. When selecting a carrier, always look for one which is made specifically for your machine and which is bolted on with as few separate brackets as possible. The universal-type carrier, with its mass of brackets and adaptor pieces, will generally prove too weak to be of any real use. A good carrier should bolt to the main frame, generally using the two suspension unit top mountings and a mudguard mounting bolt as attachment points, and have its luggage platform as low and as far forward as possible to minimise the effect of any load on the machine's stability. Look for good quality, heavy gauge tubing, good welding and good finish. Also ensure that the carrier does not prevent opening of the seat, sidepanels or tail compartment, as appropriate. When using a carrier, be very careful not to overload it. Excessive weight placed so high and so far to the rear of any motorcycle will have an adverse effect on the machine's steering and stability.

Luggage

Motorcycle luggage can be grouped under two headings: soft and hard. Both types are available in many sizes and styles and have advantages and disadvantages in use.

Soft luggage is now becoming very popular because of its lower cost and its versatility. Whether in the form of tankbags, panniers, or strap-on bags, soft luggage requires in general no brackets and no modification to the motorcycle. Equipment can be swapped easily from one motorcycle to another and can be fitted and removed in seconds. Awkwardly shaped loads can easily be carried. The disadvantages of soft luggage are that the contents cannot be secure against the casual thief, very little protection is afforded in the event of a crash, and waterproofing is generally poor. Also, in the case of panniers, carrying capacity is restricted to approximately 10 lb, although this amount will

vary considerably depending on the manufacturer's recommendation. When purchasing soft luggage, look for good quality material, generally vinyl or nylon, with strong, well-stitched attachment points. It is always useful to have separate pockets, especially on tank bags, for items which will be needed on the journey. When purchasing a tank bag, look for one which has a separate, well-padded, base. This will protect the tank's paintwork and permit easy access to the filler cap at petrol stations.

Hard luggage is confined to two types: panniers, and top boxes or tail trunks. Most hard luggage manufacturers produce matching sets of these items, the basis of which is generally that manufacturer's own heavy-duty luggage rack. Variations on this theme occur in the form of separate frames for the better quality panniers, fixed or quickly-detachable luggage, and in size and carrying capacity. Hard luggage offers a reasonable degree of security against theft and good protection against weather and accident damage. Carrying capacity is greater than that of soft luggage, around 15 – 20 lb in the case of panniers, although top boxes should never be loaded as much as their apparent capacity might imply. A top box should only be used for lightweight items, because one that is heavily laden can have a serious effect on the stability of the machine. When purchasing hard luggage look for the same good points as mentioned under fairings and windscreens, ie good quality mounting brackets and fittings, and well-finished fibreglass or ABS plastic cases. Again as with fairings, always purchase luggage made specifically for your motorcycle, using as few separate brackets as possible, to ensure that everything remains securely bolted in place. When fitting hard luggage, be careful to check that the rear suspension and brake operation will not be impaired in any way and remember that many pannier kits require re-siting of the indicators. Remember also that a non-standard exhaust system may make fitting extremely difficult.

Handlebars

The occupation of fitting alternative types of handlebar is extremely popular with modern motorcyclists, whose motives may vary from the purely practical, wishing to improve the comfort of their machines, to the purely aesthetic, where form is more important than function. Whatever the reason, there are several considerations to be borne in mind when changing the handlebars of your machine. If fitting lower bars, check carefully that the switches and cables do not foul the petrol tank on full lock and that the surplus length of cable, brake pipe, and electrical wiring are smoothly and tidily disposed of. Avoid tight kinks in cable or brake pipes which will produce stiff controls or the premature and disastrous failure of an overstressed component. If necessary, remove the petrol tank and re-route the cable from the engine/gearbox unit upwards, ensuring smooth gentle curves are produced. In extreme cases, it will be necessary to purchase a shorter brake pipe to overcome this problem. In the case of higher handlebars than standard it will almost certainly be necessary to purchase extended cables and brake pipes. Fortunately, many standard motorcycles have a custom version which will be equipped with higher handlebars and, therefore, factory-built extended components will be available from your local dealer. It is not usually necessary to extend electrical wiring, as switch clusters may be used on several different motorcycles, some being custom versions. This point should be borne in mind however when fitting extremely high or wide handlebars.

When fitting different types of handlebar, ensure that the mounting clamps are correctly tightened to the manufacturer's specifications and that cables and wiring, as previously mentioned, have smooth easy runs and do not snag on any part of the motorcycle throughout the full steering lock. Ensure that the fluid level in the front brake master cylinder remains level to avoid any chance of air entering the hydraulic system. Also check that the cables are adjusted correctly and that all handlebar controls operate correctly and can be easily reached when riding.

Crashbars

Crashbars, also known as engine protector bars, engine guards, or case savers, are extremely useful items of equipment which can contribute protection to the machine's structure if a crash occurs. They do not, as has been inferred in the US, prevent the rider from crashing, or necessarily prevent rider injury should a crash occur.

It is recommended that only the smaller, neater, engine protector type of crashbar is considered. This type will offer protection while restricting, as little as is possible, access to the engine and the machine's ground clearance. The crashbars should be designed for use specifically on your machine, and should be constructed of heavy-gauge tubing with strong, integral mounting brackets. Where possible, they should bolt to a strong lug on the frame, usually at the engine mounting bolts.

The alternative type of crashbar is the larger cage type. This type is not recommended in spite of their appearance which promises some protection to the rider as well as to the machine. The larger amount of leverage imposed by the size of this type of crashbar increases the risk of severe frame damage in the event of an accident. This type also decreases the machine's ground clearance and restricts access to the engine. The amount of protection afforded the rider is open to some doubt as the design is based on the premise that the rider will stay in the normally seated position during an accident, and the crash bar structure will not itself fail. Neither result can in any way be guaranteed.

As a general rule, always purchase the best, ie usually the most expensive, set of crashbars you an afford. The investment will be repaid by minimising the amount of damage incurred, should the machine be involved in an accident. Finally, avoid the universal type of crashbar. This should be regarded only as a last resort to be used if no alternative exists. With its usual multitude of separate brackets and spacers, the universal crashbar is far too weak in design and construction to be of any practical value.

Exhaust systems

The fitting of aftermarket exhaust systems is another extremely popular pastime amongst motorcyclists. The usual motive is to gain more performance from the engine but other considerations are to gain more ground clearance, to lose weight from the motorcycle, to obtain a more distinctive exhaust note or to find a cheaper alternative to the manufacturer's original equipment exhaust system. Original equipment exhaust systems often cost more and may well have a relatively short life. It should be noted that it is rare for an aftermarket exhaust system alone to give a noticeable increase in the engine's power output. Modern motorcycles are designed to give the highest power output possible allowing for factors such as quietness, fuel economy, spread of power, and long-term reliability. If there were a magic formula which allowed the exhaust system to produce more power without affecting these other considerations you can be sure that the manufacturers, with their large research and development facilities, would have found it and made use of it. Performance increases of a worthwhile and noticeable nature only come from well-tried and properly matched modifications to the entire engine, from the air filter, through the carburettors, port timing or camshaft and valve design, combustion chamber shape, compression ratio, and the exhaust system. Such modifications are well outside the scope of this manual but interested owners might refer to specialist books produced by the publisher of this manual which go into the whole subject in great detail.

Whatever your motive for wishing to fit an alternative exhaust system, be sure to seek expert advice before doing so. Changes to the carburettor jetting will almost certainly be required for which you must consult the exhaust system manufacturer. If he cannot supply adequately specific information it is reasonable to assume that insufficient development work has been carried out, and that particular make should be avoided. Other factors to be borne in mind are whether the exhaust system allows the use of both centre and side stands, whether it allows sufficient access to permit oil and filter changing and whether modifications are necessary to the standard exhaust system. Many two-stroke expansion chamber systems require the use of the standard exhaust pipe; this is all very well if the standard exhaust pipe and silencer are separate units but can cause problems if the two, as with so many modern two-strokes, are a one-piece unit. While the exhaust pipe can be removed easily by means of a hacksaw it is not so easy to refit the original silencer should you at any time wish to return the machine to standard trim. The same applies to several four-stroke systems.

On the subject of the finish of aftermarket exhausts, avoid black-painted systems unless you enjoy painting. As any trail-bike owner will tell you, rust has a great affinity for black exhausts and re-painting or rust removal becomes a task which must be carried out with monotonous regularity. A bright chrome finish is, as a general rule, a far better proposition as it is much easier to keep clean and to prevent rusting. Although the general finish of aftermarket exhaust systems is not always up to the standard of the original equipment the lower cost of such systems does at least reflect this fact.

When fitting an alternative system always purchase a full set of new exhaust gaskets, to prevent leaks. Fit the exhaust first to the cylinder head or barrel, as appropriate, tightening the retaining nuts or bolts by hand only and then line up the exhaust rear mountings. If the new system is a one-piece unit and the rear mountings do not line up exactly, spacers must be fabricated to take up the difference. Do not force the system into place as the stress thus imposed will rapidly cause cracks and splits to appear. Once all the mountings are loosely fixed, tighten the retaining nuts or bolts securely, being careful not to overtighten them. Where the motorcycle manufacturer's torque settings are available, these should be used. Do not forget to carry out any carburation changes recommended by the exhaust system's manufacturer.

Electrical equipment

The vast range of electrical equipment available to motorcyclists is so large and so diverse that only the most general outline can be given here. Electrical accessories vary from electric ignition kits fitted to replace contact breaker points, to additional lighting at the front and rear, more powerful horns, various instruments and gauges, clocks, anti-theft systems, heated clothing, CB radios, radio-cassette players, and intercom systems, to name but a few of the more popular items of equipment.

As will be evident, it would require a separate manual to cover this subject alone and this section is therefore restricted to outlining a few basic rules which must be borne in mind when fitting electrical equipment. The first consideration is whether your machine's electrical system has enough reserve capacity to cope with the added demand of the accessories you wish to fit. The motorcycle's manufacturer or importer should be able to furnish this sort of information and may also be able to offer advice on uprating the electrical system. Failing this, a good dealer or the accessory manufacturer may be able to help. In some cases, more powerful generator components may be available, perhaps from another motorcycle in the manufacturer's range. The second consideration is the legal requirements in force in your area. The local police may be prepared to help with this point. In the UK for example, there are strict regulations governing the position and use of auxiliary riding lamps and fog lamps.

When fitting electrical equipment always disconnect the battery first to prevent the risk of a short-circuit, and be careful to ensure that all connections are properly made and that they are waterproof. Remember that many electrical accesories are designed primarily for use in cars and that they cannot easily withstand the exposure to vibration and to the weather. Delicate components must be rubber-mounted to insulate them from vibration, and sealed carefully to prevent the entry of rainwater and dirt. Be careful to follow exactly the accessory manufacturer's instructions in conjunction with the wiring diagram at the back of this manual.

Accessories – general

Accessories fitted to your motorcycle will rapidly deteriorate if not cared for. Regular washing and polishing will maintain the finish and will provide an opportunity to check that all mounting bolts and nuts are securely fastened. Any signs of chafing or wear should be watched for, and the cause cured as soon as possible before serious damage occurs.

As a general rule, do not expect the re-sale value of your motorcycle to increase by an amount proportional to the amount of money and effort put into fitting accessories. It is usually the case that an absolutely standard motorcycle will sell more easily at a better price than one that has been modified. If you are in the habit of exchanging your machine for another at frequent intervals, this factor should be borne in mind to avoid loss of money.

Fault diagnosis

Contents

1 Introduction

This Section provides an easy reference-guide to the more common ailments that are likely to afflict your machine. Obviously, the opportunities are almost limitless for faults to occur as a result of obscure failures, and to try and cover all eventualities would require a book. Indeed, a number have been written on the subject.

Successful fault diagnosis is not a mysterious 'black art' but the application of a bit of knowledge combined with a systematic and logical approach to the problem. Approach any fault diagnosis by first accurately identifying the symptom and then checking through the list of possible causes, starting with the simplest or most obvious and progressing in stages to the most complex. Take nothing for granted, but above all apply liberal quantities of common sense.

The main symptom of a fault is given in the text as a major heading below which are listed, as Sections headings, the various systems or areas which may contain the fault. Details of each possible cause for a fault and the remedial action to be taken are given, in brief, in the paragraphs below each Section heading. Further information should be sought in the relevant Chapter.

Engine does not start when turned over

2 No fuel flow to carburettor

● Fuel tank empty or level too low. Check that the tap is turned to 'On' or 'Reserve' position as required. If in doubt, prise off the fuel feed pipe at the carburettor end and check that fuel runs from pipe when the tap is turned on.

● Tank filler cap vent obstructed. This can prevent fuel from flowing into the carburettor float bowl bcause air cannot enter the fuel tank to replace it. The problem is more likely to appear when the machine is being ridden. Check by listening close to the filler cap and releasing it. A hissing noise indicates that a blockage is present. Remove the cap and clear the vent hole with wire or by using an air line from the inside of the cap.

● Fuel tap or filter blocked. Blockage may be due to accumulation of rust or paint flakes from the tank's inner surface or of foreign matter from contaminated fuel. Remove the tap and clean it and the filter. Look also for water droplets in the fuel.

● Fuel line blocked. Blockage of the fuel line is more likely to result from a kink in the line rather than the accumulation of debris.

3 Fuel not reaching cylinder

● Float chamber not filling. Caused by float needle or floats sticking in up position. This may occur after the machine has been left standing for an extended length of time allowing the fuel to evaporate. When this occurs a gummy residue is often left which hardens to a varnish-like substance. This condition may be worsened by corrosion and crystalline deposits produced prior to the total evaporation of contaminated fuel. Sticking of the float needle may also be caused by wear. In any case removal of the float chamber will be necessary for inspection and cleaning.

● Blockage in starting circuit, slow running circuit or jets. Blockage of these items may be attributable to debris from the fuel tank by-passing the filter system or to gumming up as described in paragraph 1. Water droplets in the fuel will also block jets and passages. The carburettor should be dismantled for cleaning.

● Fuel level too low. The fuel level in the float chamber is controlled by float height. The fuel level may increase with wear or damage but will never reduce, thus a low fuel level is an inherent rather than developing condition. Check the float height, renewing the float or needle if required.

4 Engine flooding

● Float valve needle worn or stuck open. A piece of rust or other debris can prevent correct seating of the needle against the valve seat thereby permitting an uncontrolled flow of fuel. Similarly, a worn needle or needle seat will prevent valve closure. Dismantle the carburettor float bowl for cleaning and, if necessary, renewal of the worn components.

● Fuel level too high. The fuel level is controlled by the float height which may increase due to wear of the float needle, pivot pin or operating tang. Check the float height, and make any necessary adjustments. A leaking float will cause an increase in fuel level, and thus should be renewed.

● Cold starting mechanism. Check the choke (starter mechanism) for correct operation. If the mechanism jams in the 'On' position subsequent starting of a hot engine will be difficult.

● Blocked air filter. A badly restricted air filter will cause flooding. Check the filter and clean or renew as required. A collapsed inlet hose will have a similar effect. Check that the air filter inlet has not become blocked by a rag or similar item.

5 No spark at plug

● Ignition switch not on.
● Engine stop switch off.
● Spark plug dirty, oiled or 'whiskered'. Because the induction mixture of a two-stroke engine is inclined to be of a rather oily nature it is comparatively easy to foul the plug electrodes, especially where there have been repeated attempts to start the engine. A machine used for short journeys will be more prone to fouling because the engine may never reach full operating temperature, and the deposits will not burn off. On rare occasions a change of plug grade may be required but the advice of a dealer should be sought before making such a change. 'Whiskering' is a comparatively rare occurrence on modern machines but may be encountered where pre-mixed petrol and oil (petroil) lubrication is employed. An electrode deposit in the form of a barely visible filament across the plug electrodes can short circuit the plug and prevent its sparking. On all two-stroke machines it is a sound precaution to carry a new spare spark plug for substitution in the event of fouling problems.

● Spark plug failure. Clean the spark plug thoroughly and reset the electrode gap. Refer to the spark plug section and the condition guide in Chapter 4. If the spark plug shorts internally or has sustained visible damage to the electrodes, core or ceramic insulator it should be renewed. On rare occasions a plug that appears to spark vigorously will fail to do so when refitted to the engine and subjected to the compression pressure in the cylinder.

● Spark plug cap or high tension (HT) lead faulty. Check condition and security. Replace if deterioration is evident. Most spark plugs have an internal resistor designed to inhibit electrical interference with radio and television sets. On rare occasions the resistor may break down, thus preventing sparking. If this is suspected, fit a new cap as a precaution.

● Spark plug cap loose. Check that the spark plug cap fits securely over the plug and, where fitted, the screwed terminal on the plug end is secure.

● Shorting due to moisture. Certain parts of the ignition system are susceptible to shorting when the machine is ridden or parked in wet weather. Check particularly the area from the spark plug cap back to the ignition coil. A water dispersant spray may be used to dry out waterlogged components.

Recurrence of the problem can be prevented by using an ignition sealant spray after drying out and cleaning.
● Ignition or stop switch shorted. May be caused by water corrosion or wear. Water dispersant and contact cleaning sprays may be used. If this fails to overcome the problem dismantling and visual inspection of the switches will be required.
● Shorting or open circuit in wiring. Failure in any wire connecting any of the ignition components will cause ignition malfunction. Check also that all connections are clean, dry and tight.
● Ignition coil failure. Check the coil, referring to Chapter 4.
● Pulser coil, source coil or CDI unit defective. See Chapter 4.

6 Weak spark at plug

● Feeble sparking at the plug may be caused by any of the faults mentioned in the preceding Section other than those items in the first two paragraphs. Check first the spark plug, this being the most likely culprit.

7 Compression low

● Spark plug loose. This will be self-evident on inspection, and may be accompanied by a hissing noise when the engine is turned over. Remove the plug and check that the threads in the cylinder head are not damaged. Check also that the plug sealing washer is in good condition.
● Cylinder head gasket leaking. This condition is often accompanied by a high pitched squeak from around the cylinder head and oil loss, and may be caused by insufficiently tightened cylinder head fasteners, a warped cylinder head or mechanical failure of the gasket material. Re-torqueing the fasteners to the correct specification may seal the leak in some instances but if damage has occurred this course of action will provide, at best, only a temporary cure.
● Low crankcase compression. This can be caused by worn main bearings and seals and will upset the incoming fuel/air mixture. A good seal in these areas is essential on any two-stroke engine.
● Piston rings sticking or broken. Sticking of the piston rings may be caused by seizure due to lack of lubrication or overheating as a result of poor carburation or incorrect fuel type. Gumming of the rings may result from lack of use, or carbon deposits in the ring grooves. Broken rings result from over-revving, over-heating or general wear. In either case a top-end overhaul will be required.

Engine stalls after starting

8 General causes

● Improper cold start mechanism operation. Check that the operating controls function smoothly and, where applicable, are correctly adjusted. A cold engine may not require application of an enriched mixture to start initially but may baulk without choke once firing. Likewise a hot engine may start with an enriched mixture but will stop almost immediately if the choke is inadvertently in operation.
● Ignition malfunction. See Section 9. Weak spark at plug.
● Carburettor incorrectly adjusted. Maladjustment of the mixture strength or idle speed may cause the engine to stop immediately after starting. See Chapter 3.
● Fuel contamination. Check for filter blockage by debris or water which reduces, but does not completely stop, fuel flow, or blockage of the slow speed circuit in the carburettor by the same agents. If water is present it can often be seen as droplets

in the bottom of the float bowl. Clean the filter and, where water is in evidence, drain and flush the fuel tank and float bowl.
● Intake air leak. Check for security of the carburettor mounting and hose connections, and for cracks or splits in the hoses. Check also that the carburettor top is secure and that the vacuum gauge adaptor plug (where fitted) is tight.
● Air filter blocked or omitted. A blocked filter will cause an over-rich mixture; the omission of a filter will cause an excessively weak mixture. Both conditions will have a detrimental effect on carburation. Clean or renew the filter as necessary.
● Fuel filler cap air vent blocked. Usually caused by dirt or water. Clean the vent orifice.
● Choked exhaust system. Caused by excessive carbon build-up in the system, particularly around the silencer baffles. Refer to Chapter 3 for further information.
● Excessive carbon build-up in the engine. This can result from failure to decarbonise the engine at the specified interval or through excessive oil consumption. On pump-fed engines check pump adjustment.

Poor running at idle and low speed

9 Weak spark at plug or erratic firing

● Spark plug fouled, faulty or incorrectly adjusted. See Section 4 or refer to Chapter 4.
● Spark plug cap or high tension lead shorting. Check the condition of both these items ensuring that they are in good condition and dry and that the cap is fitted correctly.
● Spark plug type incorrect. Fit plug of correct type and heat range as given in Specifications. In certain conditions a plug of hotter or colder type may be required for normal running.
● Ignition timing incorrect. Check the ignition timing statically and dynamically, ensuring that the advance is functioning correctly.
● Faulty ignition coil. Partial failure of the coil internal insulation will diminish the performance of the coil. No repair is possible, a new component must be fitted.
● Pulser coil, source coil or CDI unit defective. See Chapter 4.
● Defective flywheel generator ignition source. Refer to Chapter 4 for further details on test procedures.

10 Fuel/air mixture incorrect

● Intake air leak. Check carburettor mountings and air cleaner hoses for security and signs of splitting. Ensure that carburettor top is tight and that the vacuum gauge take-off plug (where fitted) is tight.
● Mixture strength incorrect. Adjust slow running mixture strength using pilot adjustment screw.
● Carburettor synchronisation.
● Pilot jet or slow running circuit blocked. The carburettor should be removed and dismantled for thorough cleaning. Blow through all jets and air passages with compressed air to clear obstructions.
● Air cleaner clogged or omitted. Clean or fit air cleaner element as necessary. Check also that the element and air filter cover are correctly seated.
● Cold start mechanism in operation. Check that the choke has not been left on inadvertently and the operation is correct. Where applicable check the operating cable free play.
● Fuel level too high or too low. Check the float height, renewing float or needle if required. See Section 3 or 4.
● Fuel tank air vent obstructed. Obstructions usually caused by dirt or water. Clean vent orifice.

11 Compression low

● See Section 7.

Acceleration poor

12 General causes

● All items as for previous Section.
● Choked air filter. Failure to keep the air filter element clean will allow the build-up of dirt with proportional loss of performance. In extreme cases of neglect acceleration will suffer.
● Choked exhaust system. This can result from failure to remove accumulations of carbon from the silencer baffles at the prescribed intervals. The increased back pressure will make the machine noticeably sluggish. Refer to Chapter 3 for further information on decarbonisation.
● Excessive carbon build-up in the engine. This can result from failure to decarbonise the engine at the specified interval or through excessive oil consumption. Check oil pump adjustment.
● Ignition timing incorrect. Check the ignition timing as described in Chapter 4. Where no provision for adjustment exists, test the electronic ignition components and renew as required.
● Carburation fault. See Section 10.
● Mechanical resistance. Check that the brakes are not binding. On small machines in particular note that the increased rolling resistance caused by under-inflated tyres may impede acceleration.

Poor running or lack of power at high speeds

13 Weak spark at plug or erratic firing

● All items as for Section 9.
● HT lead insulation failure. Insulation failure of the HT lead and spark plug cap due to old age or damage can cause shorting when the engine is driven hard. This condition may be less noticeable, or not noticeable at all at lower engine speeds.

14 Fuel/air mixture incorrect

● All items as for Section 10, with the exception of items relative exclusively to low speed running.
● Main jet blocked. Debris from contaminated fuel, or from the fuel tank, and water in the fuel can block the main jet. Clean the fuel filter, the float bowl area, and if water is present, flush and refill the fuel tank.
● Main jet is the wrong size. The standard carburettor jetting is for sea level atmospheric pressure. For high altitudes, usually above 5000 ft, a smaller main jet will be required.
● Jet needle and needle jet worn. These can be renewed individually but should be renewed as a pair. Renewal of both items requires partial dismantling of the carburettor.
● Air bleed holes blocked. Dismantle carburettor and use compressed air to blow out all air passages.
● Reduced fuel flow. A reduction in the maximum fuel flow from the fuel tank to the carburettor will cause fuel starvation, proportionate to the engine speed. Check for blockages through debris or a kinked fuel line.

15 Compression low

● See Section 7.

Knocking or pinking

16 General causes

● Carbon build-up in combustion chamber. After high mileages have been covered large accumulations of carbon may occur.

This may glow red hot and cause premature ignition of the fuel/air mixture, in advance of normal firing by the spark plug. Cylinder head removal will be required to allow inspection and cleaning.
● Fuel incorrect. A low grade fuel, or one of poor quality may result in compression induced detonation of the fuel resulting in knocking and pinking noises. Old fuel can cause similar problems. A too highly leaded fuel will reduce detonation but will accelerate deposit formation in the combustion chamber and may lead to early pre-ignition as described in item 1.
● Spark plug heat range incorrect. Uncontrolled pre-ignition can result from the use of a spark plug the heat range of which is too hot.
● Weak mixture. Overheating of the engine due to a weak mixture can result in pre-ignition occurring where it would not occur when engine temperature was within normal limits. Maladjustment, blocked jets or passages and air leaks can cause this condition.

Overheating

17 Firing incorrect

● Spark plug fouled, defective or maladjusted. See Section 5.
● Spark plug type incorrect. Refer to the Specifications and ensure that the correct plug type is fitted.
● Incorrect ignition timing. Timing that is far too much advanced or far too much retarded will cause overheating. Check the ignition timing is correct.

18 Fuel/air mixture incorrect

● Slow speed mixture strength incorrect. Adjust pilot air screw.
● Main jet wrong size. The carburettor is jetted for sea level atmospheric conditions. For high altitudes, usually above 5000 ft, a smaller main jet will be required.
● Air filter badly fitted or omitted. Check that the filter element is in place and that it and the air filter box cover are sealing correctly. Any leaks will cause a weak mixture.
● Induction air leaks. Check the security of the carburettor mountings and hose connections, and for cracks and splits in the hoses. Check also that the carburettor top is secure and that the vacuum gauge adaptor plug (where fitted) is tight.
● Fuel level too low. See Section 3.
● Fuel tank filler cap air vent obstructed. Clear blockage.

19 Lubrication inadequate

● Oil pump settings incorrect. The oil pump settings are of great importance since the quantities of oil being injected are very small. Any variation in oil delivery will have a significant effect on the engine. Refer to Chapter 3 for further information.
● Oil tank empty or low. This will have disastrous consequences if left unnoticed. Check and replenish tank regularly.
● Transmission oil low or worn out. Check the level regularly and investigate any loss of oil. If the oil level drops with no sign of external leakage it is likely that the crankshaft main bearing oil seals are worn, allowing transmission oil to be drawn into the crankcase during induction.

20 Miscellaneous causes

● Radiator fins clogged. Accumulated debris in the radiator core will gradually reduce its ability to dissipate heat generated by the engine. It is worth noting that during the summer months dead insects can cause as many problems in this respect as road dirt and mud during the winter. Cleaning is best carried out by dislodging the debris with a high pressure hose from the back of the radiator. Once cleaned it is worth painting the matrix with a heat-dispersant matt black paint both to assist cooling and to prevent external corrosion. The fitting of some sort of mesh guard will help prevent the fins from becoming clogged, but make sure that this does not itself prevent adequate cooling.

Clutch operating problems

21 Clutch slip

● No clutch lever play. Adjust clutch lever end play according to the procedure in Chapter 1.
● Friction plates worn or warped. Overhaul clutch assembly, replacing plates out of specification.
● Steel plates worn or warped. Overhaul clutch assembly, replacing plates out of specification.
● Clutch spring broken or worn. Old or heat-damaged (from slipping clutch) springs should be replaced with new ones.
● Clutch release not adjusted properly. See the adjustments section of Chapter 1.
● Clutch inner cable snagging. Caused by a frayed cable or kinked outer cable. Replace the cable with a new one. Repair of a frayed cable is not advised.
● Clutch release mechanism defective. Worn or damaged parts in the clutch release mechanism could include the shaft, cam, actuating arm or pivot. Replace parts as necessary.
● Clutch hub and outer drum worn. Severe indentation by the clutch plate tangs of the channels in the hub and drum will cause snagging of the plates preventing correct engagement. If this damage occurs, renewal of the worn components is required.
● Lubricant incorrect. Use of a transmission lubricant other than that specified may allow the plates to slip.

22 Clutch drag

● Clutch lever play excessive. Adjust lever at bars or at cable end if necessary.
● Clutch plates warped or damaged. This will cause a drag on the clutch, causing the machine to creep. Overhaul clutch assembly.
● Clutch spring tension uneven. Usually caused by a sagged or broken spring. Check and replace springs.
● Transmission oil deteriorated. Badly contaminated transmission oil and a heavy deposit of oil sludge on the plates will cause plate sticking. The oil recommended for this machine is of the detergent type, therefore it is unlikely that this problem will arise unless regular oil changes are neglected.
● Transmission oil viscosity too high. Drag in the plates will result from the use of an oil with too high a viscosity. In very cold weather clutch drag may occur until the engine has reached operating temperature.
● Clutch hub and outer drum worn. Indentation by the clutch plate tangs of the channels in the hub and drum will prevent easy plate disengagement. If the damage is light the affected areas may be dressed with a fine file. More pronounced damage will necessitate renewal of the components.
● Clutch housing seized to shaft. Lack of lubrication, severe wear or damage can cause the housing to seize to the shaft. Overhaul of the clutch, and perhaps the transmission, may be necessary to repair damage.
● Clutch release mechanism defective. Worn or damaged release mechanism parts can stick and fail to provide leverage. Overhaul clutch cover components.
● Loose clutch hub nut. Causes drum and hub misalignment, putting a drag on the engine. Engagement adjustment continually varies. Overhaul clutch assembly.

Gear selection problems

23 Gear lever does not return

● Weak or broken centraliser spring. Renew the spring.
● Gearchange shaft bent or seized. Distortion of the gearchange shaft often occurs if the machine is dropped heavily on the gear lever. Provided that damage is not severe straightening of the shaft is permissible.

24 Gear selection difficult or impossible

● Clutch not disengaging fully. See Section 22.
● Gearchange shaft bent. This often occurs if the machine is dropped heavily on the gear lever. Straightening of the shaft is permissible if the damage is not too great.
● Gearchange arms, pawls or pins worn or damaged. Wear or breakage of any of these items may cause difficulty in selecting one or more gears. Overhaul the selector mechanism.
● Gearchange shaft centraliser spring maladjusted. This is often characterised by difficulties in changing up or down, but rarely in both directions. Adjust the centraliser anchor bolt as described in Chapter 1.
● Gearchange drum stopper cam or detent plunger damaged. Failure, rather than wear of these items may jam the drum thereby preventing gearchanging or causing false selection at high speed.
● Selector forks bent or seized. This can be caused by dropping the machine heavily on the gearchange lever or as a result of lack of lubrication. Though rare, bending of a shaft can result from a missed gearchange or false selection at high speed.
● Selector fork end and pin wear. Pronounced wear of these items and the grooves in the gearchange drum can lead to imprecise selection and, eventually, no selection. Renewal of the worn components will be required.
● Structural failure. Failure of any one component of the selector rod and change mechanism will result in improper or fouled gear selection.

25 Jumping out of gear

● Detent plunger assembly worn or damaged. Wear of the plunger and the cam with which it locates and breakage of the detent spring can cause imprecise gear selection resulting in jumping out of gear. Renew the damaged components.
● Gear pinion dogs worn or damaged. Rounding off the dog edges and the mating recesses in adjacent pinion can lead to jumping out of gear when under load. The gears should be inspected and renewed. Attempting to reprofile the dogs is not recommended.
● Selector forks, gearchange drum and pinion grooves worn. Extreme wear of these interconnected items can occur after high mileages especially when lubrication has been neglected. The worn components must be renewed.
● Gear pinions, bushes and shafts worn. Renew the worn components.
● Bent gearchange shaft. Often caused by dropping the machine on the gear lever.
● Gear pinion tooth broken. Chipped teeth are unlikely to cause jumping out of gear once the gear has been selected fully; a tooth which is completely broken off, however, may cause problems in this respect and in any event will cause transmission noise.

26 Overselection

● Pawl spring weak or broken. Renew the spring.
● Detent plunger worn or broken. Renew the damaged items.

● Stopper arm spring worn or broken. Renew the spring.
● Gearchange arm stop pads worn. Repairs can be made by welding and reprofiling with a file.
● Selector limiter claw components (where fitted) worn or damaged. Renew the damaged items.

Abnormal engine noise

27 Knocking or pinking

● See Section 16.

28 Piston slap or rattling from cylinder

● Cylinder bore/piston clearance excessive. Resulting from wear, or partial seizure. This condition can often be heard as a high, rapid tapping noise when the engine is under little or no load, particularly when power is just beginning to be applied. Reboring to the next correct oversize should be carried out and a new oversize piston fitted.
● Connecting rod bent. This can be caused by over-revving, trying to start a very badly flooded engine (resulting in a hydraulic lock in the cylinder) or by earlier mechanical failure. Attempts at straightening a bent connecting rod from a high performance engine are not recommended. Careful inspection of the crankshaft should be made before renewing the damaged connecting rod.
● Gudgeon pin, piston boss bore or small-end bearing wear or seizure. Excess clearance or partial seizure between normal moving parts of these items can cause continuous or intermittent tapping noises. Rapid wear or seizure is caused by lubrication starvation.
● Piston rings worn, broken or sticking. Renew the rings after careful inspection of the piston and bore.

29 Other noises

● Big-end bearing wear. A pronounced knock from within the crankcase which worsens rapidly is indicative of big-end bearing failure as a result of extreme normal wear or lubrication failure. Remedial action in the form of a bottom end overhaul should be taken; continuing to run the engine will lead to further damage including the possibility of connecting rod breakage.
● Main bearing failure. Extreme normal wear or failure of the main bearings is characteristically accompanied by a rumble from the crankcase and vibration felt through the frame and footrests. Renew the worn bearings and carry out a very careful examination of the crankshaft.
● Crankshaft excessively out of true. A bent crank may result from over-revving or damage from an upper cylinder component or gearbox failure. Damage can also result from dropping the machine on either crankshaft end. Straightening of the crankshaft is not be possible in normal circumstances; a replacement item should be fitted.
● Engine mounting loose. Tighten all the engine mounting nuts and bolts.
● Cylinder head gasket leaking. The noise most often associated with a leaking head gasket is a high pitched squeaking, although any other noise consistent with gas being forced out under pressure from a small orifice can also be emitted. Gasket leakage is often accompanied by oil seepage from around the mating joint or from the cylinder head holding down bolts and nuts. Leakage results from insufficient or uneven tightening of the cylinder head fasteners, or from random mechanical failure. Retightening to the correct torque figure will, at best, only provide a temporary cure. The gasket should be renewed at the earliest opportunity.

● Exhaust system leakage. Popping or crackling in the exhaust system, particularly when it occurs with the engine on the overrun, indicates a poor joint either at the cylinder port or at the exhaust pipe/silencer connection. Failure of the gasket or looseness of the clamp should be looked for.

Abnormal transmission noise

30 Clutch noise

● Clutch outer drum/friction plate tang clearance excessive.
● Clutch outer drum/spacer clearance excessive.
● Clutch outer drum/thrust washer clearance excessive.
● Primary drive gear teeth worn or damaged.
● Clutch shock absorber assembly worn or damaged.

31 Transmission noise

● Bearing or bushes worn or damaged. Renew the affected components.
● Gear pinions worn or chipped. Renew the gear pinions.
● Metal chips jammed in gear teeth. This can occur when pieces of metal from any failed component are picked up by a meshing pinion. The condition will lead to rapid bearing wear or early gear failure.
● Engine/transmission oil level too low. Top up immediately to prevent damage to gearbox and engine.
● Gearchange mechanism worn or damaged. Wear or failure of certain items in the selection and change components can induce mis-selection of gears (see Section 24) where incipient engagement of more than one gear set is promoted. Remedial action, by the overhaul of the gearbox, should be taken without delay.
● Chain snagging on cases or cycle parts. A badly worn chain or one that is excessively loose may snag or smack against adjacent components.

Exhaust smokes excessively

32 White/blue smoke (caused by oil burning)

● Cylinder cracked, worn or scored. These conditions may be caused by overheating, lack of lubrication, component failure or advanced normal wear. The cylinder barrel should be renewed and, if necessary, a new piston fitted.
● Oil pump settings incorrect. Check and reset the oil pump as described in Chapter 2.
● Crankshaft main bearing oil seals worn. Wear in the main bearing oil seals, often in conjunction with wear in the bearings themselves, can allow transmission oil to find its way into the crankcase and thence to the combustion chamber. This condition is often indicated by a mysterious drop in the transmission oil level with no sign of external leakage.
● Accumulated oil deposits in exhaust system. If the machine is used for short journeys only it is possible for the oil residue in the exhaust gases to condense in the relatively cool silencer. If the machine is then taken for a longer run in hot weather, the accumulated oil will burn off producing ominous smoke from the exhaust.

33 Black smoke (caused by over-rich mixture)

● Air filter element clogged. Clean or renew the element.
● Main jet loose or too large. Remove the float chamber to check for tightness of the jet. If the machine is used at high altitudes rejetting will be required to compensate for the lower atmospheric pressure.

● Cold start mechanism jammed on. Check that the mechanism works smoothly and correctly and that, where fitted, the operating cable is lubricated and not snagged.

● Fuel level too high. The fuel level is controlled by the float height which can increase as a result of wear or damage. Remove the float bowl and check the float height. Check also that floats have not punctured; a punctured float will lose buoyancy and allow an increased fuel level.

● Float valve needle stuck open. Caused by dirt or a worn valve. Clean the float chamber or renew the needle and, if necessary, the valve seat.

Poor handling or roadholding

34 Directional instability

● Steering head bearing adjustment too tight. This will cause rolling or weaving at low speeds. Re-adjust the bearings.

● Steering head bearing worn or damaged. Correct adjustment of the bearing will prove impossible to achieve if wear or damage has occurred. Inconsistent handling will occur including rolling or weaving at low speed and poor directional control at indeterminate higher speeds. The steering head bearing should be dismantled for inspection and renewed if required. Lubrication should also be carried out.

● Bearing races pitted or dented. Impact damage caused, perhaps, by an accident or riding over a pot-hole can cause indentation of the bearing, usually in one position. This should be noted as notchiness when the handlebars are turned. Renew and lubricate the bearings.

● Steering stem bent. This will occur only if the machine is subjected to a high impact such as hitting a curb or a pot-hole. The lower yoke/stem should be renewed; do not attempt to straighten the stem.

● Front or rear tyre pressures too low.

● Front or rear tyre worn. General instability, high speed wobbles and skipping over white lines indicates that tyre renewal may be required. Tyre induced problems, in some machine/tyre combinations, can occur even when the tyre in question is by no means fully worn.

● Swinging arm bearings worn. Difficulties in holding line, particularly when cornering or when changing power settings indicates wear in the swinging arm bearings. The swinging arm should be removed from the machine and the bearings renewed.

● Swinging arm flexing. The symptoms given in the preceding paragraph will also occur if the swinging arm fork flexes badly. This can be caused by structural weakness as a result of corrosion, fatigue or impact damage, or because the rear wheel spindle is slack.

● Wheel bearings worn. Renew the worn bearings.

● Loose wheel spokes. The spokes should be tightened evenly to maintain tension and trueness of the rim.

● Tyres unsuitable for machine. Not all available tyres will suit the characteristics of the frame and suspension, indeed, some tyres or tyre combinations may cause a transformation in the handling characteristics. If handling problems occur immediately after changing to a new tyre type or make, revert to the original tyres to see whether an improvement can be noted. In some instances a change to what are, in fact, suitable tyres may give rise to handling deficiences. In this case a thorough check should be made of all frame and suspension items which affect stability.

35 Steering bias to left or right

● Rear wheel out of alignment. Caused by uneven adjustment of chain tensioner adjusters allowing the wheel to be askew in the fork ends. A bent rear wheel spindle will also misalign the wheel in the swinging arm.

● Wheels out of alignment. This can be caused by impact damage to the frame, swinging arm, wheel spindles or front forks. Although occasionally a result of material failure or corrosion it is usually as a result of a crash.

● Front forks twisted in the steering yokes. A light impact, for instance with a pot-hole or low curb, can twist the fork legs in the steering yokes without causing structural damage to the fork legs or the yokes themselves. Re-alignment can be made by loosening the yoke pinch bolts, wheel spindle and mudguard bolts. Re-align the wheel with the handlebars and tighten the bolts working upwards from the wheel spindle. This action should be carried out only when there is no chance that structural damage has occurred.

36 Handlebar vibrates or oscillates

● Tyres worn or out of balance. Either condition, particularly in the front tyre, will promote shaking of the fork assembly and thus the handlebars. A sudden onset of shaking can result if a balance weight is displaced during use.

● Tyres badly positioned on the wheel rims. A moulded line on each wall of a tyre is provided to allow visual verification that the tyre is correctly positioned on the rim. A check can be made by rotating the tyre; any misalignment will be immediately obvious.

● Wheels rims warped or damaged. Inspect the wheels for runout as described in Chapter 6.

● Swinging arm bearings worn. Renew the bearings.

● Wheel bearings worn. Renew the bearings.

● Steering head bearings incorrectly adjusted. Vibration is more likely to result from bearings which are too loose rather than too tight. Re-adjust the bearings.

● Loosen fork component fasteners. Loose nuts and bolts holding the fork legs, wheel spindle, mudguards or steering stem can promote shaking at the handlebars. Fasteners on running gear such as the forks and suspension should be check tightened occasionally to prevent dangerous looseness of components occurring.

● Engine mounting bolts loose. Tighten all fasteners.

37 Poor front fork performance

● Damping fluid level incorrect. If the fluid level is too low poor suspension control will occur resulting in a general impairment of roadholding and early loss of tyre adhesion when cornering and braking. Too much oil is unlikely to change the fork characteristics unless severe overfilling occurs when the fork action will become stiffer and oil seal failure may occur.

● Damping oil viscosity incorrect. The damping action of the fork is directly related to the viscosity of the damping oil. The lighter the oil used, the less will be the damping action imparted. For general use, use the recommended viscosity of oil, changing to a slightly higher or heavier oil only when a change in damping characteristic is required. Overworked oil, or oil contaminated with water which has found its way past the seals, should be renewed to restore the correct damping performance and to prevent bottoming of the forks.

● Damping components worn or corroded. Advanced normal wear of the fork internals is unlikely to occur until a very high mileage has been covered. Continual use of the machine with damaged oil seals which allows the ingress of water, or neglect, will lead to rapid corrosion and wear. Dismantle the forks for inspection and overhaul.

● Weak fork springs. Progressive fatigue of the fork springs, resulting in a reduced spring free length, will occur after extensive use. This condition will promote excessive fork dive under braking, and in its advanced form will reduce the at-rest extended length of the forks and thus the fork geometry. Renewal of the springs as a pair is the only satisfactory course of action.

● Bent stanchions or corroded stanchions. Both conditions will prevent correct telescoping of the fork legs, and in an advanced state can cause sticking of the fork in one position. In a mild form corrosion will cause stiction of the fork thereby increasing the time the suspension takes to react to an uneven road surface. Bent fork stanchions should be attended to immediately because they indicate that impact damage has occurred, and there is a danger that the forks will fail with disastrous consequences.

38 Front fork judder when braking (see also Section 41)

● Wear between the fork stanchions and the fork legs. Renewal of the affected components is required.
● Slack steering head bearings. Re-adjust the bearings.
● Warped brake disc or drum. If irregular braking action occurs fork judder can be induced in what are normally serviceable forks. Renew the damaged brake components.

39 Poor rear suspension performance

● Rear suspension unit damper worn out or leaking. The damping performance of most rear suspension units falls off with age. This is a gradual process, and thus may not be immediately obvious. Indications of poor damping include hopping of the rear end when cornering or braking, and a general loss of positive stability.
● Weak rear spring. If the suspension unit spring fatigues it will promote excessive pitching of the machine and reduce the ground clearance when cornering. It is probable that if spring fatigue has occurred the damper units will also require renewal.
● Swinging arm flexing or bearings worn. See Sections 34 and 36.
● Bent suspension unit damper rod. This is likely to occur only if the machine is dropped or if seizure of the piston occurs. If either happens the suspension unit should be renewed.

Abnormal frame and suspension noise

40 Front end noise

● Oil level low or too thin. This can cause a 'spurting' sound and is usually accompanied by irregular fork action.
● Spring weak or broken. Makes a clicking or scraping sound. Fork oil will have a lot of metal particles in it.
● Steering head bearings loose or damaged. Clicks when braking. Check, adjust or replace.
● Fork clamps loose. Make sure all fork clamp pinch bolts are tight.
● Fork stanchion bent. Good possibility if machine has been dropped. Repair or replace tube.

41 Rear suspension noise

● Fluid level too low. Leakage of a suspension unit, usually evident by oil on the outer surfaces, can cause a spurting noise. The suspension unit should be renewed.
● Defective rear suspension unit with internal damage. Renew the suspension unit.

Brake problems

42 Brakes are spongy or ineffective – disc brakes

● Air in brake circuit. This is only likely to happen in service due to neglect in checking the fluid level or because a leak has developed. The problem should be identified and the brake system bled of air.
● Pad worn. Check the pad wear against the wear lines provided and renew the pads if necessary.
● Contaminated pads. Cleaning pads which have been contaminated with oil, grease or brake fluid is unlikely to prove successful; the pads should be renewed.
● Pads glazed. This is usually caused by overheating. The surface of the pads may be roughened using glass-paper or a fine file.
● Brake fluid deterioration. A brake which on initial operation is firm but rapidly becomes spongy in use may be failing due to water contamination of the fluid. The fluid should be drained and then the system refilled and bled.
● Master cylinder seal failure. Wear or damage of master cylinder internal parts will prevent pressurisation of the brake fluid. Overhaul the master cylinder unit.
● Caliper seal failure. This will almost certainly be obvious by loss of fluid, a lowering of fluid in the master cylinder reservoir and contamination of the brake pads and caliper. Overhaul the caliper assembly.
● Brake lever improperly adjusted. Adjust the clearance between the lever end and master cylinder plunger to take up lost motion, as recommended in Routine maintenance.

43 Brakes drag – disc brakes

● Disc warped. The disc must be renewed.
● Caliper piston, caliper or pads corroded. The brake caliper assembly is vulnerable to corrosion due to water and dirt, and unless cleaned at regular intervals and lubricated in the recommended manner, will become sticky in operation.
● Piston seal deteriorated. The seal is designed to return the piston in the caliper to the retracted position when the brake is released. Wear or old age can affect this function. The caliper should be overhauled if this occurs.
● Brake pad damaged. Pad material separating from the backing plate due to wear or faulty manufacture. Renew the pads. Faulty installation of a pad also will cause dragging.
● Wheel spindle bent. The spindle may be straightened if no structural damage has occurred.
● Brake lever or pedal not returning. Check that the lever or pedal works smoothly throughout its operating range and does not snag on any adjacent cycle parts. Lubricate the pivot if necessary.
● Twisted caliper support bracket. This is likely to occur only after impact in an accident. No attempt should be made to re-align the caliper; the bracket should be renewed.

44 Brake lever or pedal pulsates in operation – disc brakes

● Disc warped or irregularly worn. The disc must be renewed.
● Wheel spindle bent. The spindle may be straightened provided no structural damage has occurred.

45 Disc brake noise

● Brake squeal. This can be caused by the omission or incorrect installation of the anti-squeal shim fitted to the rear of one pad. Squealing can also be caused by dust on the pads, usually in combination with glazed pads, or other contamination from oil, grease, brake fluid or corrosion. Persistent squealing which cannot be traced to any of the normal causes can often be cured by applying a thin layer of high temperature silicone grease to the rear of the pads. Make absolutely certain that no grease is allowed to contaminate the braking surface of the pads.
● Glazed pads. This is usually caused by high temperatures or

contamination. The pad surfaces may be roughened using glass-paper or a fine file. If this approach does not effect a cure the pads should be renewed.

● Disc warped. This can cause a chattering, clicking or intermittent squeal and is usually accompanied by a pulsating brake lever or uneven braking. The disc must be renewed.

● Brake pads fitted incorrectly or undersize. Longitudinal play in the pads due to omission of the locating springs (where fitted) or because pads of the wrong size have been fitted will cause a single tapping noise every time the brake is operated. Inspect the pads for correct installation and security.

46 Brakes are spongy or ineffective – drum brakes

● Brake cable deterioration. Damage to the outer cable by stretching or being trapped will give a spongy feel to the brake lever. The cable should be renewed. A cable which has become corroded due to old age or neglect of lubrication will partially seize making operation very heavy. Lubrication at this stage may overcome the problem but the fitting of a new cable is recommended.

● Worn brake linings. Determine lining wear using the external brake wear indicator on the brake backplate, or by removing the wheel and withdrawing the brake backplate. Renew the shoe/lining units as a pair if the linings are worn below the recommended limit.

● Worn brake camshaft. Wear between the camshaft and the bearing surface will reduce brake feel and reduce operating efficiency. Renewal of one or both items will be required to rectify the fault.

● Worn brake cam and shoe ends. Renew the worn components.

● Linings contaminated with dust or grease. Any accumulations of dust should be cleaned from the brake assembly and drum using a petrol dampened cloth. Do not blow or brush off the dust because it is asbestos based and thus harmful if inhaled. Light contamination from grease can be removed from the surface of the brake linings using a solvent; attempts at removing heavier contamination are less likely to be successful because some of the lubricant will have been absorbed by the lining material which will severely reduce the braking performance.

47 Brake drag – drum brakes

● Incorrect adjustment. Re-adjust the brake operating mechanism.

● Drum warped or oval. This can result from overheating or impact or uneven tension of the wheel spokes. The condition is difficult to correct, although if slight ovality only occurs, skimming the surface of the brake drum can provide a cure. This is work for a specialist engineer. Renewal of the complete wheel hub is normally the only satisfactory solution.

● Weak brake shoe return springs. This will prevent the brake lining/shoe units from pulling away from the drum surface once the brake is released. The springs should be renewed.

● Brake camshaft, lever pivot or cable poorly lubricated. Failure to attend to regular lubrication of these areas will increase operating resistance which, when compounded, may cause tardy operation and poor release movement.

48 Brake lever or pedal pulsates in operation – drum brakes

● Drums warped or oval. This can result from overheating or impact or uneven spoke tension. This condition is difficult to correct, although if slight ovality only occurs skimming the surface of the drum can provide a cure. This is work for a specialist engineer. Renewal of the hub is normally the only satisfactory solution.

49 Drum brake noise

● Drum warped or oval. This can cause intermittent rubbing of the brake linings against the drum. See the preceding Section.

● Brake linings glazed. This condition, usually accompanied by heavy lining dust contamination, often induces brake squeal. The surface of the linings may be roughened using glass-paper or a fine file.

50 Brake induced fork judder

● Worn front fork stanchions and legs, or worn or badly adjusted steering head bearings. These conditions, combined with uneven or pulsating braking as described in Sections 44 and 48 will induce more or less judder when the brakes are applied, dependent on the degree of wear and poor brake operation. Attention should be given to both areas of malfunction. See the relevant Sections.

Electrical problems

51 Battery dead or weak

● Battery faulty. Battery life should not be expected to exceed 3 to 4 years, particularly where a starter motor is used regularly. Gradual sulphation of the plates and sediment deposits will reduce the battery performance. Plate and insulator damage can often occur as a result of vibration. Complete power failure, or intermittent failure, may be due to a broken battery terminal. Lack of electrolyte will prevent the battery maintaining charge.

● Battery leads making poor contact. Remove the battery leads and clean them and the terminals, removing all traces of corrosion and tarnish. Reconnect the leads and apply a coating of petroleum jelly to the terminals.

● Load excessive. If additional items such as spot lamps, are fitted, which increase the total electrical load above the maximum alternator output, the battery will fail to maintain full charge. Reduce the electrical load to suit the electrical capacity.

● Rectifier failure.

● Alternator generating coils open-circuit or shorted.

● Charging circuit shorting or open circuit. This may be caused by frayed or broken wiring, dirty connectors or a faulty ignition switch. The system should be tested in a logical manner. See Section 54.

52 Battery overcharged

● Regulator faulty. Overcharging is indicated if the battery becomes hot or it is noticed that the electrolyte level falls repeatedly between checks. In extreme cases the battery will boil causing corrosive gases and electrolyte to be emitted through the vent pipes.

● Battery wrongly matched to the electrical circuit. Ensure that the specified battery is fitted to the machine.

53 Total electrical failure

● Fuse blown. Check the main fuse. If a fault has occurred, it must be rectified before a new fuse is fitted.

● Battery faulty. See Section 51.

● Earth failure. Check that the frame main earth strap from the battery is securely affixed to the frame and is making a good contact.

● Ignition switch or power circuit failure. Check for current flow through the battery positive lead (red) to the ignition switch. Check the ignition switch for continuity.

54 Circuit failure

● Cable failure. Refer to the machine's wiring diagram and check the circuit for continuity. Open circuits are a result of loose or corroded connections, either at terminals or in-line connectors, or because of broken wires. Occasionally, the core of a wire will break without there being any apparent damage to the outer plastic cover.

● Switch failure. All switches may be checked for continuity in each switch position, after referring to the switch position boxes incorporated in the wiring diagram for the machine. Switch failure may be a result of mechanical breakage, corrosion or water.

● Fuse blown. Refer to the wiring diagram to check whether or not a circuit fuse is fitted. Replace the fuse, if blown, only after the fault has been identified and rectified.

55 Bulbs blowing repeatedly

● Vibration failure. This is often an inherent fault related to the natural vibration characteristics of the engine and frame and is, thus, difficult to resolve. Modifications of the lamp mounting, to change the damping characteristics, may help.

● Intermittent earth. Repeated failure of one bulb, particularly where the bulb is fed directly from the generator, indicates that a poor earth exists somewhere in the circuit. Check that a good contact is available at each earthing point in the circuit.

● Reduced voltage. Where a quartz-halogen bulb is fitted the voltage to the bulb should be maintained or early failure of the bulb will occur. Do not overload the system with additional electrical equipment in excess of the system's power capacity and ensure that all circuit connections are maintained clean and tight.

YAMAHA RD/DT 125 LC

Check list

Pre-ride checks

1   Check for correct operation of the front and rear brake
2   Check the operation of the clutch
3   Ensure that the throttle cable operates smoothly
4   Check the level of coolant in the radiator
5   Check the engine oil level in the tank
6   Check the transmission oil level
7   Ensure that the final drive chain is correctly adjusted
8   Inspect the wheels and tyres for damage and under-inflation of the tyres
9   Ensure that all accessories and fittings are securely fastened
10 Check that the lights and electrical system function correctly
11 Check that there is sufficient petrol in the tank to complete your journey

Weekly or every 200 miles (300 km)

1   Top up the engine oil tank
2   Check the level of coolant in the reservoir and top up if necessary
3   Check the final drive chain tension
4   Adjust the brakes
5   Check the tyre pressures
6   Check the level of electrolyte in the battery
7   Check the operation of the front and rear suspension
8   Check the security of all fastenings and lubricate exposed portions of control cables

Monthly or every 1000 miles (1500 km)

1   Check the transmission oil level and top up if necessary
2   Remove, clean and lubricate the final drive chain
3   Remove and clean the air filter element – DT model
4   Examine the battery terminals for corrosion

Three monthly or every 2000 miles (3000 km)

1   Clean and reset the spark plug gap
2   Remove and clean the air filter element – RD model
3   Check the clutch free play and adjust if necessary
4   Change the transmission oil
5   Adjust the carburettor and throttle cable
6   Check the oil pump setting
7   Check the security of the exhaust system and for leakage around the gasket
8   Check the degree of brake wear

Six monthly or every 4000 miles (6000 km)

1   Decarbonize the engine
2   Overhaul the cooling system
3   Renew the spark plug
4   Check the ignition timing
5   Clean the petrol tank filter
6   Overhaul the brakes
7   Lubricate the controls and stand pivots
8   Lubricate the control and instrument cables
9   Grease and adjust the steering head bearings
10 Renew the front fork oil
11 Grease the wheel bearings and speedometer drive gearbox
12 Grease the sub-frame pivot and suspension unit mountings

Additional maintenance items

1   Renew the radiator coolant
2   Renew the hydraulic brake fluid
3   Renew the hydraulic brake caliper seals
4   Renew the hydraulic brake hose
5   Cleaning the machine

Adjustment data

Tyre pressures	Front	Rear
DT model		
Solo	18 psi (1.26 kg/cm²)	22 psi (1.54 kg/cm²)
Pillion	22 psi (1.54 kg/cm²)	26 psi (1.82 kg/cm²)
RD model		
Solo	26 psi (1.82 kg/cm²)	28 psi (1.96 kg/cm²)
Solo – continuous high speed riding	26 psi (1.82 kg/cm²)	32 psi (2.24 kg/cm²)
Pillion	26 psi (1.82 kg/cm²)	32 psi (2.24 kg/cm²)
Pillion – continuous high speed riding	26 psi (1.82 kg/cm²)	40 psi (2.80 kg/cm²)

Spark plug type   NGK BR8ES

Spark plug gap   0.7 – 0.8 mm (0.028 – 0.031 in)

Idle speed
RD model   1300 ± 50 rpm
DT model   1350 ± 50 rpm

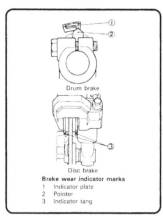

Drum brake

Disc brake

Brake wear indicator marks
1   Indicator plate
2   Pointer
3   Indicator tang

Recommended lubricants

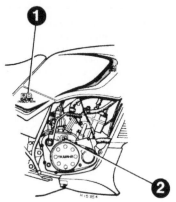

Component	Quantity	Type/viscosity
1 Engine RD model	1.1 lit (1.94 Imp pt)	Two-stroke oil
DT model	1.0 lit (1.76 Imp pt)	
2 Transmission	550 cc (0.97 Imp pt)	SAE 10W/30 SE engine oil
3 Cooling system	800 cc (1.41 Imp pt)	50% distilled water/ 50% corrosion-inhibited ethylene glycol antifreeze
4 Front forks RD model	147 cc (5.17 fl oz)	SAE 10 fork oil
DT model	304 cc (10.7 fl oz)	SAE 10W/30 SE engine oil
5 Wheel bearings	As required	High melting point grease
6 Steering head bearings	As required	High melting point grease
7 Sub-frame pivot bolt	As required	High melting point grease
8 Final drive chain	As required	Aerosol chain lubricant
9 Hydraulic front brake	As required	SAE J1703 or DOT 3
10 Control cables	As required	Light machine oil
11 Pivot points	As required	General purpose grease

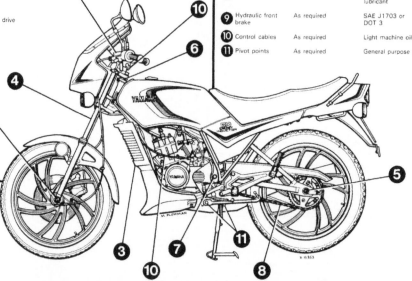

ROUTINE MAINTENANCE GUIDE

Routine maintenance

Periodic routine maintenance is a continuous process which should commence immediately the machine is used. The object is to maintain all adjustments and to diagnose and rectify minor defects before they develop into more extensive, and often more expensive, problems.

It follows that if the machine is maintained properly, it will both run and perform with optimum efficiency, and be less prone to unexpected breakdowns. Regular inspection of the machine will show up any parts which are wearing, and with a little experience, it is possible to obtain the maximum life from any one component, renewing it when it becomes so worn that it is liable to fail.

Regular cleaning can be considered as important as mechanical maintenance. This will ensure that all the cycle parts are inspected regularly and are kept free from accumulations of road dirt and grime.

Cleaning is especially important during the winter months, despite its appearance of being a thankless task which very soon seems pointless. On the contrary, it is during these months that the paintwork, chromium plating, and the alloy casings suffer the ravages of abrasive grit, rain and road salt. A couple of hours spent weekly on cleaning the machine will maintain its appearance and value, and highlight small points, like chipped paint, before they become a serious problem.

The various maintenance tasks are described under their respective mileage and calendar headings, and are accompanied by diagrams and photographs where pertinent.

It should be noted that the intervals between each maintenance task serve only as a guide. As the machine gets older, or if it is used under particularly arduous conditions, it is advisable to reduce the period between each check.

For ease of reference, most service operations are described in detail under the relevant heading. However, if further general information is required, this can be found under the pertinent Section heading and Chapter in the main text.

Although no special tools are required for routine maintenance, a good selection of general workshop tools is essential. Included in the tools must be a range of metric ring or combination spanners, a selection of crosshead screwdrivers, and two pairs of circlip pliers, one external opening and the other internal opening. Additionally, owing to the extreme tightness of most casing screws on Japanese machines, an impact screwdriver, together with a choice of large or small cross-head screw bits, is absolutely indispensable. This is particularly so if the engine has not been dismantled since leaving the factory.

Pre-operation checks

Yamaha recommend that the checks shown in the accompanying table should be carried out each time the machine is used. Whilst this is sound advice it will obviously depend upon the use to which the machine is put. Few owners will feel inclined to carry out the complete sequence on every occasion. It is suggested that the list is tailored to individual usage but that all items are checked with reasonable frequency and before any long journeys.

(a) Front brake. Check brake operation, free play and fluid level.
(b) Rear brake. Check brake operation and free play.
(c) Clutch. Check clutch operation and lever free play.
(d) Throttle. Check for full and free operation of the twist grip.
(e) Coolant. Check coolant level.
(f) Engine oil. Check the oil level in the oil tank.
(g) Transmission oil. Check the transmission oil level.
(h) Final drive chain. Check the chain adjustment and lubrication.
(i) Wheels and tyres. Check the wheels for damage and the tyres for damage or wear. Check the tyre pressures.
(j) Fittings and fasteners. Check all items for security and damage.
(k) Lights. Check that all illumination and warning lights are operating correctly.
(l) Petrol. Check that the level is sufficient to complete your journey or to get to a filling station.

Weekly or every 200 miles (300 km)

This is where the proper procedure of routine maintenance begins. The daily checks serve to ensure that the machine is safe and legal to use, but contribute little to maintenance other than to give the owner an accurate picture of what item needs attention. However, if done conscientiously, they will give early warning, as has been stated, of any faults which are about to appear. When performing the following weekly maintenance tasks, therefore, carry out the daily checks first.

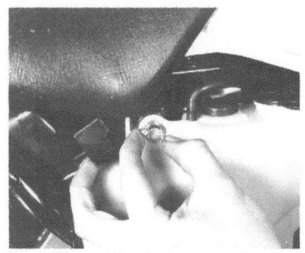

Release wing bolt and hinge oil tank outwards RD125

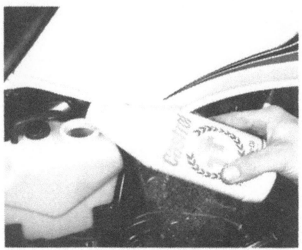

Top-up using high-quality two stroke oil

DT125 expansion tank is in front of oil tank ...

... while RD125 expansion tank is next to radiator

1 Topping up the oil tank

Engine lubrication is by pump injection fed by oil carried in a frame-mounted oil tank fitted behind the right-hand side panel. Although the machine is equipped with a low oil level warning lamp which will indicate the need for refilling it is better to check the oil level regularly and top up as required. Remove the plastic side panel to expose the oil tank. The tank is constructed from a translucent white plastic which allows the oil level to be seen quite easily. To top up, release the single wing nut and allow the tank to pivot outwards (RD125 LC models only). On DT125 LC models the seat must be removed to gain access to the filler cap. Pull off the filler cap and add oil as required. Fill the tank to the base of the filler neck and no further, to prevent spillage, using only good quality oil designed for two-stroke injection systems. Ensure that the filler cap is refitted securely.

2 Checking the coolant level

The cooling system is of the semi-sealed type and is unlikely to require frequent topping up. This does not mean however, that regular checks should be neglected. The system employs an expansion tank located next to the oil tank on DT125 LC models, and next to the radiator on RD125 LC models. This allows room for expansion when the engine becomes hot, the displaced water being drawn back into the radiator when it cools.

To check the level, remove the right-hand side panel or radiator cover, as applicable. The reservoir tank has upper and lower level lines marked on its side, and the coolant level must be between these marks with the engine cold.

The mixture is a mixture of 50% soft or distilled water and 50% ethylene glycol antifreeze with corrosion inhibitors for use in aluminium engines. For topping up purposes, distilled water is best, or soft water where this is available. Those living in hard water areas should avoid the use of hard tap water because of the risk of scaling in the cooling system. **Clean** rainwater may be used as another alternative. When topping up note that while the tank has a capacity of 250 cc (0.44 pint) on RD125 LC models and 110 cc (0.19 pint) on DT125 LC models, the amount required to raise the level from the 'Low' to the 'Full' mark is only 180 cc (0.32 pint) on RD125 LC models and 90 cc (0.16 pint) on DT125 LC models. The appropriate amount of distilled water should be kept prepared in a clean container so that it is ready for use. Remember that the level must be checked with the engine cold, it will vary with the temperature of the engine.

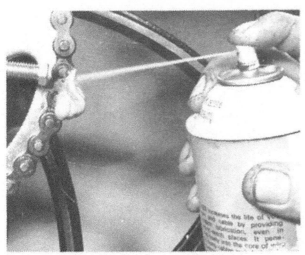

Use aerosol chain lubricant at frequent intervals

Check chain tension in position shown – note tensioner pressed away from chain – DT125 only

Use index marks to preserve wheel alignment or ...

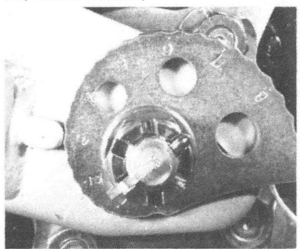

... on DT125 machines, use the number provided for the same purpose

3   Final drive chain

Check that the final drive chain is properly lubricated and adjusted. Although it is not likely to be needed at each weekly check, the full procedure is given here for easy reference. For any further details see Chapter 6, Section 19.

Lubrication is most effectively accomplished by the use of a special chain grease such as Linklyfe or Chainguard. This is, however, a long and potentially messy process which should be made at intervals of 500 – 1000 miles depending on the use to which the machine is put. A better solution for daily maintenance is the use of one of the many proprietary chain greases applied with an aerosol can. This can be applied very quickly, while the chain is in place on the machine, and makes very little mess. It should be applied at least once a week, and daily if the machine is used in wet weather conditions. If the roller surfaces look dry, then they need lubrication. Engine oil can be used for this task, but remember that it is flung off the chain far more easily than grease, thus making the rear end of the machine unnecessarily dirty, and requires more frequent application if it is to perform its tasks adequately. Also remember that surplus oil will eventually find its way on to the tyre, with quite disastrous consequences.

To check the chain tension, wedge the tensioner blade clear of the chain (DT125 LC models only), push the machine off its stand and along the ground, testing the chain tension at points all along the length of the chain until the tightest spot is found. The check is made on the lower run of the chain, midway between the two sprockets. When the chain's tightest spot is at this point, measure the total amount of vertical free play in the chain. There should be 35 – 40 mm (1.4 – 1.6 in) on RD125 LC models, and 45 – 55 mm (1.8 – 2.2 in) on DT125 LC models. If adjustment is necessary, remove the split-pin from the wheel spindle nut and slacken the nut by just enough to permit the spindle to be moved. Draw the spindle back the amount required by tightening the drawbolt adjusters on RD125 LC models or by rotating the snail cams on DT125 LC models. Use the reference marks or numbers provided to ensure that the wheel spindle is moved back by exactly the same amount on each side, thus preserving accurate wheel alignment. Tighten the rear wheel spindle nut to a torque setting of 8.5 kgf m and fit a new split-pin, spreading its ends securely. Check and reset if necessary the rear brake adjustment and the stop lamp rear switch height and check that all nuts and bolts are securely fastened and that the rear wheel is free to revolve easily. On DT125 LC models, remember to free the chain tensioner so that it springs up to bear against the chain, and to examine the tensioner pad, the guide roller, and the chain guides for wear. As described in Section 19 of Chapter 6, these components must be renewed if seriously worn to protect the chain and to prolong its life as much as possible.

4   Adjusting the brakes

The hydraulic front brake of the RD125 LC model requires no adjustment. Check that the fluid level in the master cylinder reservoir is above the 'Lower' level mark. Remember that in a hydraulic brake, the level will sink only gradually, as the pads wear and more fluid is needed to maintain pressure.

A rapid drop in the fluid level is indicative of a leak somewhere in the system; immediate attention should be given to curing the leak. Refer to Chapter 6, Sections 6 – 12 as appropriate. For the purposes of routine maintenance, it will suffice to keep the fluid level topped up above the 'Lower' mark on the reservoir. Never overfill the reservoir, as fluid spillage will inevitably result. Top up using only hydraulic fluid of the correct specification; this is SAE J1703 (UK) or DOT 3 (US). The drum brake fitted to the DT125 LC model is adjusted at the cable adjuster on the front wheel. Slacken the locknut and turn the adjuster nut as necessary to give 5 – 8 mm (0.20 – 0.30 in) of free play, measured between the butt end of the lever and its handlebar clamp when the brake is firmly applied. Use the handlebar adjuster for quick adjustments only.

The rear drum brake fitted to both models is adjusted by means of a single nut at the rear end of the brake operating rod. Turn the nut clockwise to reduce the free play, if necessary, to measure 20 – 30 mm ($\frac{3}{4}$ – $1\frac{1}{4}$ in) at the brake pedal tip. Check that the rear wheel rotates freely and that the stop lamp is functioning properly. Remember that the stop lamp switch height must be adjusted every time the rear brake adjustment is altered. To adjust the switch height, turn its plastic sleeve nut as required until the stop lamp bulb lights when the brake pedal free play has been taken up and the rear brake shoes are just beginning to engage the brake drum.

Complete brake maintenance by oiling all lever pivot points, all exposed lengths of cable, cable nipples and the rear brake linkage with a few drops of oil from a can. Remember not to allow excessive oil on to the operating linkage, in case any surplus should find its way into the brake drum or onto the tyre.

5   Checking the tyre pressures

It is essential that the tyres are kept inflated to the correct pressure at all times. Under- or over-inflated tyres can lead to accelerated rates of wear, and more importantly, can render the machine inherently unsafe. Whilst this may not be obvious during normal riding, it can become painfully and expensively so in an emergency situation, as the tyres' adhesion limits will be greatly reduced.

Check the tyre pressures with a pressure gauge that is known to be accurate. Always check the pressure when the tyres are cold. If the machine has travelled a number of miles, the tyres will have become hot and consequently the pressure will have increased. A false reading will therefore result.

It is recommended that a small pocket gauge is purchased and carried on the machine, as the readings on garage forecourt gauges can vary and may often be inaccurate.

The pressures given are those recommended for the tyres fitted as original equipment. If replacement tyres are purchased, the pressure settings may vary. Any reputable tyre distributor will be able to give this information.

Tyre pressures (cold)

	RD125 LC	DT125 LC
Front – solo	26 psi (1.82 kg/cm²)	18 psi (1.26 kg/cm²)
Rear – solo	28 psi (1.96 kg/cm²)	22 psi (1.54 kg/cm²)
Front – pillion	26 psi (1.82 kg/cm²)	22 psi (1.54 kg/cm²)
Rear – pillion	32 psi (2.24 kg/cm²)	26 psi (1.82 kg/cm²)

For continuous high speed riding on RD125 LC models only, the rear tyre pressure should be increased to:

| Rear – solo | 32 psi | (2.24 kg/cm²) |
| Rear – pillion | 40 psi | (2.80 kg/cm²) |

Fluid level can be seen in rear face of master cylinder

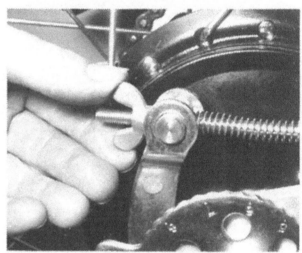

Rear brake is adjusted via nut at rear of brake rod

Check tyre pressures with an accurate gauge

Check tread wear and remove any stones or sharp objects

Electrolyte levels can be checked via transparent case

At the same time, examine the tyres themselves. Look for damage, especially splitting of the sidewalls. Remove any small stones or other road debris caught between the treads. Any such debris will work its way into the tyre and may penetrate the inner tube, thus causing rapid deflation and a resultant loss of control. The tendency to pick up road debris is much more pronounced in the rear tyre and this, therefore, should be checked particularly carefully. The depth of tread remaining should also be measured in view of both the legal and safety aspects. It is vital to keep the tread depth above the UK legal limit of 1 mm of depth over three-quarters of the tread breadth around the entire circumference of the tyre. Many riders however, consider nearer 2 mm to be the limit for secure roadholding, traction, and braking, especially in adverse weather conditions.

6 Checking the battery

The battery is located behind the left-hand sidepanel on RD125 LC models and behind the right-hand sidepanel on DT125 LC models. Remove the appropriate panel and check that the electrolyte levels are between the upper and lower level marks on the battery casing. Top up to the upper level if necessary. Check that the connections are clean and tight and that the vent tube is free from kinks or blockages.

7 Checking the front and rear suspension

Ensure that the front forks operate smoothly and progressively by pumping them up and down whilst the front brake is held on. Any faults revealed by this check should be investigated further, because any deterioration in the handling of the machine can have serious consequences if left unremedied. Check the condition of the fork stanchions on RD125 LC models. As with most current production machines, the fork stanchions are left exposed in the interests of fashion, and are thus prone to damage from stone chips or abrasion. Any damage to the stanchions will lead to rapid wear of the fork seals and can only be cured by renewing the stanchions. This is both costly and time consuming, so it is worth checking that the area below each dust seal is kept clean and greased. Remove any abrasive grit which may have accumulated around the dust seal lip. The above problems can be eliminated by fitting fork gaiters, these being available from most accessory stockists. On DT125 LC models, check that the gaiters are undamaged and securely retained. At regular intervals, the gaiter should be pulled off the top of the fork lower leg and the stanchion checked for corrosion. Remove any dirt that may have got inside the gaiter and apply a smear of grease to the stanchion surface to prevent the onset of corrosion. Check that the rear

suspension works smoothly, that all nuts and bolts are securely fastened, and that there are no traces of free play at the sub-frame pivot.

8 General checks and lubrication

Check around the machine, looking for loose nuts, bolts or screws, retightening them as necessary. Check the stand and lever pivots for security and lubricate them with light machine oil or engine oil. Make sure that the stand springs are in good condition.

It is advisable to lubricate the handlebar switches and stop lamp switches with WD40 or a similar water dispersant lubricant. This will keep the switches working properly and prolong their life, especially if the machine is used in adverse weather conditions. Apply a few drops of engine oil to the exposed inner portion of each control cable. This will prevent the cables drying up between the more thorough lubrications that should be carried out at each 4000 mile/6 monthly service.

Monthly or every 1000 miles (1500 km)

Complete the operations listed under the weekly/200 mile service heading, then carry out the following:

1 Checking the transmission oil

Although the transmission oil level is unlikely to drop between normal oil changes, a regular check at this interval will ensure that any sudden drop is noticed before damage occurs. It is worth noting that a loss of oil can be due only to external leakage, which should be immediately obvious, or to oil being drawn into the crankcase via the crankshaft right-hand main bearing oil seal. Any such loss of oil should be investigated promptly.

To check the oil level, the machine must be thoroughly warmed up to normal operating temperature, preferably taking it on a short trip. Stand the machine on its wheels on level ground and support it so that it is absolutely vertical. The level of the oil can be seen in the sight glass set in the crankcase right-hand cover immediately beneath the kickstart lever, and should fall between the maximum and minimum level marks in the side of the glass. Add oil as necessary through the filler plug orifice set in the top of the crankcase right-hand cover, using only a good quality SAE 10W 30 SE engine oil.

Note that if the machine is used only for short journeys, light grey-brown deposits will appear in the sight glass due to condensation forming in the oil which cannot warm up properly under these conditions. A journey of approximately 50 – 100 miles will remove the deposits by allowing the oil to warm up thoroughly and restore a clear brown appearance to the oil in the sight glass. If this does not work and is accompanied by a steady loss of coolant, particularly on a high mileage machine, the water pump should be dismantled for the pump seal to be renewed as described in the relevant Sections of Chapter 2, as the two symptoms together indicate that coolant is finding its way into the transmission oil.

2   Lubricating the chain

As stated elsewhere in this Manual, it is recommended that the drive chain be lubricated using a special chain grease at regular intervals, backed up by frequent applications of lubricant from an aerosol. If Linklyfe, Chainguard, or similar is to be employed for regular lubrication, remove the chain by disconnecting it at its split link, wash it thoroughly in a petrol/paraffin mixture and allow it to dry. Heat the grease according to the manufacturer's instructions and immerse the chain in it. Swill the chain gently around to allow the grease fully to penetrate the rollers and withdraw it removing as much of the surplus as possible. When refitting the chain, ensure that the connecting link spring clip is replaced with its closed end facing the direction of normal travel of the chain. Note that this need only be done at intervals of 500 – 1000 miles, not every week. The actual interval is completely dependent on the use to which the machine is put, and the weather conditions in which it is used. If in doubt, remember that a tin of chain grease and some time spent applying it cost considerably less than a new chain and sprockets.

3   Cleaning the air filter – DT125 LC only

It is recommended that if a DT125 LC model is used regularly on off-road trips, the air filter is cleaned at far more frequent intervals than normal. It is difficult to state a specific mileage/time interval for air filter cleaning, since the task is entirely dependent upon the use to which the machine is put. For example, if the machine is taken for a very dusty trail riding session in mid summer, the air filter should be cleaned as soon as possible after returning home. On the other hand, if the machine is ridden principally on the road, the air filter cleaning interval can be extended to the same as the RD125 LC model. For this reason the operation is included at this point as a reminder to DT125 LC owners, but the description of the operation is included at the 2000 mile/3 monthly service.

4   Checking the battery

On both of the machines described in this Manual, it is essential that the battery is maintained in excellent condition to prolong its life. In addition to the weekly check of the electrolyte level, the condition of the terminals should be examined. The exposed terminals employed on the battery fitted to these machines are prone to corroding, producing a variety of faults in the electrical system if allowed to go unchecked. Clean away all traces of dirt and corrosion, scraping the terminals and connections with a knife and using emery cloth to finish off. Remake the connections while the joint is still clean, and then smear the assembly with petroleum jelly (not grease) to prevent recurrence of the corrosion. Finish off by checking that the battery is securely clamped in its mountings and that the vent tube is quite clean and free from kinks or blockages.

Three monthly or every 2000 miles (3000 km)

Complete the operations listed under the previous mileage/time headings and then carry out the following:

1   Cleaning and resetting the spark plug gap

Remove and examine the spark plug, comparing its general condition with the   photographs in Chapter 4 to gain an impression of general running conditions. The standard plug is

an NGK BR8ES, this being considered the best type for general use. If a lot of hard riding is done the standard plugs may show signs of overheating, with the porcelain insulator nose appearing blistered and the electrodes heavily eroded. Assuming that the ignition and carburation systems are set accurately, change the plug for the next coldest grade of that type, NGK BR9ES or equivalent.

Conversely, if the machine is normally used for short trips or is ridden gently there may be some inclination towards plug fouling giving the plug a sooty or oily appearance. Again the fuel and ignition systems can affect this aspect, but if all is in order change to the next hottest grade, NGK BR7ES or equivalent.

If the plug grade is correct it will now be necessary to clean and adjust the electrode gap. Note that if the electrodes show signs of excessive wear the plugs should be renewed, otherwise the opposing faces of the electrodes can be cleaned up and burnished using fine abrasive paper. It is inadvisable to leave abrasive particles on the plug end in case they find their way into the bores and cause scoring. To this end the plug should be washed in petrol to remove residual oil and the electrodes cleaned with the plug facing downwards.

Check the electrode gap using feeler gauges and if necessary adjust it to 0.7 – 0.8 mm (0.028 – 0.032 in) by bending the outer (earth) electrode. On no account attempt to bend the centre electrode – this will only result in a cracked insulator nose. Check that the plug threads are clean and wipe a trace of graphite or molybdenum grease on them to prevent their sticking in the cylinder head threads. Fit the plugs by hand, then tighten them just enough to obtain an effective seal. The specified torque figure is 2.0 kgf m (14.5 lbf ft).

2   Cleaning the air filter element

The air filter element on the Yamaha LC models is of the oiled foam type and is housed in a moulded plastic casing beneath the fuel tank on RD125 LC machines, and behind the left-hand side-panel on the DT125 LC. To gain access to the filter element on DT125 LC machines, remove the sidepanel which is retained by a single screw in the middle of the lower edge, by a prong at the front of the lower edge, and by a clip at the upper rear edge which engages around the frame tube. When the sidepanel has been removed the air filter casing will be seen immediately below the exhaust pipe. Slacken and remove the single screw which retains the element cover, then withdraw the cover, the plastic supporting frame, and the two parts of the element itself, noting that the orange fibrous part must be refitted facing outwards, and must be outside the grey

Oiled foam element can be removed for cleaning – RD125

Remove left-hand sidepanel to reach air filter – DT125

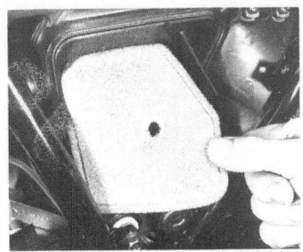

DT125 is fitted with two-piece filter element. Do not omit either part

foam element. On RD125 LC models, the sidepanels and the seat and tank must be removed as follows. Unlock the seat and lift it clear of the frame. Turn the fuel tap to the 'Off' position and pull off the fuel pipe. Remove the single rubber mounted bolt at the rear of the tank. Raise the rear of the tank slightly and then ease the tank backwards to disengage it from its mounting rubbers at the steering head. Remove the air filter cover by unscrewing its three retaining screws. The plastic frame flat foam element can now be lifted away.

Wash the element in clean petrol to remove the old oil and any dust which has been trapped. When it is clean, wrap it in some clean rag and gently squeeze out the remaining petrol. The element should now be left for a while to allow any residual petrol to evaporate. Soak the cleaned element in engine oil and then squeeze out any excess to leave the foam damp but not dripping. Refit the element, ensuring that the cover seals properly and using a thin smear of grease to help prevent the entry of dirt. The sidepanel (DT125 LC only) or the petrol tank, seat, and sidepanels (RD125 LC only) can then be refilled.

Note that a damaged element must be renewed immediately. Apart from the risk of damage from ingested dust, the holed filter will allow a much weaker mixture and may lead to overheating or seizure. It follows that the machine must never be used without the filter in position.

3 Checking the clutch adjustment

While the clutch can be adjusted at two points, at the operating mechanism and at the cable itself, it will suffice for the purposes of Routine Maintenance to regard the operating mechanism as set and to make all normal adjustments using the cable length adjusters. The clutch is adjusted correctly when the pointed end of the operating mechanism lever aligns exactly with the raised index mark on the crankcase top when the handlebar lever is released, and when there is 2 – 3 mm (0.08 – 0.12 in) of free play in the cable at the same time, the free play being measured between the butt end of the clutch handlebar lever and its handlebar clamp.

Check that the operating mechanism is aligned correctly and reset the cable free play, if necessary, using the adjusters provided. Note that while the DT125 LC model is fitted with two cable adjusters, the in-line adjuster must be used first and the handlebar lever adjuster used only to achieve the correct setting.

In the event that adjustment is no longer possible with the cable adjusters, the crankcase right-hand cover must be removed as described in Section 8 of Chapter 1 and the operating mechanism reset as described in Section 35 of the

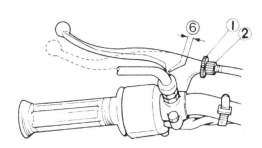

Handlebar lever adjustment

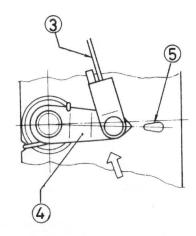

Operating mechanism adjustment

Clutch adjustment

1	Lock nut	4	Operating lever
2	Handlebar lever adjuster	5	Raised index mark
3	Operating cable	6	Lever free play

same Chapter. Screw in the cable adjusters to achieve maximum free play in the cable, then unscrew them to bring the pointed end of the operating lever into exact alignment with the raised index mark on the crankcase top, thus achieving the initial setting for the adjustment of the operating mechanism. If the clutch cable outer should become so compressed that correct clutch adjustment can no longer be made, the cable must be renewed to ensure correct and safe operation of the clutch itself.

4 Changing the transmission oil

Before the old oil is drained, the machine must be fully warmed up to normal operating temperature so that the oil is thinned and will drain faster, taking any particles of swarf and dirt with it, thus making the oil change far more effective. Support the machine upright on level ground, place a container of suitable size underneath the engine/gearbox unit, then remove the filler and drain plugs. Allow the oil to drain as completely as possible.

Check the condition of the drain plug sealing washer, renewing it if necessary, then refit the drain plug, tightening it to a torque setting of 2.0 kgf m (14.5 lbf ft). Add 550 cc (0.97 pint) of good quality SAE10W 30 SE engine oil through the filler plug orifice then refit the filler plug and start the engine. Allow it to warm up then stop it and leave the machine for a few minutes to allow the oil level to settle. With the machine standing on its wheels, vertically upright on level ground, the oil level should fall between the maximum and minimum level marks on the periphery of the sight glass. Add more oil or drain off the surplus to achieve this.

Once the oil level is known to be correct, refit the filler plug and check that both plugs are securely tightened. Remove any surplus oil and check subsequently that no oil leaks appear.

5 Checking carburettor adjustment

If rough running of the engine has developed, some adjustment of the carburettor pilot setting and tick-over speed may be required. If this is the case refer to Chapter 3, Section 8 for details. Do not make these adjustments unless they are obviously required, there is little to be gained by unwarranted attention to the carburettor. Complete carburettor maintenance by removing the drain plug on the float chamber, turning the

petrol on, and allowing a small amount of fuel to drain through, thus flushing any water or dirt from the carburettor. Refit the drain securely and switch the petrol off.

Once the carburettor has been checked and reset if necessary, the throttle cable free play can be checked. Open and close the throttle several times, allowing it to snap shut under its own pressure. Ensure that it is able to shut off quickly and fully at all handlebar positions. Check that there is 4 – 7 mm ($\frac{1}{8}$ – $\frac{1}{4}$ in approx) free play measured around the circumference of the inner flange of the rubber twistgrip. If not, use the adjuster at the twistgrip to achieve the correct setting, completing the operation, if necessary, with the adjuster on the carburettor top. Open and close the throttle again to settle the cable and to check that adjustment is not disturbed.

6 Checking the oil pump settings

The engine depends upon the Autolube pump to deliver the required amount of oil at any given throttle setting. It follows that it is important that the pump adjustment is maintained

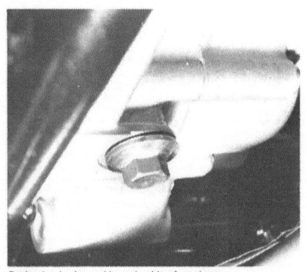

Drain plug is situated in underside of crankcase

Use good quality engine oil to fill gearbox ...

... to correct level

within the specified limits if it is wished to avoid the untimely demise of the engine. The pump settings should be checked **after** the throttle cable free play and carburettor adjustment have been adjusted as described above. The oil pump checks can be divided into two parts: checking the minimum stroke setting, and checking that the oil pump is synchronised correctly with the carburettor. The first step in each case is to remove the oil/water pump cover from the front of the crankcase right-hand cover.

To check the minimum stroke adjustment, start the engine and allow it to idle. Observe the front end of the pump unit, where it will be noticed that the pump adjustment plate moves in and out. When the plate is out to its fullest extent, stop the engine and measure the gap between the plate and the raised boss of the pump pulley using feeler gauges. Do not force the feeler gauge into the gap – it should be a light sliding fit. Make a note of the reading, then repeat the procedure several times. The largest gap is indicative that the pump is at its minimum stroke position. If the pump is set up correctly, the gap found should be 0.20 – 0.25 mm (0.008 – 0.010 in). If the gap is found to be incorrect it will be necessary to remove the nut from the end of the pump and to withdraw the spring washer and the adjustment plate so that the clearance can be reset with shims. If necessary shims can be purchased from Yamaha dealers in thicknesses of 0.3, 0.5 and 1.0 mm (0.0118, 0.0197 and 0.0394 in). Add or subtract shims as required, refit the adjustment plate, spring washer, and the retaining nut and recheck the minimum stroke measurement. In practice, this part of the oil pump setting is unlikely to require frequent adjustment.

When the pump minimum stroke setting is known to be correct, the oil pump cable adjustment should be checked. This applies to RD125 LC models only, as the DT125 LC is fitted with a throttle cable junction box that compensates automatically for wear or stretch in the cable and renders unnecessary all the usual inspection and adjustment. It is, however, recommended that the throttle is opened and closed while the oil pump pulley is watched closely to see that it revolves smoothly and easily throughout its full movement. If any stiffness or other sign of trouble is noticed, the cable and junction box should be examined closely as described in Section 2 of Chapter 3. If the movement is satisfactory, apply a few drops of oil to the exposed length of the inner cable, and to the moving parts of the oil pump itself, then refit the outer cover, taking especial care to check the routing and condition of the oil feed pipes as described in Section 17 of Chapter 3. Be careful to carry out the modification described in the last two paragraphs of that Section if it is found to be necessary.

For RD125 LC models only, the oil pump cable adjustment is checked by observing the relationship between the pump plunger pin and the pulley when all traces of free play have been eliminated from the cables. There are various marks on the outer face of the pulley, one of which is a round indentation drilled in the pulley at about 90° from the cable end nipple (see accompanying photograph). This is the pulley alignment mark to be used; all others should be ignored. Slowly rotate the twistgrip until all free play in the cables has just been taken up, and check that the pump plunger pin aligns exactly with the pulley alignment mark. Adjust the oil pump cable as necessary by rotating the cable adjuster set in the crankcase right-hand cover. Tighten the adjuster locknut, replace the rubber sleeve over the adjuster and open and close the throttle fully to settle the cables. Check that the adjustment has not altered and reset it if necessary. Lubricate the exposed length of the oil pump cable inner and the moving parts of the oil pump with a few drops of oil, then refit the oil/water pump cover, taking care that the oil feed pipes are correctly routed and are not trapped or kinked.

7 Checking the exhaust system

While removable baffles are not fitted to either of the machines described in this Manual, it will be advisable to check

Alignment mark should coincide with plunger pin at idle (see text)

the exhaust system at regular intervals so that the build-up of carbon deposits and of oily sludge can be dealt with as soon as possible.

First check that the exhaust system is securely fastened and that there are no exhaust leaks. Renew the exhaust port gasket and, on DT125 LC models only, the rubber seal between the two parts of the system, if leaks are found at either of these points. Some idea of the condition of the exhaust can be gained by looking at the tail pipe. If the mixture and oil pump settings are correct, and the machine is ridden normally, there should be a thin film of sooty black carbon with a slight trace of oiliness. If the machine is ridden hard, the deposits will tend to be a lighter colour, almost grey, and there will be no traces of oil. If any of the above is found, and the spark plug electrode colour, carburettor settings, air filter condition, and the oil pump settings are known to be satisfactory from the previous Routine Maintenance operations, then the carbon deposits inside the exhaust system will be kept to a minimum and will take a long time to build up to the point where a full decarbonising operation is necessary. On the other hand, if the rear of the exhaust is excessively oily and the carbon deposits rather thicker than those described above, then either one of the settings mentioned above is incorrect, causing the engine to run inefficiently so that it produces too much waste in the form of carbon, or the machine is being ridden too slowly which produces the same symptoms. If the engine settings are known to be satisfactory thanks to the previous Routine Maintenance operations and the carbon build-up is caught at an early stage, then the simplest and most satisfying method of cleaning the exhaust is to take the machine on a good hard run until the exhaust is too hot to touch and the excessive smoke produced by the dispersal of the carbon/oil build-up has disappeared. This is a well known trick employed by many mechanics to restore lost power, and can have quite dramatic results. If the build-up has been allowed to develop too much, however, the complete exhaust must be removed from the machine and decarbonised using one of the methods described in Section 15 of Chapter 3.

8 Checking brake wear

In addition to the weekly adjustment of the brakes, they must be checked to assess the amount of friction material remaining on the brake pads or brake shoes. In all cases the task is simplified by the provision of external wear indicators.

The brake pads fitted to the front brake of RD125 LC models can be checked by looking at the caliper from behind,

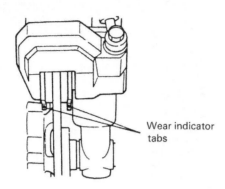

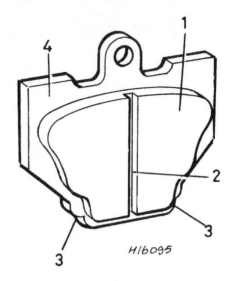

Front brake pad wear check

Front brake pad

1 *Friction material*	3 *Wear indicator tabs*
2 *Central groove*	4 *Metal backplate*

and then from in front of the fork right-hand lower leg. The wear indicator tabs formed by the outermost corners of the metal backing of each pad will indicate the need for pad renewal when they touch the disc, although for the reasons given in Section 7 of Chapter 6, it is recommended that the central groove of each pad be used instead of the tabs. When the friction material is worn down to the point that the central groove disappears, the pads should be considered worn out and renewed. If at any time a build-up of dirt appears around the pads or a pronounced squealing noise on application of the brake indicates that foreign matter is embedded in the friction material, the pads must be removed for cleaning. The task of removing and refitting the pads, whether for cleaning or for renewal, is described in Section 7 of Chapter 6.

In the case of the drum brakes fitted to the rear wheel of RD125 LC models or to both wheels of DT125 LC models, a pointer is fixed to the brake camshaft to serve as a wear indicator in conjunction with an arc cast in the brake backplate; the indicator being used as follows. Apply the brake hard and look at the pointer; if it is pointing at the end of the arc, or beyond it, the brake shoes must be considered worn out and renewed immediately, as described in the relevant Sections of Chapter 6.

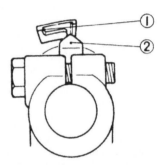

Drum brake wear indicator plate

1 *Wear limit*	2 *Indicator*

Six monthly or every 4000 miles (6000 km)

Complete the tasks listed under the previous service headings and then carry out the following:

1   Decarbonising the engine

Decarbonisation, or 'decoking' as it is often called, is an essential periodic maintenance task on all two-stroke engines. Because oil is deliberately burnt during combustion, the rate of carbon build up is greater than in a four-stroke engine and in time performance is impaired. If the carbon is not removed it may ultimately build up to the point where localised over-heating during combustion causes seizure or a holed piston.

The decarbonisation operation is rather more complicated on the LC models than on conventional air-cooled two-strokes because it is necessary to drain the cooling system. Although Yamaha recommend that it is carried out at 4000-mile intervals it has been found in practice that these models run cleaner than many two-strokes given that excessive short town journeys are avoided. It may be advisable to reduce the frequency of decarbonisation somewhat in view of the amount of work involved. It should also be noted that many Yamaha Dealers feel

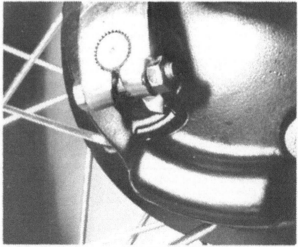

Brake shoes are worn out when pointer reaches outer end of arc

that this job should only be undertaken by experienced mechanics, so novices may be better advised to leave this work to the local Yamaha Dealer. For the more ambitious, proceed as follows.

Before starting work make sure that new cylinder head, cylinder base and exhaust port gaskets are to hand. It is also worth checking the hoses and purchasing replacements for any that appear perished or cracked. A large bowl or bucket will be needed to catch the coolant when this is drained, but note that if this is close to two years old it should be discarded and fresh coolant added during reassembly.

Place the machine on its stand, leaving adequate room to work on both sides. Slacken and remove the screw which retains the radiator shroud to the radiator. Lift the shroud away so that the radiator cap can be removed. Check that the engine is quite cold, then slowly remove the cap.

It will now be necessary to drain the system via the two drain plugs. Obtain a plastic bowl or bucket to catch the coolant and make up some sort of chute to keep the coolant off the engine casings. A piece of stiff card will suffice for this purpose. Remove each of the drain plugs in turn and allow the water to drain thoroughly, then disconnect the radiator bottom hose and drain off any residual coolant. It is important to ensure that none remains in the system so that there is no risk of water finding its way into the engine when the cylinder head is removed.

Pull off the temperature gauge sender lead and disconnect the radiator top hose from the adaptor stub. Release the two screws which secure the coolant pipe to the cylinder barrel, then carefully lift the coolant pipe out of the crankcase right-hand cover, plugging the waterway with rag to prevent dirt getting into the water pump. With reference to the relevant Sections of Chapter 3 where necessary, remove the seat, the sidepanels, the petrol tank, and the exhaust system, then remove the carburettor, the YEIS chamber and the reed valve block.

The cylinder head is retained by four chromed dome nuts, each with a plain washer underneath it. Slacken the nuts progressively by about one turn at a time, working in a diagonal sequence until all are removed. Before attempting to remove the cylinder head have some tissue or absorbent rag nearby to soak up any spilt coolant.

Break the seal between the cylinder barrel and the head by tapping around the latter with a soft-faced mallet or by judicious levering. Lift the cylinder head clear of the barrel and wipe up any drops of coolant, especially those which find their way into the bore. Remove any residual coolant by inverting and shaking the cylinder head.

Remove the carbon deposits from the inside of the cylinder head using a hardwood or soft metal scraper to avoid scoring the soft aluminium alloy casting. Although not strictly essential it is a good idea to get the smoothest possible finish on the inside of the combustion chamber because this will improve gas flow and reduce the rate at which carbon builds up in the future. If it is felt worthwhile, some time spent with fine grades of abrasive paper and then metal polish can produce a mirror finish.

To clean the piston and exhaust port properly, the cylinder barrel should be removed. Partially remove the barrel until there is sufficient room to pack some clean rag around the connecting rod to catch any debris which might drop down as the piston emerges from the bore. Finish removing the barrel, supporting the piston as it emerges from its bore.

Examine the piston rings, piston, and bore for signs of damage, taking remedial action as required (see Section 18 and 19 of Chapter 1 for details). If all seems well, proceed as follows.

Carefully scrape away any carbon deposits from the exhaust port, taking care not to burr the edge of the bore – any nicks at this point could catch and break the piston rings. Try to prevent any of the carbon debris from entering the inlet or transfer ports. This is best accomplished by packing the cylinder bore with clean rag.

Remove the carbon from the piston crown using a hardwood or soft metal scraper. Like the cylinder head, the piston will benefit from a polished finish, but this is by no means essential. Check that the piston rings are free in their grooves, and if necessary remove the rings and clean them. When decarbonisation is complete, clean the top of the crankcase and the rag around the connecting rods.

When reassembling the engine, make sure that the gasket faces are clean and dry and always use new gaskets. If the old gaskets are re-used it is likely that leakage will develop, and in some cases this will allow coolant to be forced into the crankcase. Not surprisingly, the main and big-end bearings will not respond well to water cooling and will soon corrode and wear out. The piston should be coated with oil prior to insertion in the bore. Check that the ring ends are properly located by the pins in the ring grooves, then feed each one into the bore by hand. A tapered lead in makes this operation reasonably easy. Once the rings are engaged in the bore, remove the rag from the crankcase mouth and push the barrel carefully down to rest on the crankcase. Refit the clutch cable bracket on the cylinder left-hand rear mounting stud, then fit the four flanged nuts and tighten them in a diagonal sequence and in progressive stages to a torque setting of 2.8 kgf m (20 lbf ft).

Refit the coolant pipe and reed valve block, using new gaskets and seals where appropriate, then fit a new cylinder head gasket followed by the cylinder head itself. Tighten the four cylinder head nuts, again progressively and in a diagonal sequence, to a torque setting of 2.2 kgf m (16 lbf ft).

Complete the remainder of the reassembly work by reversing the instructions followed on dismantling. Remember to refill the cooling system and to bleed the oil pump to carburettor feed pipe, if necessary, then check that there are no coolant or oil leaks once the engine has been started again. Use a torque wrench to check that the cylinder barrel and head nuts are tightened to their correct respective torque settings once the engine has been warmed up and then allowed to cool down again. Finally, remember that the disturbed components will need time to settle down again, even if none have been renewed. Use the engine gently for the first 50 miles or so to allow this bedding in to take place properly.

2 Overhauling the cooling system

The cooling system should be checked for signs of leakage or damage and any suspect hoses renewed. This is best undertaken in conjunction with decarbonisation when many of the parts concerned will have been removed. The used coolant can be put back into the system if it is kept clean, but should be renewed every two years as a matter of course. For further information. refer to Chapter 2.

3 Renewing the spark plug

It is usually recommended that the spark plug is renewed as a precautionary measure at this stage. Always ensure that a plug of the correct type and heat range is fitted, and that the gap is set to the prescribed 0.7 – 0.8 mm (0.028 – 0.032 in) prior to installation. If the old plug is in reasonable condition, it can be cleaned and re-gapped and carried as an emergency spare in the toolbox.

4 Checking the ignition timing

It is recommended by the manufacturer that the ignition timing is checked at intervals of 4000 miles (6000 km). While this is commendable as a safety measure, it is felt that this is very much the counsel of perfection and is not necessary unless a loss of power or other symptoms developed by the engine indicates that a check of the ignition timing would be warranted. Should this be the case, the ignition timing checking procedure is given for both models in Section 9 of Chapter 4.

5   Cleaning the petrol tap filter

Switch the petrol tap to the 'Off' position and remove the filter level from the bottom of the petrol tap by unscrewing it with a close-fitting ring spanner, then pick out the sealing O-ring and the filter gauze. Remove all traces of dirt from the filter and filter bowl and check the condition of the sealing O-ring, renewing it if it is worn or damaged. Refit the filter gauze, the O-ring, and the filter bowl, tightening the bowl by just enough to nip the O-ring tight.

Note that if excessive traces of dirt or water appear in the filter bowl, the petrol tank must be drained, removed from the machine and flushed out, as described in Section 2 of Chapter 3, to prevent blockages in the fuel system. If such dirt or water gets through to the carburettor, it should be removed from the machine and cleaned out, as described in Sections 5 and 6 of Chapter 3.

Finally, examine the petrol feed pipe, checking that it is securely fastened at the tap and at the carburettor unions. Any signs of splits or hardening will mean that the pipe must be renewed to prevent any chance of petrol leaks.

6   Overhauling the brakes

If for any reason the brakes have not been stripped for examination, this must now be done. If no leaks are apparent in the hydraulic brake system of the RD125 LC model then it will suffice to remove the pads for cleaning and examination, this procedure being described in Chapter 6, Section 7. Check that full lever pressure is maintained. If there appears to be some sponginess present the hydraulic system must first be bled to ensure that any air is removed. This procedure is fully described in Section 12 of Chapter 6. If it is necessary to bleed the system again after a short interval, this indicates that air is getting into the system somewhere. If this is not indicated by a fluid leak, the master cylinder must be overhauled, followed by the caliper if necessary. This is described in Sections 10 and 11 of Chapter 6.

Overhauling the drum brakes fitted with either of the machines described in this Manual must be preceded by the removal of the wheel concerned, as described in the relevant parts of Sections 4 or 15 of Chapter 6. The brake components can then be dismantled, cleaned, checked for wear, and reassembled following the instructions given in Sections 13 and 17 of Chapter 6. It is important that moving parts such as the brake camshaft are lubricated with a smear of high melting-point grease on reassembly. Refer to the maintenance tasks described in Sections 9, 10, 11 and 12 of this service before refitting the wheels.

7   Greasing the controls and stand pivots

It is essential that the smooth, safe, operation of the controls is ensured by regular lubrication, which will serve two purposes. The first, and most important, is that a layer of grease will prevent the onset of corrosion which would cause the controls to become stiff and jerky in operation, and the second is that lubrication will minimise the wear that will inevitably occur on any bearing surface. In this case the wear would cause the controls to be sloppy in operation, rendering them awkward and imprecise to use.

Dismantle, clean and grease the handlebar lever pivots, the twistgrip, the rear brake pedal pivot, and, on RD125 LC models only, the gearchange linkage. For further information refer to the relevant Sections of Chapters 1, 3 and 5. Do not forget, on DT 125 LC models only, to apply a few drops of oil to the pivoting tips of the brake pedal and gearchange levers.

Similarly, grease the stand pivots, referring to the relevant Sections of Chapter 5, and remember to check that the footrest pivots are well lubricated.

8   Lubricating the control cables and instrument cables

In addition to the general lubrication made at each weekly service, the control cables must be examined thoroughly and

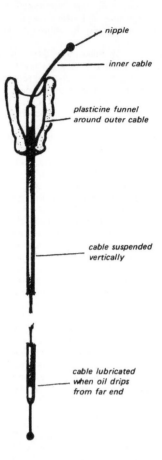

nipple

inner cable

plasticine funnel around outer cable

cable suspended vertically

cable lubricated when oil drips from far end

Oiling a control cable

properly lubricated to ensure that they remain in good order. Check the outer cables for signs of damage, then examine the exposed portions of inner cables. Any signs of kinking or fraying will indicate that renewal is required. To obtain maximum life and reliability from the cables they should be thoroughly lubricated. To do the job properly and quickly use one of the hydraulic cable oilers available from most motorcycle shops. Free one end of the cable and assemble the cable oiler as described by the manufacturer's instructions. Operate the oiler until oil emerges from the lower end, indicating that the cable is lubricated throughout its length. This process will expel any dirt or moisture and will prevent its subsequent ingress.

If a cable oiler is not available, an alternative is to remove the cable from the machine. Hang the cable upright and make up a small funnel arrangement using plasticine or by tapping a plastic bag around the upper end. Fill the funnel with oil and leave it overnight to drain through. Note that where nylon-lined cables are fitted, they should be used dry or lubricated with a silicone-based lubricant suitable for this application. On no account use ordinary engine oil because this will cause the liner to swell, pinching the cable. On DT125 LC models only, remember to inspect and to lubricate, if necessary, the throttle/oil pump cable junction box as described in Sectin 20 of Chapter 3.

Similarly, to ensure the correct and accurate operation of the speedometer, and the tachometer (where fitted), the drive cables must be removed from the machine as described in Section 17 of Chapter 5 and greased. Remember to grease all but the upper six inches of the cable to prevent damage to the instrument head.

9 Adjusting and greasing the steering head bearings

The handling and roadholding characteristics are to a great degree dependent upon the condition of the steering and suspension components. Although any sudden change will be immediately apparent to the rider, the gradual deterioration in performance caused by general wear can often be overlooked and deficiencies may not be apparent unless the machine is ridden close to its limits. A regular and systematic check can, therefore, avoid unpleasant surprises on the road.

Roll the machine backwards and forwards with the front brake held on, noting the action of the front forks. These should operate smoothly and progressively and should show no signs of bouncing due to ineffective damping. Any signs of knocking felt through the handlebar or heard are often traced to slack or worn steering head bearings.

To check for wear, place the machine on its centre stand and place blocks or a crate beneath the crankcase so that the front wheel is raised clear of the ground. Persuade a friend to grasp the lower end of the fork legs and push and pull them. If play is present in the bearings it may prove easier to feel than see, by placing a finger between the frame headstock and the bearing shroud.

To adjust the steering head bearings, slacken the top bolt at the centre of the upper fork yoke. Using a C-spanner, slacken the adjustment nut by about a half turn, then slowly tighten it until a slight resistance is felt. It is very easy to overtighten the bearings, and this can cause stiff steering and rapid bearing wear. It is necessary to **just** eliminate free play and no more. When adjustment is complete check that the steering moves easily from lock to lock, then secure the top bolt to lock the adjustment. Any play remaining in the front end will probably be due to wear in the fork legs, and if found these should be overhauled as described in Chapter 5, Sections 2, 5 and 6.

If the steering head bearings have not been dismantled for any other reason, they should be stripped for examination and greasing at this interval. This task would fit in conveniently with the fork oil change, especially on RD125 LC models, and is described in Chapter 5, Sections 2, 3 and 4.

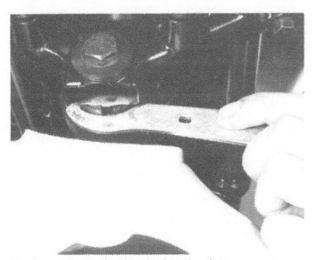

Use C-spanner to adjust steering head bearings

10 Changing the fork oil

This is an important task which must be carried out to ensure the continuing stability and safety of the machine on the road. Fork oil gradually degenerates as it loses viscosity and is contaminated by water and dirt, which produces a very gradual loss of damping. This can occur over a long period of time, thus being completely unnoticed by the rider until the machine is in a dangerous condition. Regular changes of the fork oil will

eliminate this possibility. Refer to Chapter 5, Sections 2, 5 and 6 for details of removal and refitting. Note that as RD125 LC machines are not fitted with front fork drain plugs, each fork leg must be removed from the machine and inverted to drain the oil. This does, however, offer an excellent opportunity to examine the forks for wear or damage and fits in well with the other maintenance operations that are necessary at this interval.

On DT125 LC machines only, drain plugs are provided. To use these, place a suitable container at the side of the front wheel, some distance from the drain plug to be removed and lay a sheet of cardboard or newspaper against the wheel to prevent oil getting on to the brake or tyre. Remove the drain plug and pump gently on the forks to eject the oil by applying the front brake and leaning on the handlebars. When all the oil has been pumped out, repeat the process on the other leg. Leave the machine for a while to allow as much oil as possible to drain the bottom of the fork legs, then repeat the pumping action to expel the remainder. Refit the drain plugs, tightening them carefully, then remove the fork top plugs and the fork springs as described in Section 6 of Chapter 5. Take care to support the machine so that it cannot topple over when the springs are removed. Have the full amount of fork oil ready for each leg but remember that as there will be some oil left in the fork leg, a certain amount will be surplus to requirements and that it is more important to have the oil level correct.

Using the dipstick described in Section 6 of Chapter 5, add oil until it is 168.4 mm (6.63 in) from the top of the stanchion when the stanchion is fully compressed and the fork springs are removed. Gently pump the forks up and down to distribute the oil around each fork leg, then recheck the oil level, adding oil if necessary. When the level is correct, raise the front of the machine to extend fully the forks, then refit the springs and top plugs as described.

11 Greasing the wheel bearings and speedometer drive

While the wheels are removed for the other maintenance tasks described at this service interval, the opportunity should be taken to examine the wheel bearings. Remove them as described in Sections 5 and 16 of Chapter 6, for cleaning and examination, renew any that are worn or damaged, then refit them, packing each with high melting-point grease.

Examine also the speedometer drive mechanism and pack it with grease as described in Section 18 of Chapter 5.

12 Greasing the sub-frame pivot and suspension unit mountings

If the regular weekly check reveals wear in the sub-frame pivot bearings at any time, the sub-frame must be removed for new components to be fitted, the necessary work being described in Section 10 of Chapter 5. It is also necessary to grease the bearings at regular intervals to minimise wear and to prevent the onset of corrosion.

Support the machine securely in an upright position so that the rear wheel is clear of the ground, then wedge a wooden block or similar under the rear tyre so that the weight is taken off the sub-frame pivot. Remove the rear brake pedal (DT125 LC only), then slacken and remove the pivot bolt retaining nut. Using a hammer and a long metal drift, tap out the pivot bolt. If sufficient care is exercised, the sub-frame will not be disturbed.

Thoroughly clean the pivot bolt, removing all traces of dirt and corrosion, then smear it with grease. Pack liberal quantities of grease through the frame apertures and into the pivot bearings, then refit the pivot bolt. It is pushed through from right to left on RD125 LC models, and from left to right on DT125 LC models. Place a finger firmly over the frame aperture on the opposite side to the pivot bolt, thus trapping the grease in the pivot bearings as the bolt is pushed through and filling all the tiny cracks where water or road dirt might gather to start the process of corrosion. Refit the pivot bolt retaining nut and its spring or plain washer, then tighten the nut securely to a torque setting of 4.3 kgf m (31 lbf ft).

Referring to Section 11 of Chapter 5 where necessary, carry out a similar procedure to grease the mounting bolt and pin of the rear suspension unit. Always use a new split-pin to secure the suspension unit (bottom) mounting pin.

Additional routine maintenance

Certain aspects of routine maintenance make it impossible to place operations under specific mileage or calendar headings, or may necessitate modification of these headings. A good example of the latter is the effect of a dusty environment on certain maintenance operations, notably cleaning the air filter element. Similarly a machine ridden over rough or dirty roads will require more frequent attention to the cycle parts, ie suspension components and wheels, and to the chain. A machine ridden at constant high speeds will need attention to the brakes and engine/transmission components far more than a machine ridden at excessively slow speeds. The latter will, however, need more frequent decarbonising. The problem of how to achieve the correct balance between too little maintenance, which will result in premature and expensive damage to the machine, and too much, is a delicate one which is unfortunately only resolved by personal experience. This experience is best gained by strict adherence to the specified mileage/time headings until the owner feels qualified to alter them to suit his own machine.

Some tasks are recommended for safety reasons due to the fact that materials can deteriorate through old age alone, irrespective of usage or mileage; four such tasks are given below.

1   Renewing the coolant

It is recommended that the cooling system is drained, flushed out with clean water, and refilled with a fresh mixture of coolant every two years. This is to minimise the build-up of mineral deposits and dirt in the various waterways. The task of draining, flushing and refilling the cooling system is described in Sections 2, 3 and 4 of Chapter 2.

2   Renewing the brake fluid

If the brake fluid is not completely changed during the course of routine maintenance, it should be changed at least every two years. Brake fluid is hydroscopic, which means that it absorbs moisture from the air. Although the system is sealed, the fluid will gradually deteriorate and must be renewed before contamination lowers its boiling point to an unsafe level.

Before starting work, obtain a full can of new DOT 3 or SAE J1703 hydraulic fluid and read Chapter 6, Section 12. Prepare the clear plastic tube and glass jar in the same way as for bleeding the hydraulic system, open the bleed nipple by unscrewing it $\frac{1}{4} - \frac{1}{2}$ a turn with a spanner and apply the front brake lever gently and repeatedly. This will pump out the old fluid. Keep the master cylinder reservoir topped up at all times, otherwise air may enter the system and greatly lengthen the operation. The old brake fluid is invariably much darker in colour than the new, making it much easier to see when the old fluid has been pumped out and the new fluid has completely replaced it.

When the new fluid appears in the clear plastic tubing with no traces of old fluid contaminating it, close the bleed nipple,

remove the plastic tubing and refit the rubber dust cap on the nipple. Top the master cylinder reservoir up to the raised level mark protruding from the inside face of the reservoir. Carefully dry the diaphragm with a clean lint-free cloth, fold it into its compressed state, and refit the diaphragm and the reservoir cover, tightening securely the two retaining screws. Wash off any surplus fluid with fresh water and check for any fluid leaks which may subsequently appear. Remember to check that full brake pressure is restored and that the front brake is working properly before taking the machine out on the road.

3   Renewing the caliper seals

The manufacturer recommends that the two seals fitted in the caliper body around the piston be renewed as a safety precaution against their sudden failure. While this is commendable, it should be pointed out that the seals are only sold with the piston as Yamaha replacement parts, thus making the exercise quite costly. It is therefore advised that the owner makes regular and conscientious examinations of all the components of the hydraulic system and takes instant action as described in Chapter 6, whenever the slightest sign of a fluid leak appears, or if lever pressure and the brake efficiency is reduced in any way, indicating the presence of a potentially serious fault.

4   Renewing the brake hose

As mentioned in Section 8 of Chapter 6 which discusses the brake hose in detail, although the hose is very strongly constructed, it is under considerable strain and can deteriorate with age. For this reason the manufacturer recommends that the hose be renewed every four years, regardless of its apparent condition. This does not mean that the hose can be forgotten until its time is up, it must be inspected at regular intervals and renewed at the slightest sign of damage or deterioration.

5   Cleaning the machine

Keeping the motorcycle clean should be considered as an important part of the routine maintenance, to be carried out whenever the need arises. A machine cleaned regularly will not only succumb less speedily to the inevitable corrosion of external surfaces, and hence maintain its market value, but will be far more approachable when the time comes for maintenance or service work. Furthermore, loose or failing components are more readily spotted when not partially obscured by a mantle of road grime and oil.

Surface dirt should be removed using a sponge and warm, soapy water; the latter being applied copiously to remove the particles of grit which might otherwise cause damage to the paintwork and polished surfaces.

Oil and grease is removed most easily by the application of a cleaning solvent such as 'Gunk' or 'Jizer'. The solvent should be applied when the parts are still dry and worked in with a stiff brush. Large quantities of water should be used when rinsing off, taking care that water does not enter the carburettors, air cleaners or electrics.

If desired, a polish such as Solvol Autosol can be applied to the aluminium alloy parts to restore the original lustre. This does not apply in instances, much favoured by Japanese manufacturers, where the components are lacquered. Application of a wax polish to the cycle parts and a good chrome cleaner to the chrome parts will also give a good finish. Always wipe the machine down if used in the wet, and make sure the chain is well oiled. There is less chance of water getting into control cables if they are regularly lubricated, which will prevent stiffness of action.

Castrol Lubricants

Castrol Engine Oils
Castrol Grand Prix

Castrol Grand Prix 10W/40 four stroke motorcycle oil is a superior quality lubricant designed for air or water cooled four stroke motorcycle engines, operating under all conditions.

Castrol Super TT Two Stroke Oil

Castrol Super TT Two Stroke Oil is a superior quality lubricant specially formulated for high powered Two Stroke engines. It is readily miscible with fuel and contains selective modern additives to provide excellent protection against deposit induced pre-ignition, high temperature ring sticking and scuffing, wear and corrosion.
Castrol Super TT Two Stroke Oil is recommended for use at petrol mixture ratios of up to 50:1.

Castrol R40

Castrol R40 is a castor-based lubricant specially designed for racing and high speed rallying, providing the ultimate in lubrication. Castrol R40 should never be mixed with mineral-based oils, and further additives are unnecessary and undesirable. A specialist oil for limited applications.

Castrol Gear Oils
Castrol Hypoy EP90

An SAE 90 mineral-based extreme pressure multi-purpose gear oil, primarily recommended for the lubrication of conventional hypoid differential units operating under moderate service conditions. Suitable also for some gearbox applications.

Castrol Hypoy Light EP 80W

A mineral-based extreme pressure multi-purpose gear oil with similar applications to Castrol Hypoy but

an SAE rating of 80W and suitable where the average ambient temperatures are between 32°F and 10°F. Also recommended for manual transmissions where manufacturers specify an extreme pressure SAE 80 gear oil.

Castrol Hypoy B EP80 and B EP90

Are mineral-based extreme pressure multi-purpose gear oils with similar applications to Castrol Hypoy, operating in average ambient temperatures between 90°F and 32°F. The Castrol Hypoy B range provides added protection for gears operating under very stringent service conditions.

Castrol Greases
Castrol LM Grease

A multi-purpose high melting point lithium-based grease suitable for most automotive applications, including chassis and wheel bearing lubrication.

Castrol MS3 Grease

A high melting point lithium-based grease containing molybdenum disulphide. Suitable for heavy duty chassis application and some CV joints where a lithium-based grease is specified.

Castrol BNS Grease

A bentone-based non melting high temperature grease for ultra severe applications such as race and rally car front wheel bearings.

Other Castrol Products
Castrol Girling Universal Brake and Clutch Fluid

A special high performance brake and clutch fluid with an advanced vapour lock performance. It is the only fluid recommended by

Girling Limited and surpasses the performance requirements of the current SAE J1703 Specification and the United States Federal Motor Vehicle Safety Standard No. 116 DOT 3 Specification.
In addition, Castrol Girling Universal Brake and Clutch fluid fully meets the requirements of the major vehicle manufacturers.

Castrol Fork Oil

A specially formulated fluid for the front forks of motorcycles, providing excellent damping and load carrying properties.

Castrol Chain Lubricant

A specially developed motorcycle chain lubricant containing non-drip, anti corrosion and water resistant additives which afford excellent penetration, lubrication and protection of exposed chains.

Castrol Everyman Oil

A light-bodied machine oil containing anti-corrosion additives for both household use and cycle lubrication.

Castrol DWF

A de-watering fluid which displaces moisture, lubricates and protects against corrosion of all metals. Innumerable uses in both car and home. Available in 400gm and 200gm aerosol cans.

Castrol Easing Fluid

A rust releasing fluid for corroded nuts, locks, hinges and all mechanical joints. Also available in 250ml tins.

Castrol Antifreeze

Contains anti-corrosion additives with ethylene glycol. Recommended for the cooling system of all petrol and diesel engines.

Chapter 1 Engine, clutch and gearbox

For modifications, and information relating to later models, see Chapter 8

Contents

Specifications

Note: *Specifications for DT125 LC model are given only where they differ from those of RD125 LC model*

	RD125 LC	DT125 LC
Engine		
Type	Water-cooled, single cylinder, two-stroke	
Bore	56 mm (2.20 in)	
Stroke	50 mm (1.97 in)	
Capacity	123 cc (7.5 cu in)	
Compression ratio	6.4 : 1	7.2 : 1
Power output:		
Unrestricted model	21.1 bhp @ 9500 rpm	16.2 bhp @ 7000 rpm
Restricted model	12.2 bhp @ 7500 rpm	12.2 bhp @ 6000 rpm

Maximum torque
    Unrestricted model .... 1.6 kgf m @ 9250 rpm   1.7 kgf m @ 7000 rpm
    Restricted model .... 1.2 kgf m @ 7000 rpm   1.4 kgf m @ 6000 rpm

Cylinder head
    Type .... Cast aluminium. with integral water passages
    Maximum warpage .... 0.03 mm (0.0012 in)

Cylinder barrel
    Type .... Cast aluminium, with cast iron liner
    Standard bore size .... 56.000 − 56.020 mm (2.205 − 2.206 in)
    Wear limit .... 56.10 mm (2.209 in)
    Maximum taper .... 0.08 mm (0.003 in)
    Maximum ovality .... 0.05 mm (0.002 in)

Piston rings
Top:
    Type .... Keystone
    Thickness .... 1.2 mm (0.0472 in)
    Width .... 2.2 mm (0.0866 in)
    End gap − installed .... 0.30 − 0.45 mm (0.012 − 0.018 in)
    Ring/groove clearance .... 0.03 − 0.05 mm (0.001 − 0.002 in)
Second:
    Type .... Plain, with expander
    Thickness .... 1.2 mm (0.0472 in)
Width:
    RD125 LC .... 1.85 mm (0.0728 in)
    DT125 LC .... 2.20 mm (0.0866 in)
End gap − installed:
    RD125 LC .... 0.30 − 0.50 mm (0.012 − 0.020 in)
    DT125 LC .... 0.30 − 0.45 mm (0.012 − 0.018 in)
Ring groove clearance:
    RD125 LC .... 0.03 − 0.05 mm (0.001 − 0.002 in)
    DT125 LC .... 0.03 − 0.07 mm (0.001 − 0.003 in)

Piston
    Standard OD .... 55.94 − 56.00 mm (2.202 − 2.205 in)
    Piston/cylinder clearance .... 0.050 − 0.055 mm (0.0020 − 0.0022 in)

Crankshaft
    Big-end bearing deflection − at small-end .... 2 mm (0.08 in)
    Big-end side clearance .... 0.2 − 0.7 mm (0.008 − 0.028 in)
    Maximum runout .... 0.03 mm (0.001 in)
    Width across flywheels .... 55.90 − 55.95 mm (2.201 − 2.203 in)

Primary drive
    Type .... Helical gear
    Reduction ratio .... 3.227 : 1 (71/22T)

Clutch
    Type .... Wet, multi-plate
    Number of friction plates 6
    Number of plain plates 5
    Number of springs 4
    Friction plate thickness .... 3.0 mm (0.12 in)
    Wear limit .... 2.7 mm (0.11 in)
    Plain plate thickness .... 1.2 mm (0.05 in)
    Plain plate warpage limit .... 0.05 mm (0.002 in)
    Spring free length .... 34.5 mm (1.36 in)
    Wear limit .... 33.5 mm (1.32 in)
    Push rod maximum runout .... 0.15 mm (0.006 in)
    Outer drum thrust clearance .... 0 − 0.5 mm (0 − 0.02 in)

Gearbox

	RD125 LC	DT125 LC
Type	6 speed, constant mesh	6-speed, constant mesh
Gear ratios:		
1st	2.833 : 1 (34/12T)	3.500 : 1 (35/10T)
2nd	1.812 : 1 (29/16T)	2.214 : 1 (31/14T)
3rd	1.368 : 1 (26/19T)	1.555 : 1 (28/18T)
4th	1.142 : 1 (24/21T)	1.190 : 1 (25/21T)
5th	1.000 : 1 (23/23T)	1.000 : 1 (23/23T)
6th	0.916 : 1 (22/24T)	0.840 : 1 (21/25T)

Final drive

Type .... Chain and sprocket

Reduction ratio:

Unrestricted model	2.875 : 1 (46/16T)	3.266 : 1 (49.15T)
Restricted model	2.813 : 1 (45/16T)	3.266 : 1 (49.15T)

Torque settings

Component	kgf m	lbf ft
Spark plug	2.0	14.5
Cylinder head nuts	2.2	16.0
Cylinder barrel nuts	2.8	20.0
Cylinder head and barrel studs	1.0	7.0
Coolant drain plugs	1.0	7.0
Temperature gauge sender unit	1.0	7.0
Generator rotor nut	5.0	36.0
Water pump cover screws	1.0	7.0
Oil pump mounting screws	0.5	3.5
Reed valve assembly mounting bolts	0.8	6.0
Exhaust pipe front mounting nuts	1.8	13.0
Crankcase fastening screws	0.8	6.0
Crankcase cover screws	1.0	7.0
Transmission oil drain plug	2.0	14.5
Primary drive gear nut	6.5	47.0
Clutch centre nut	6.5	47.0
Clutch spring bolts	1.0	7.0
Balancer shaft gear nut	6.5	47.0
Oil seal retainer screw	1.6	11.5
Bearing retainer screws	1.0	7.0
Kickstart lever retaining nut	6.5	47.0
Gearbox sprocket mounting bolts	1.0	7.0
Detent arm bolt	1.5	11.0
Coolant pipe mounting screws	1.2	9.0
Engine mounting bolts	2.5	18.0
Gearchange linkage mounting bolt	1.5	11.0
Selector cam retaining screw	1.4	10.0
Clutch adjuster locknut	0.8	6.0

1   General description

The Yamaha RD125 and DT125 LC models employ a water-cooled, single-cylinder, two-stroke engine built in unit with the primary drive, clutch and gearbox. The cylinder head and barrel castings are of light alloy construction and incorporate cast-in passages for the coolant. The cylinder barrel has a cast iron liner. The crankshaft is of conventional design, having needle roller bearings at the connecting rod big- and small-ends, and is supported by two ball journal main bearings. A gear driven single shaft primary balance is fitted to counter the imbalance in all single-cylinder engines.

Primary drive is by helical-cut gears to the wet multi-plate clutch mounted on the end of the gearbox input shaft. The gearbox is of the six-speed constant mesh type and is lubricated by oil bath shared with the primary drive, the oil being contained in a reservoir formed by the main crankcase castings. The engine oil and water pumps are mounted on the crankcase right-hand cover and are driven from the crankshaft via the primary drive pinion and nylon idler gears.

Although the use of water-cooling is normally associated with increased weight and complexity, Yamaha have, by dint of careful design and maximum use of lightweight materials, kept the weight of the power unit down to the point where it is over 2lb (1kg) lighter than the air-cooled unit it supersedes. The penalty of increased complexity is more than offset by the more consistent performance and increased longevity which can be gained from water-cooled engine units.

2   Operations with the engine/gearbox unit in the frame

The items listed below can be overhauled with the engine/gearbox unit in place. When a number of these operations need to be undertaken simultaneously it is usually worthwhile taking the unit out of the frame to gain better access and more comfortable working conditions. Engine removal is fairly straightforward and can be expected to take about one hour.

a) Cylinder head, barrel and piston
b) Clutch assembly and primary gears
c) Oil pump
d) Water pump
e) Kickstart mechanism
f) Ignition pickup
g) Generator assembly
h) Gear selector mechanism (external components only)
i) Final drive sprocket

3   Operations with the engine/gearbox unit removed from the frame

To gain access to the items listed below it is essential that the engine unit is removed from the frame and the crankcase halves separated:

a) Crankshaft assembly
b) Gearbox components
c) Gear selector drum and forks

4   Removing the engine/gearbox unit from the frame

1   Before commencing any dismantling work it will be necessary to drain the transmission oil. Because the oil will flow easier when hot, and therefore drain faster and more completely, draining of the oil is best carried out when the machine has just completed a journey and the engine/gearbox

unit is fully warmed up to normal operating temperature. Place a suitable container below the engine into which to drain the transmission oil. Remove the filler plug from the top of the crankcase right-hand cover, then slacken and remove the drain plug which is situated on the underside of the crankcase unit. It will be necessary to use a socket spanner and extension or a long box spanner to unscrew the drain plug as it can only be reached via a large aperture in the bashplate of the DT125 LC model or via the gap between the trailing edge of the belly fairing and the exhaust pipe of the RD125 LC model. When the plug has been removed, allow the oil to drain. If time permits, the machine should be left to stand overnight so that the oil can drain completely and the machine can cool down. Once the oil has drained fully, clean the drain plug and examine the sealing washer, renewing it unless it is in good condition. Refit the drain plug, tightening it to 2.0 kgf m (14.5 lbf ft), and refit also the filler plug.

2   The machine must now be placed on a stand so that it is held securely in an upright position. When working on an RD125 LC model this is achieved easily by the use of the centre stand and a block of wood placed ahead of the front wheel to prevent the machine falling forwards off the stand. The DT125 LC model, however, presents problems as it is fitted only with a prop stand. The usual solution is to place a stout wooden box underneath the engine/gearbox unit so that both wheels are clear of the ground, but it should be noted that this does restrict access to some components, and it is therefore recommended that a metal 'paddock'-type stand is either fabricated, or purchased from one of the many suppliers who advertise regularly in the national motorcycle press. It is useful, but by no means essential, to raise the machine to a convenient working height by placing it, on its stand, on a strong wooden bench or similar support. If this course of action is adopted, great care must be taken to ensure that the machine is securely tied down so that it cannot fall.

3   It is now necessary to remove all those components which will restrict access to the various items to be withdrawn during the course of engine/gearbox unit removal. On the RD125 LC model, these components consist of the sidepanels, the seat, the petrol tank, the belly fairing and the radiator cover. Both sidepanels are each retained at three points, by a prong at the top leading edge which engages in a rubber grommet set in the petrol tank, and by two clips which engage on rubber grommets placed over lugs on the frame, one clip being at the bottom edge and the other at the extreme rear of the sidepanel moulding. Be very careful in pulling the sidepanels away from their mountings, to avoid damaging the moulding. The seat is retained by two catches at the rear, one on each side, and by a moulded prong at the front. Unlock the helmet/seat lock, press down both catches and lift the seat up at the rear, then pull carefully backwards to release the seat from its front mounting. The petrol tank is retained by a single bolt at the rear and is supported by two rubber mountings at the front. Turn the petrol tap to the 'Off' position and use a suitable pair of pliers to release the wire petrol pipe retaining clip. Slide the clip down the petrol pipe until it is clear of the petrol tap spigot and carefully pull the petrol pipe away from the tap. Slacken and remove the tank rear mounting bolt, then lift the tank up at the rear and pull it backwards. The belly fairing is retained by two bolts at its rear edge which thread into lugs on each of the frame cradle tubes. Slacken and remove these two bolts and release the clips which secure the fairing to the lower edge of the radiator cover. The radiator cover itself is retained by a single screw at its upped edge, and by two clips at the bottom which engage on rubber grommets placed over lugs on the frame. Slacken and remove the single screw, then release the two bottom clips to remove the radiator cover. Carefully check all mounting rubbers and grommets, making a note to renew any that are lost or damaged, and place all the items removed to one side where they cannot be damaged.

4   When working on the DT125 LC model, the items to be removed are the sidepanels, the seat, the petrol tank, the crankcase bashplate and the radiator cover. The sidepanels are retained by prongs at their lower front edges which engage in rubber grommets set in the frame, by single screws in the middle of the lower edge, and by a clip at the upper rear edge which engages around the frame tube on the left-hand side, and which engages on a rubber grommet placed over a lug on the frame on the right-hand side. Slacken the single retaining screw on each sidepanel and pull the moulding carefully away from its two retaining mountings, taking care not to damage it. Unlock the helmet/seat lock and release the seat catch. Raise the seat at the rear and pull it backwards to disengage it from its mounting underneath the petrol tank. Turn the petrol tap to the 'Off' position and use a suitable pair of pliers to release the wire petrol pipe retaining clip. Slide the clip down the petrol pipe until it is clear of the petrol tap spigot and carefully pull the petrol pipe away from the tap. Slacken and remove the single tank rear mounting bolt, then lift the tank up at the rear and pull it backwards off its front mountings. Slacken and remove the three bolts which secure the bashplate to the frame bottom rails and withdraw the bashplate. The radiator cover is fastened to the radiator mounting bracket at three points. Slacken and remove the single screw which is set horizontally in the cover's right-hand side, then pull the cover carefully forwards to disengage the two prongs which are held in rubber grommets set in the top and the bottom rails of the radiator mounting bracket. Take great care not to damage the moulding. Carefully check all mounting rubbers and grommets, making a note to renew any that are lost or damaged, and place all the components removed to one side where they cannot be lost or damaged.

5   The battery should now be disconnected to prevent any possibility of damage due to short circuits when the various electrical components are disconnected. Unplug the snap connector set in the battery positive (+ ve) lead next to the fuse casing assembly. Note that if it is anticipated that the machine is to be out of service for some time, the battery should be removed and arrangements made to give it a refresher charge every month or so, as described in the relevant Sections of Chapter 7.

6   It will now be necessary to drain the cooling system, but note that this should only be done when the engine has cooled down to avoid any risk of scalding when the radiator pressure cap is released. Place a thick rag, such as a piece of old towel, over the cap, and twist the cap slowly anti-clockwise until the stop is reached. The cap must then be pushed down and twisted further to clear the stop, whereupon it can be lifted away. If at any time a hissing sound is heared, stop rotating the cap until the residual pressure is released and the hissing stops. The cooling system can now be drained. A suitable container, such as a plastic bowl or bucket can be used to catch the coolant. If the coolant is fairly new (less than two years old) it can be re-used, in which case make sure that the container is clean. Slacken and remove the four screws which fasten the oil and water pump cover to the front of the crankcase right-hand cover. Note that an impact drive should be used to slacken any tightly-fastened screws, but also that in this case extreme caution must be exercised as the cover is only a light plastic moulding and can be easily cracked. Ease the cover away from the oil feed pipe grommet.

7   Two drain plugs are provided for the cooling system, one on the left-hand side of the cylinder barrel, and the other in the water pump casing. Because the drain plugs are situated inboard of the engine casings it will be necessary to make up some sort of chute to direct the coolant into the drain container. This need not be elaborate; a piece of stiff card will suffice. Remove each drain plug in turn and allow the coolant to drain into the container. If any coolant is spilled it should be washed off immediately if the painted finish of the crankcases is not to be marked. When the coolant is fully drained, inspect the drain plug washers, renewing them if there is any doubt about their condition, and refit the drain plugs, tightening them to a torque setting of 1.0 kgf m (7lbf ft).

8   The radiator hoses should be disconnected at their lower ends, ie at the crankcase right-hand cover and cylinder head

unions. Use a heavy pair of pliers to release each hose retaining clip by squeezing together the protruding ends of the clip. Slide each clip up the hose until the clip is clear of the union, then carefully pull or twist the hose off the union. Be prepared to catch or mop up any residual coolant which might be released by disconnecting the hoses. Although it is not strictly necessary to remove the radiator during the course of engine/gearbox removal, if left in place the hoses will hinder operations to a certain extent. To avoid this, slacken the clip at the radiator end of each hose and rotate the hose right round until it is pointing away from the engine/gearbox unit. Note that in the case of the RD125 LC model, the radiator bottom hose is nearly straight and therefore cannot be moved out of the way in this manner. In this case, it will be necessary to release completely the clip and to remove the hose from the radiator as described above. Take care not to damage the radiator, especially if the hose is a particularly tight fit.

9   The exhaust system is the next item to be removed. On the RD125 LC model, this is a straightforward operation which involves only the slackening and removal of the two exhaust front mounting nuts and the single rear mounting bolt which passes through the centre of the right-hand footrest mounting bracket. Be careful not to damage the painted finish of the exhaust system as it is withdrawn from the machine. Replace the two nuts on their cylinder barrel studs and refit the bolt in the footrest bracket to prevent the loss of these components. When working on the DT125 LC model, note that the two-piece exhaust system is fastened to the frame at four points, in addition to the cylinder barrel mounting. Start by slackening fully the two clips which secure the rubber sleeve sealing the joint between the two parts of the system. Slacken and remove the two bolts which fasten the rear section to the frame and carefully pull the assembly backwards to separate the two parts. Then slacken and remove the two front mounting nuts and the two mounting bolts to release the front part of the exhaust. Carefully manoeuvre the front section forwards clear of the frame and withdraw it from the machine taking great care not to damage the paintwork of either the exhaust or of the machine itself. Refit the mounting nuts and bolts in their correct respective locations to avoid loss or damage. Inspect all the rubber mounting components, renewing any that are worn or damaged, and put the exhaust system to one side to await cleaning and/or reassembly.

10   Disconnect the YEIS hose at its union on the inlet stub using the same method as given above for radiator hose removal, and taking great care not to damage the stub itself. Slacken and remove the single bolt securing the YEIS chamber to the frame and remove the complete YEIS assembly. Take this opportunity to inspect the assembly for cracks, splits, or other damage, renewing any component which is found to be faulty. Remove the spark plug cap from the spark plug, secure the plug cap and HT lead clear of the engine/gearbox unit, then remove the spark plug itself. Pull the temperature gauge sender unit wire off the terminal at the top of the sender unit and push the wire up into a convenient point in the steering head area so that it is also secured clear of the engine/gearbox unit.

11   Slacken and remove the large nut which retains the kickstart lever on its shaft, holding the lever with one hand to prevent rotation of the shaft. Pull the lever off the shaft and put it to one side with the nut. On RD125 LC models only, use a suitable pair of pliers to release the large knurled, threaded ring which secures the tachometer cable to the drive assembly, then pull the cable upwards out of the drive assembly. Tape the cable to a frame top tube to secure it out of the way and ensure that the oil seal which fits around the tachometer inner cable is stored in a safe place.

12   Rotate the oil pump pulley by hand to the full throttle position to gain maximum free play in the cable, then disengage the cable inner wire from the pulley as far as possible. It will be noted that on some models the pulley return spring outer end has a long extension which passes through the pulley on the outside of the cable inner wire, thus serving as a cable retainer. Some models are fitted with a separate spring clip to serve the

same purpose. Using a small, thin-bladed electrical screwdriver, gently prise the spring away from the pulley until the cable can be slipped out of the pulley track, or gently pull away the spring clip. Allow the spring extension to return to its normal position or refit the spring clip and disengage the cable end nipple from the pulley. Pull the complete cable up through the crankcase right-hand cover. Using a suitable pair of pliers, release the wire clip which secures the larger diameter oil feed pipe to its oil pump union. Pull the feed pipe carefully off the union, placing a finger over the pipe end temporarily to stop the flow of oil. The pipe may then be plugged using a screw or bolt of suitable size provided care is taken not to damage the end of the pipe itself. Using a small flat-bladed screwdriver, carefully displace the tubular clips fitted at each end of the smaller diameter, oil pump/carburettor, oil feed pipe. The clips should be eased down the length of the pipe until they are clear of the union spigots on the carburettor and on the oil pump. Disconnect the pipe at both ends, but be careful to pull off only one end at a time and to plug it immediately with a suitably-sized screw or wooden dowel to avoid the loss of oil and the entry of dirt or air. In some cases, the oil tank/oil pump feed pipe is secured to the crankcase top surface by a clamp which is fastened by one of the crankcase right-hand cover screws. In such cases, the pipe must be freed from the clamp by slackening the screw and raising the clamp. Tuck both oil feed pipes out of the way beneath the oil tank.

13   Slacken the clamps which secure the air filter to the carburettor and the carburettor to the inlet stub. Tilt the carburettor to one side, unscrew the carburettor top and withdraw the throttle valve assembly. Wrap the throttle valve and needle in a clean rag to prevent damage, then tape the throttle cable and oil pump cable together to a convenient point on a frame top tube so that both are well out of harm's way. Carefully remove the carburettor body from the machine, disengaging first the air filter hose. The task of carburettor removal is extremely awkward and requires care and patience if damage to the components is to be avoided.

14   Moving round to the left-hand side of the machine, slacken the locknut of the clutch cable adjuster fitted to the handlebar lever and, for DT125 LC models only, the locknut of the in-line cable adjuster fitted near the handlebar end of the cable. Screw the adjuster, or for DT125 LC models, the adjusters, in as far as possible to gain the maximum amount of free play in the cable. Using a thin-bladed electrical screwdriver or a pair of needle-nose pliers, bend back the thin strip of metal which prevents the cable lower end nipple from jumping out of its slot in the trunnion on the end of the clutch actuating assembly. Compress the rubber sleeve which protects the cable inner, press the lever assembly towards the front of the machine and use a pair of needle-nose pliers to slip the cable end nipple out of the trunnion slot. Disengage the cable from its bracket at the base of the cylinder barrel and allow it to hang down in front of the engine/gearbox unit.

15   Remove the gear lever fitted to the DT125 LC model after slackening and removing the pinch bolt which clamps the lever to the shaft. The position of the lever should be marked to aid correct refitting on reassembly, before the lever is pulled off the shaft splines. When working on the RD125 LC model, slacken and remove first the bolt which secures the linkage at its pivot point to the frame, then the pinch bolt which clamps the linkage to the gearchange shaft. When both bolts have been removed the linkage can be withdrawn as a complete assembly. Inspect all the component parts of the linkage, making a note to renew any that are worn or damaged. The crankcase left-hand cover fitted to the DT125 LC model is secured by five screws, but that fitted to the RD125 LC model is secured by six. Slacken and remove these screws, noting that an impact screwdriver should be used to slacken any that are particularly tight, but also that, as in the case of the oil/water pump cover, the crankcase left-hand cover is only a light plastic moulding and may be cracked easily by the over-enthusiastic or inexpert use of such a tool. When the screws have been removed withdraw the cover.

16   Slacken and remove the two bolts which secure the

gearbox sprocket, applying the rear brake, if necessary, to prevent sprocket rotation. Rotate the sprocket retaining plate until it can be pulled off the shaft, then pull away the sprocket itself. If this is difficult to achieve, depress the chain tensioner fitted to the DT125 LC model or slacken the spindle nut and push the rear wheel of the RD125 LC model forward to gain the necessary slack in the chain. When the sprocket is removed, disengage it from the chain and allow the chain to hang over the rear sub-frame pivot. Starting from the sealing grommet set in the crankcase top surface, trace the main lead from the generator stator plate up to the connectors by which it is joined to the main wiring loom, freeing the lead from any guides or cable clamps which fasten it to the frame. On the RD125 LC model, the lead passes up the frame left-hand front downtube to the horizontal bracing tube immediately below the steering head, ending in two multi-pin block connectors and two snap connectors. The lead fitted to the DT125 LC model is somewhat longer, running across the top of the engine/gearbox unit and up the frame right-hand rear downtube, terminating in a single multi-pin block connector and two snap connectors in the vicinity of the rear suspension unit top mounting. Uncouple the connectors and coil the lead around the cylinder barrel so that it cannot foul any other component as the engine/gearbox unit is withdrawn from the frame.

17 The last operation in the preliminary dismantling sequence concerns the DT125 LC model only. To gain access to the rear sub-frame pivot bolt and its retaining nut, and to provide sufficient clearance for the engine/gearbox unit to be withdrawn, the rear brake pedal assembly must be removed. Slacken off as far as possible the rear brake adjusting nut and use a small flat-bladed screwdriver to remove the circlip which retains the brake pedal on its pivot. Unhook the pedal return spring and the stop lamp rear switch extension rod from the pedal, then pull the pedal assembly off its pivot. Remove the pedal return spring and switch extension rod and put them to one side so that they are not lost or damaged.

18 On both models described in this Manual, the engine/gearbox unit is retained in the frame at three points: the front mounting is a bolt which is fitted from left to right and is fastened at its right-hand end by a self-locking nut and a plain washer, the rear lower mounting is a plain bolt which is threaded into a captive nut and is also fitted from left to right, and the rear upper mounting is the rear sub-frame pivot. Slacken and remove the rear sub-frame pivot bolt retaining nut and the engine front mounting bolt retaining nut, then remove the plain or spring washer fitted under each. Slacken and remove the engine lower rear mounting bolt. Tap the rear sub-frame pivot bolt back through until its threaded end is just flush with the frame. Find a bolt, a metal rod, or a length of wooden dowelling of the same outside diameter as the pivot bolt and

use this as a drift to tap the pivot bolt through the frame and through the sub-frame pivot immediately behind the frame. Note that the sub-frame pivot bolt is fitted from right to left on the RD125 LC model, and from left to right on the DT125 LC model; also that if corrosion has formed on the pivot bolt it will be necessary to use a hammer and a long metal drift to tap the bolt out of the frame and the pivot bushes so that it can be cleaned, greased, and refitted.

19 The engine/gearbox unit is now ready for removal. Make a final check to ensure that all components have been removed or disconnected and that nothing will impede the lifting out of the unit. Using the drift described in the previous paragraph, tap the sub-frame pivot bolt out until the drift has passed through the first sub-frame pivot and is just clear of the crankcase lug which fits between the two sub-frame pivots. Lay a thick piece of rag over the frame cradle tubes to protect the paintwork and pull out the pivot bolt by hand until it is withdrawn from the crankcase lug and the engine/gearbox unit is felt to drop. It is essential that the sub-frame is supported by the drift on one side and by the pivot bolt on the other, as it would otherwise drop at an awkward angle and, by jamming the crankcase lug, make removal of the engine very difficult. Once the engine/gearbox unit rear upper mounting is free, tap out the front mounting bolt with a hammer and a long metal drift. Raise the unit at the back to clear the rear mountings and lift the complete assembly out to the right of the machine. Place the unit on a workbench ready for further dismantling.

4.1 Drain plug is reached through bashplate aperture on DT125

4.3a Release petrol pipe retaining clip to remove pipe

4.3b Slacken and remove bolts to release belly fairing – RD125

4.4 Crankcase bashplate is secured by three bolts – DT125

4.5 Disconnect battery to prevent short circuits

4.7 Release coolant drain plugs and allow coolant to drain

4.8 Use heavy pliers to release hose clips

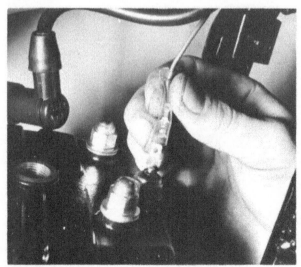

4.10 Disconnect temperature gauge sender unit wire

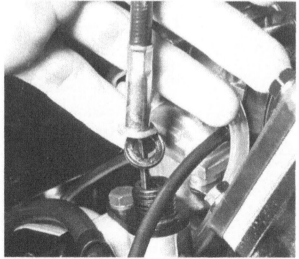

4.11 Do not forget oil seal fitted to RD125 tachometer drive cable

4.12a Note projection of return spring (arrowed) retaining inner cable – other models have spring clips

4.12b Use screws or bolts to plug oil pipes

4.13a Withdraw carburettor throttle valve assembly ...

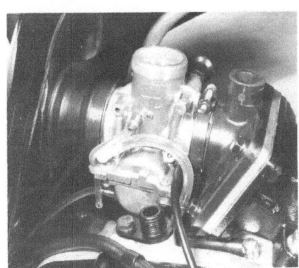

4.13b ... then disengage carburettor from stubs

4.15a Gearchange linkage must be removed ...

4.15b ... before crankcase cover can be withdrawn – RD125 only

4.16a Gearbox sprocket is retained by two bolts and a locking plate

4.16b Allow chain to hang over sub-frame pivot

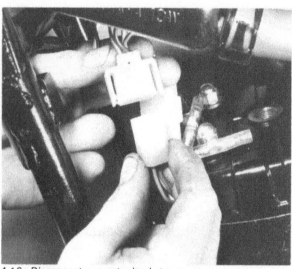

4.16c Disconnect generator lead at connectors

4.17a Gently prise out circlip retaining DT125 brake pedal

4.17b Do not lose or forget stop lamp switch components – DT125

4.18a The three engine mounting points – RD125 (arrowed)

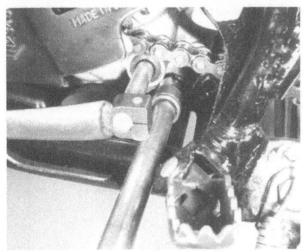

4.18b Engine rear lower mounting bolt – DT125

4.18c Note how sub-frame pivot bolt is also an engine mounting

5 Dismantling the engine/gearbox unit: preliminaries

1   Before any dismantling work is undertaken, the external surfaces of the unit should be thoroughly cleaned and degreased. This will prevent the contamination of the engine internals, and will also make working a lot easier and cleaner. A high flash point solvent, such as paraffin (kerosene) can be used, or better still, a proprietary engine degreaser such as Gunk. Use old paintbrushes and toothbrushes to work the solvent into the various recesses of the engine casings. Take care to exclude solvent or water from the electrical components and inlet and exhaust ports. The use of petrol (gasoline) as a cleaning medium should be avoided, because the vapour is explosive and can be toxic if used in a confined space.
2   When clean and dry, arrange the unit on the workbench, leaving a suitable clear area for working. Gather a selection of small containers and plastic bags so that parts can be grouped together in an easily identifiable manner. Some paper and a pen should be on hand to permit notes to be made and labels attached where necessary. A supply of clean rag is also required.

3   Before commencing work, read through the appropriate section so that some idea of the necessary procedure can be gained. When removing the various engine components it should be noted that great force is seldom required, unless specified. In many cases, a component's reluctance to be removed is indicative of an incorrect approach or removal method. If in any doubt, re-check with the text.

6 Dismantling the engine/gearbox unit: removing the cylinder head, barrel and piston

1   This operation can be undertaken with the engine unit in or out of the frame. In the former case it will first be necessary to drain the cooling system and to disconnect the carburettor and YEIS hose, the exhaust system, spark plug cap, and water temperature sender lead. The whole dismantling sequence is described in Routine Maintenance in the section relating to decarbonisation.
2   When the cylinder head and barrel are removed care must be taken to prevent any residual coolant from finding its way into the engine. To this end, lift the above components away carefully, keeping them level to avoid spillage, and have some absorbent cloth or paper to hand so that any spillage can be mopped up.
3   Detach the radiator hose union from the cylinder head by unscrewing its single fixing screw and by pulling it out of the cylinder head casting with a twisting action, as if unscrewing it. Remove the temperature gauge sender unit and the spark plug. The cylinder head is retained by four chromed dome nuts, each with a plain washer underneath it. Slacken the nuts progressively, by about one turn at a time, in a diagonal sequence, to avoid any possibility of the casting being warped by the uneven release of pressure. The cylinder head can be lifted away when the nuts and washers have all been removed. If the head proves to be stuck to the barrel, tap around the joint with a soft-faced mallet to break the seal. Once the joint is broken, lift the head clear and wipe up any spilt coolant, especially if this has entered the cylinder bore. Remove the head gasket.
4   Rotate the crankshaft until the piston reaches the highest point in the cylinder bore. Progressively slacken, and then remove, the four bolts which retain the reed valve case and inlet stub to the cylinder. Tap the joint with a soft-faced mallet and lift the inlet stub and reed valve assembly away. Slacken and remove the two screws which secure the coolant pipe to the cylinder and lift the coolant pipe out of its housing in the crankcase right-hand cover, noting that some residual coolant may be released by the action. If the pipe is a tight fit in the crankcase cover, use a long-bladed screwdriver or a long, thin piece of wood to lever it straight up from its housing. Mop up any spilt coolant. Slacken the four flanged nuts which secure the cylinder barrel to the crankcase, noting that these nuts must also be slackened progressively and in a diagonal sequence as described above for the cylinder head nuts. Remove the nuts and the clutch cable bracket which is clamped on the left-hand rear cylinder stud. If work is being carried out with the engine in the frame, tie the cable clear of the cylinder barrel so that it cannot impede progress.
5   Gently tap around the cylinder base joint with a soft-faced mallet to break the seal, then lift the barrel far enough to expose the bottom of the piston skirt. Pack some clean rag firmly into the crankcase mouth to catch any dirt or debris which may drop as the piston emerges from the bore, then lift the barrel off the piston and place it to one side in the inverted position to ensure that all the residual coolant is drained away.
6   The piston should be removed with the rag in position in case a circlip is dropped. Remove the circlip from the piston by prising it free with an electrical screwdriver. Displace the gudgeon pin by pushing it through from the opposite side until it clears the small-end eye and the piston can be lifted clear. If the pin is a tight fit, it may be necessary to warm the piston so

that the grip on the gudgeon pin is released. A rag soaked in boiling water will suffice, if it is placed on the piston crown.

7   If the gudgeon pin is still a tight fit after warming the piston it can be lightly tapped out of position with a hammer and soft metal drift. **Do not** use excess force and make sure the connecting rod is supported during this operation, or there is a risk of its bending.

8   When the piston is free of the connecting rod, remove the gudgeon pin completely and also the second circlip. Place the piston, rings and gudgeon pin aside for further attention, but discard the circlips. They should never be re-used: new circlips must be obtained and fitted during rebuilding. Remove the

small-end bearing and place it inside the piston for safe keeping.

9   The piston rings are removed by holding the piston in both hands and prising gently the ring ends apart with the thumbnails until the rings can be lifted out of their grooves and on to the piston lands, one side at a time. The rings can then be slipped off the piston and put to one side for cleaning and examination. The bottom ring is supported by a very thin expander which must be prised carefully from the groove with a sharp-pointed instrument. If the rings are stuck in their grooves by excessive carbon deposits, use three strips of thin metal sheet to remove them, as shown in the accompanying illustration.

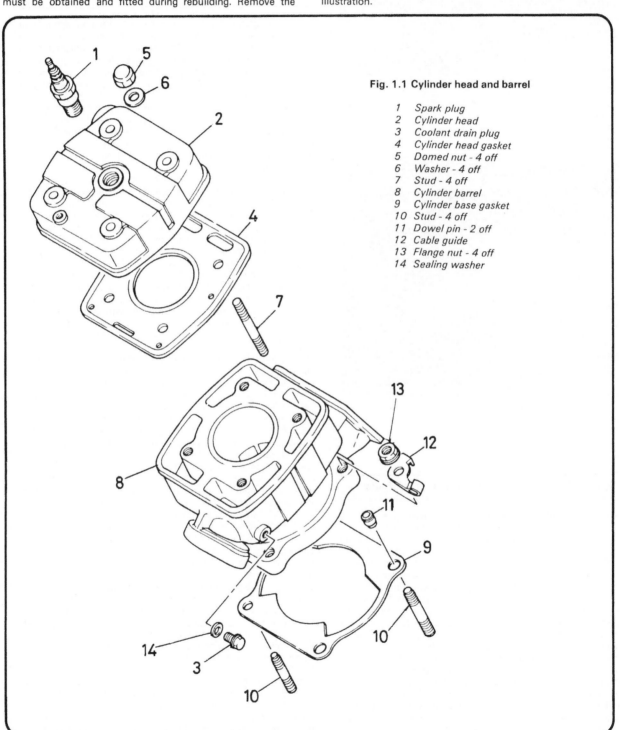

Fig. 1.1 Cylinder head and barrel

1   Spark plug
2   Cylinder head
3   Coolant drain plug
4   Cylinder head gasket
5   Domed nut - 4 off
6   Washer - 4 off
7   Stud - 4 off
8   Cylinder barrel
9   Cylinder base gasket
10 Stud - 4 off
11 Dowel pin - 2 off
12 Cable guide
13 Flange nut - 4 off
14 Sealing washer

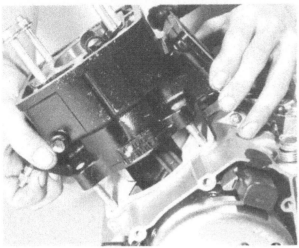

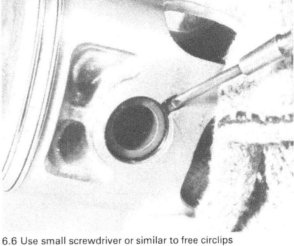

6.5 Remember to pack crankcase mouth with clean rag before removing barrel from piston

6.6 Use small screwdriver or similar to free circlips

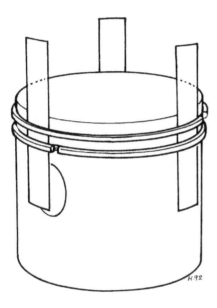

Fig. 1.2 Method of removing and replacing piston rings

7 Dismantling the engine/gearbox unit: removing the flywheel generator

1 The flywheel generator rotor and stator may be removed with the engine/gearbox unit on the workbench or installed in the frame, although in the latter case it will be necessary first to remove the gearchange pedal linkage, if working on an RD125 LC model, and to remove the crankcase left-hand cover of both models. It is essential that the correct Yamaha rotor puller, Part number 90890-01189, is available to remove the rotor safely. If the tool cannot be hired or bought from a local Yamaha dealer, note that there are pattern versions produced which are suitable for use on the majority of small-capacity Japanese motorcycles. These pattern flywheel pullers should be available at a very reasonable price at any good motorcycle dealer, and it was one of these which was used on the machines featured in the photographs accompanying the text of this Manual.

2 Note that due to the proximity of the crankcase wall to the rotor, and to the design of the rotor itself, no other method of rotor removal is recommended. If a flywheel puller of the type described and shown in the photographs is not available, take the machine to a local Yamaha dealer for the work to be carried out. Remember that it is better to pay a small labour charge than to have to pay for a complete new generator assembly or a new crankshaft as a result of damage.

3 The first step in removing the generator assembly is to lock the crankshaft so that the rotor retaining nut can be slackened. If the engine is in the frame, this can be achieved by selecting top gear and applying the back brake, thus locking the crankshaft through the transmission. If the engine is on the bench, the rotor can be held by the use of a strap wrench. On RD125 LC models care must be taken not to damage the raised ignition magnet on the periphery of the rotor itself. If the cylinder head, barrel and piston have been removed, however, the simplest and most effective method of locking the crankshaft is to pass a close-fitting metal bar through the small-end eye of the connecting rod. The bar can then be supported on two clean wooden blocks placed on the crankcase mouth so that the crankcase castings are not damaged. When the crankshaft has been locked by any of the above methods, slacken and remove the generator retaining nut and pick out the spring washer and plain washer behind it.

4 To remove the rotor, fully unscrew the rotor puller centre bolt and thread the outer body of the puller into the thread in the centre of the rotor. Note that a left-hand thread is employed. Tighten the tool body firmly so that it is screwed as far as possible into the rotor, then tighten the centre bolt firmly against the crankshaft end. Tap the centre bolt smartly on its head with a hammer to shock the rotor free. If this does not work at the first attempt, tighten the tool centre bolt further and tap it again. While this method almost always works at the first or second attempt, cases have been known of extreme stubbornness on the part of the rotor and if such a case is suspected, the work should be entrusted to a Yamaha dealer who will have the experience necessary to overcome the problem. Before this is done, remove the puller and check that there is no obvious reason for the difficulty, such as the retaining nut washers having been left in place. Repeat the operation but do not use excessive force either in tightening the puller centre bolt or in hitting the bolt head with the hammer. As previously stated, it would be better to incur a small labour charge than to risk damaging the rotor or crankshaft due to inexpert or over-enthusiastic use of the rotor puller.

5 Before the stator is removed it is advisable to mark

accurately its position in relation to the crankcase to ensure that the ignition timing is approximately correct on reassembly, although this will need to be checked in any case. Using a hammer and an old flat-bladed screwdriver or a sharp-pointed instrument, punch or scratch a line between the stator plate and one of the mounting bosses on the crankcase itself. Note that in the case of the DT125 LC model such a mark is already stamped at about the 10 o'clock position and can be used as a datum. When the original position of the stator plate has been marked, disconnect the neutral indicator switch by slackening the screw on the switch outer end and pulling the wire away. Slacken and remove the two stator plate mounting screws, disengage the generator lead sealing grommet from the crankcase wall, and pull the stator assembly away. Prise the Woodruff key out of the crankshaft keyway and store it with the rotor for safe keeping.

8 Dismantling the engine/gearbox unit: removing the crankcase right-hand cover

1   Although the crankcase right-hand cover can be removed with the engine/gearbox unit in or out of the frame, if the task is to be undertaken with the unit installed in the frame, a considerable amount of preliminary dismantling will be necessary. Note that although the oil and water pumps are fitted to the right-hand cover, they may be left in situ and need not be disturbed when removing the cover.

2   The operations to be carried out before the cover can be removed with the engine/gearbox unit in the frame are as follows:

    a)   Drain the transmission oil.
    b)   Remove the oil/water pump cover and radiator cover.
    c)   Drain the coolant, then disconnect the radiator bottom hose and either rotate it out of the way (DT125 LC) or remove it completely (RD125 LC).
    d)   Remove the right-hand cover/cylinder barrel coolant pipe.
    e)   Disconnect, plug, and remove both oil feed pipes, then disconnect the oil pump cable.
    f)   Remove the kickstart lever.
    g)   Remove the rear brake pedal assembly (DT125 LC only).

All the above operations are described in full in Section 4 of this Chapter with the exception of (d) which is described in Section 6.

3   The cover is now ready to be removed. Slacken all eight screws around the periphery of the cover, taking care not to lose the oil feed pipe clamp and, on the DT125 LC only, the stop lamp switch extension rod bracket. Ensure that all the screws are fully released and withdraw the cover, using a soft-faced mallet to break the seal of the gasket, should this be necessary.

4   The cover is located by two large dowel pins which may work loose and fall clear as the cover is removed. If such is the case, replace them immediately in the crankcase so that they are not lost. Similarly, a thrust washer is fitted over the shaft of the white nylon tachometer drive gear of the RD125 LC model; check that this is correctly in place and has not stuck to the cover or dropped away. Mop up any residual oil or coolant that may have spilt as the cover was withdrawn, and remove the cover gasket. The latter should be renewed before the cover is refitted.

5   If the cover screws are removed from the cover at any time, ensure that each screw is refitted in its original position as they are different in length.

7.4 Pattern version of rotor puller being used to remove generator rotor

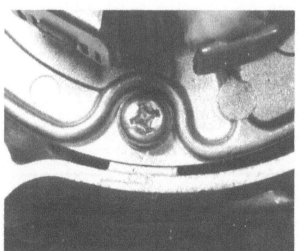

7.5a Mark stator plate before removal – RD125 ...

7.5b ... but note that there is an alignment mark on DT125

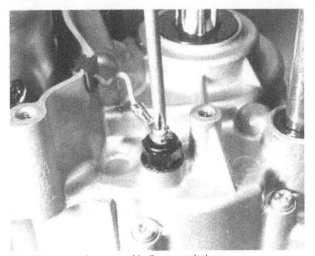

7.5c Disconnecting neutral indicator switch

9 Dismantling the engine/gearbox unit: removing the clutch assembly

1   As previously stated, the clutch can be withdrawn with the engine/gearbox unit in or out of the frame, although in either case the crankcase right-hand cover must be removed first, as described in the previous Section. Also, if it is wished to examine the clutch lifting mechanism with the engine in the frame, the cable must be disconnected to allow the operating lever assembly to be lifted out.

2   Before the clutch is removed there is one more consideration which should be remembered. If the work is being undertaken with the engine in the frame the nut which secures the primary drive pinion must be slackened at this stage if it is wished to remove the pinion. To prevent crankshaft rotation as the nut is removed, select top gear and apply the rear brake. Once the nut has been loosened the clutch can be dismantled.

3   Slacken and remove the clutch pushrod locknut, using a screwdriver to hold the pushrod, and remove also the plain washer behind the locknut. Slacken and remove the four bolts which secure the clutch springs, releasing them evenly by about one turn at a time in a diagonal sequence until they are free of spring pressure and can be unscrewed. Lift out the bolts and springs and withdraw the pressure plate, noting for future reference the cast arrow which aligns with the stamped circular mark in the clutch centre. Pull out the pushrod and refit immediately the locknut and washer so that neither is lost.

4   Remove the clutch friction and plain plates from the clutch assembly noting for future reference the way in which each is fitted. This applies particularly to the metal plain plates. Lift the plates out one by one and put them to one side to await examination and reassembly.

5   If the engine/gearbox unit is to be carefully dismantled, or if it is wished to examine the clutch lifting mechanism, the components of the mechanism must be removed. The simplest method is to tip the engine/gearbox unit on its side so that the ball and second pushrod will slide out of the input shaft centre. If this is not possible, or if it does not work correctly, find a long, slim, steel rod which will fit inside the input shaft. Magnetise the rod by stroking it gently across the magnets of the generator rotor and push it into the input shaft. As the rod is withdrawn gently it should bring the ball and then the pushrod with it. When the ball and pushrod have been removed, disengage the clutch actuating lever return spring from the crankcase top and lift the lever away, complete with its return spring and the plain washer.

6   The clutch centre nut must now be removed, and to enable this to be done, the clutch centre must be prevented from rotating. If the engine/gearbox unit is in the frame the transmission should be placed in top gear and the rear brake applied hard while the tab washer is flattened and the nut is slackened. If, however, the engine/gearbox unit is out of the frame, there are two alternatives. Yamaha produce a clutch holding tool and this can be ordered through Yamaha dealers as Part number 90890-01024. The tool is effectively a plain plate with a handle welded to it, and this can be improvised quite easily if an unwanted plate can be acquired.

7   The second alternative, which is the solution employed in the photographs accompanying the text, is to improvise a tool which locks the clutch centre via the gearbox sprocket and the transmission and uses a few readily available components. Knock back the tab of the locking washer, place the transmission in gear, and replace temporarily the gearbox sprocket over the splines of the output shaft. Find a length of used final drive chain, a metal tube large enough to accept two runs of the chain, and a metal drift which is slim enough to pass between the links of the chain. Engage the centre of the chain length around the sprocket and pass both free ends down the tube so that they can protrude at the other end. Ensure that the tube is jammed hard against the chain ends as they leave the sprocket, pull the chain tight, and push the drift through a convenient link in both runs of the chain so that the complete assembly is locked up and forms the locking tool shown in the photographs. Once the clutch centre is locked, the retaining nut can be slackened by one person alone, but the task is made easier if the aid of an assistant is enlisted to hold the locking tool and to steady the engine/gearbox unit while the nut is slackened.

8   Remove the clutch centre retaining nut and its tab washer. The clutch centre can then be pulled off the input shaft, followed by a spacer and the clutch outer drum. Behind the clutch outer drum is a spacer which is keyed in place on the input shaft. Slide the spacer off the shaft, pick out the key with a pair of needle-nose pliers, and remove the dished thrust washer behind the key, noting that the thrust washer is fitted with its convex side towards the crankcase wall. Store all the clutch components together in a clean container to avoid losing or damaging them.

9.7 Tool fabricated to lock transmission during dismantling and reassembly

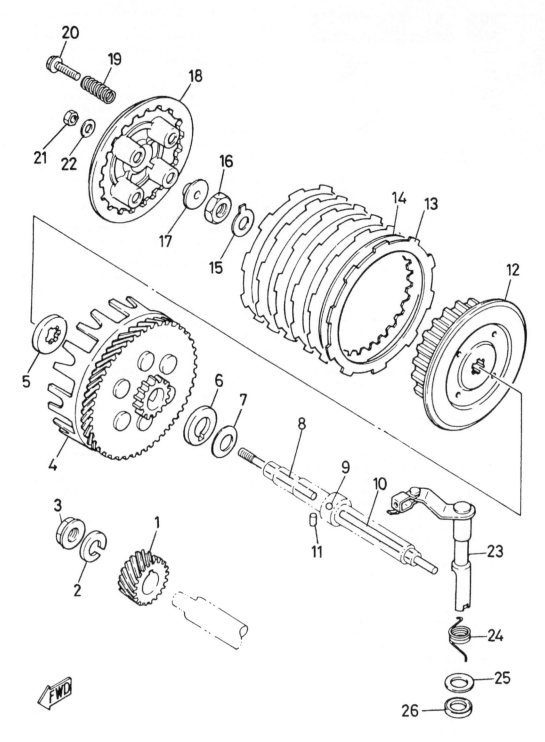

Fig. 1.3 Clutch

1	Primary drive gear	8	Pushrod - 1st	15	Tab washer	21	Lock nut
2	Spring washer	9	Ball	16	Nut	22	Washer
3	Nut	10	Pushrod - 2nd	17	Push rod end	23	Actuating lever
4	Outer drum	11	Key	18	Pressure plate	24	Return spring
5	Spacer	12	Clutch centre	19	Spring - 4 off	25	Washer
6	Spacer	13	Friction plate - 6 off	20	Bolt - 4 off	26	Oil seal
7	Dished thrust washer	14	Plain plate - 5 off				

10 Dismantling the engine/gearbox unit: removing the kickstart assembly and idler gear

1   The kickstart assembly and idler gear can be removed with the engine/gearbox unit in or out of the frame, once the crankcase right-hand cover has been withdrawn as described in Section 8 of this Chapter. Note that while the kickstart shaft can be removed with the clutch assembly in situ, if the kickstart idler gear is to be removed, the clutch must be withdrawn first.

2   To remove the kickstart assembly, use a suitable pair of pliers to unhook the outer end of the kickstart return spring from its locating hole in the crankcase wall, then allow the spring to return slowly to its relaxed position. Care is required at this point as the spring is under moderately strong tension. The kickstart assembly may then be lifted out of the crankcase.

3   Slide the white nylon spring guide down the length of the shaft and remove it. Disengage the inner end of the spring from the hole in the kickstart shaft and remove the spring, making a note of the way in which it is fitted. Slide off the large plain washer and the kickstart pinion, the latter complete with its friction clip.

4   If removal of the kickstart idler gear is required, this can be done only after the clutch has been removed. Using a pair of circlip pliers, remove the circlip from the output shaft right-hand end, then withdraw the first thrust washer, the idler gear itself, and the second thrust washer. Lastly, use the circlip pliers to remove the inner retaining circlip from the output shaft end. Because both circlips and both thrust washers are identical components, there is no need to differentiate between the inner and outer parts. Store all the kickstart components together in a clean container so that nothing is lost or damaged.

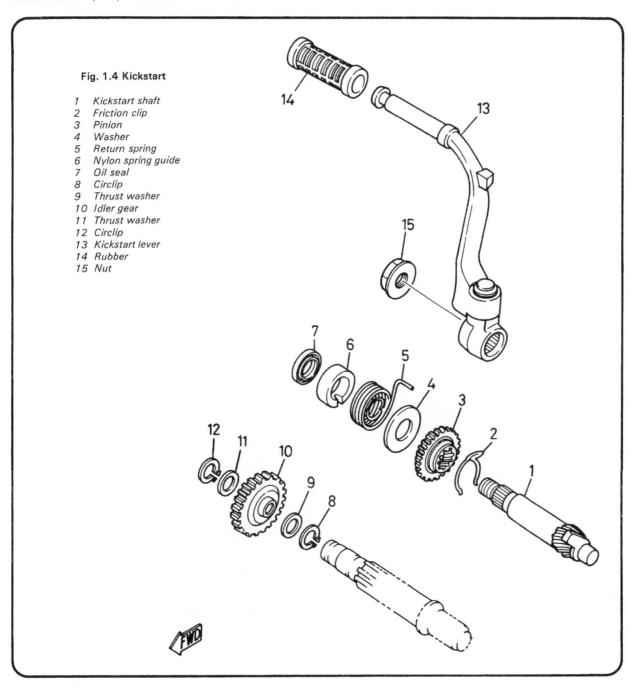

Fig. 1.4 Kickstart

1   Kickstart shaft
2   Friction clip
3   Pinion
4   Washer
5   Return spring
6   Nylon spring guide
7   Oil seal
8   Circlip
9   Thrust washer
10 Idler gear
11 Thrust washer
12 Circlip
13 Kickstart lever
14 Rubber
15 Nut

11 Dismantling the engine/gearbox unit: removing the primary drive gear, balancer drive gear and tachometer drive gear

1   All three of the components mentioned in the heading can be removed with the engine/gearbox unit in or out of the frame but it will be necessary to remove first the crankcase right-hand cover and the clutch as described in Sections 8 and 9 of this Chapter. It would be advisable, however, first to devote some thought to the problem of which method of locking the crankshaft is to be adopted so that the primary drive and balancer shaft gear retaining nuts can be slackened.

2   The manufacturer's recommended method is to lock the crankshaft by wedging a tightly-wadded piece of rag between the teeth of the primary drive gear and the clutch outer drum. A similar procedure is adopted for the balancer shaft gear once the clutch outer drum has been removed. This is, however, a risky and somewhat imprecise method for any but the most experienced mechanic to adopt. An alternative, if the generator rotor should still be fitted to the crankshaft left-hand end, is to hold the rotor with a strap wrench so that the crankshaft is locked. This method, however, will require the aid of an assistant to hold the strap wrench, and on RD 125 LC models, care will be necessary to avoid damaging the ignition trigger on the periphery of the generator rotor. A further possibility assumes that the clutch has not yet been dismantled and was mentioned in Section 9 of this Chapter to forewarn the owner. In such a case, it is possible to place the transmission in top gear and to apply the rear brake, if the work is being carried out in the frame, or to assemble and use the gearbox sprocket locking tool if the engine/gearbox unit is on the bench. Using this method, the primary drive gear nut is easy to slacken but the balancer shaft gear nut can only be reached with an open-ended spanner due to the proximity of the clutch outer drum, and so requires extreme care if damage is to be avoided.

3   Having considered the above alternatives, the method adopted when dismantling the machine featured in this Manual is much the simplest and safest method, but presupposes that the cylinder head, barrel, and piston have been removed. Once the clutch has been withdrawn as described in section 9 of this Manual, bend back the raised tab of the lock washer which secures the balancer shaft gear retaining nut and pass a close fitting metal bar through the small-end eye of the connecting rod. Support the bar on two clean wooden blocks placed over the crankcase mouth to prevent damage to the castings. The crankshaft is then locked and both nuts can be slackened and removed.

4   To remove the primary drive gear, withdraw the large spring washer which lies behind the retaining nut, then slide the gear itself off the crankcase right-hand end. Before removing the larger diameter balancer shaft drive gear which is also on the crankshaft note the stamped dot on its periphery which should align with a similar mark on the balancer shaft gear when the piston is at TDC. If these marks are indistinct they should be picked out with a blob of paint or a spirit-based felt marker before the balancer timing is disturbed. Once this point has been attended to, slide the balancer shaft drive gear off the crankshaft and prise out the long, square-section Woodruff key from the crankshaft keyway.

5   To remove the balancer shaft gear, lift off first the tab washer, then the gear itself. Prise the short, square-section Woodruff key from the balancer shaft keyway. Store the two groups of components, those from the crankshaft and those from the balancer shaft, in separate containers to avoid any possibility of confusion on reassembly.

6   The tachometer fitted to the RD 125 LC model only is driven from the clutch outer drum, the drive being transmitted via a white nylon drive gear to a vertically-mounted driven gear which is machined at the top to accept the drive cable inner. The whole assembly can be removed, once the crankcase right-hand cover is withdrawn, whether the clutch is in situ or not. The drive gear and its shaft can be pulled out easily by hand, but

note that there are two small, identical, thrust washers which are fitted one on each side of the drive gear for the purpose of location of the gear. These must not be lost or allowed to drop clear. Remove the drive gear and place the gear, thrust washers, and shaft together in a clean container. The remainder of the tachometer drive assembly is removed by unscrewing the single retaining bolt set in the crankcase top surface and by pulling away the black plastic housing and vertical driven gear. Note that if the housing is a particularly tight fit in the crankcase, it should be pulled out with a twisting motion, as if unscrewing it. Do not attempt to lever it out as the plastic moulding would only be damaged.

12 Dismantling the engine/gearbox unit: removing the external gear selector components

1   The components of the gear selector mechanism which can be removed without separating the crankcase halves consist of the gearchange shaft, incorporating the selector claw assembly, and the selector detent arm. Either component may be removed with the engine/gearbox unit in the frame or on the workbench, but in each case it will be necessary first to remove the crankcase right-hand cover and the clutch assembly, as described in Sections 8 and 9 of this Chapter. If the gearchange shaft is to be removed it will be necessary, of course, to release the gearchange pedal from the shaft left-hand end.

2   When the preliminary dismantling has been carried out and the selector claw assembly of the gearchange shaft is exposed, asses the state of the shaft return spring by applying pressure first downwards, and then upwards, to the claw end of the shaft. In both cases, release the shaft and allow it to return under spring pressure to the central position before applying pressure in the opposite direction. The shaft should centre itself quickly and positively, with no trace of free play in the centre position or of weakness in the spring. If any fault is apparent the spring must be renewed on reassembly. This test is quick and easy to carry out, and must be considered essential if the engine/gearbox unit is being stripped to find a fault in the selector mechanism.

3   The gearchange shaft can be pulled from the crankcase by hand. If the return spring is to be renewed, clamp the shaft in a vice fitted with padded jaws and use a pair of pliers or a small-bladed screwdriver to remove the large circlip at the shaft right-hand end. Note carefully the way in which the spring is fitted,

12.1 External gear selector components

then prise its ends off the protruding tang of the selector claw plate. The spring can then be removed from the shaft.

4 The selector detent arm is tensioned by a heavy spring pivoted on a shouldered bolt. Note carefully the way in which the spring is fitted, then gradually slacken the bolt until the spring pressure can be released by allowing the detent arm to ride over the selector cam. Slacken fully the bolt and remove the bolt, the arm, and the spring as a complete assembly.

13 Dismantling the engine/gearbox unit: separating the crankcase halves

1 The engine/gearbox unit must be removed from the frame and all preliminary dismantling operations carried out before the crankcase halves can be separated. These operations are described in Sections 4-12 of this Chapter. When the engine/gearbox unit has been dismantled as described, make a final check to ensure that nothing has been omitted which might obstruct operations. Rotate the selector drum so that the protruding ears on the selector cam are aligned with the cut-outs in the crankcase wall. Place two clean wooden blocks on the workbench and support the engine/gearbox unit on them so that the crankcase left-hand side is uppermost.

2 Before the twin crankcase fastening screws are removed, there is a simple device which can be used to ensure that none are lost and that each screw is refitted in its correct position. Find a sheet of cardboard and draw on it a rough outline of the crankcase left-hand side, marking the position of each of the screws. Punch holes in the cardboard at each screw position, then as each screw is removed it can be pushed into its correct corresponding hole in the cardboard and kept there until needed on reassembly.

3 The screws are likely to be tight and will require the use of an impact drive and a hammer to shock them free. Starting at the outside and working inwards in a diagonal sequence, slacken each screw by one turn at a time until the pressure is released and the screws can be fully removed. Ensure that any breather tube clamps are retained with their mounting screws on the cardboard template.

4 When all the crankcase fastening screws have been removed, invert the engine/gearbox unit and support it on the two wooden blocks so that the crankcase right-hand half is uppermost. The manufacturer's recommended method of separating the crankcase halves involves the use of a special tool which is used as follows. Refit the primary drive gear retaining nut on to the crankshaft right-hand end, screwing it down until the nut is flush with the crankshaft end. Using an impact screwdriver and a hammer, slacken and remove the single screw which retains the crankshaft right-hand main bearing oil seal retaining plate. Wind the tool centre bolt out as far as possible and fit the complete tool to the crankcase, screwing its two mounting bolts as far as possible into the tapped holes on each side of the right-hand main bearing boss. Tighten the centre bolt on to the crankshaft right-hand end and check that the tool is positioned absolutely square. It may be necessary slightly to slacken either of the tool mounting bolts so that the tool can be positioned accurately and is not pressing on the crankshaft at an awkward angle. Draw the crankcase right-hand half off the left-hand half by tightening the tool centre bolt and by using a soft-faced mallet to tap gently around the crankcase joint area and on the right-hand ends of the gearbox shafts. Check carefully at frequent intervals that the crankcase halves are parting evenly so that there is no possibility of the casting lifting at an angle, thus causing one or more of the gearbox shafts to stick in their respective right-hand bearings, and remember to withdraw the tool centre bolt far enough to permit removal of the crankshaft nut when the initial pressure is released because the nut will otherwise prevent the crankshaft from being pressed through the right-hand main bearing. The nut is merely a precaution to prevent damage to the crankshaft end caused by the initial pressure required to separate the castings.

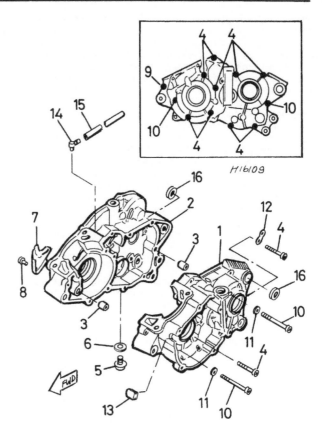

H16109

Fig. 1.5 Crankcases

1	Left-hand crankcase half	9	Screw
2	Right-hand crankcase half	10	Screw - 2 off
3	Dowel pin - 2 off	11	Washer - 2 off
4	Screw - 9 off	12	Pipe guide
5	Drain plug	13	Grommet
6	Sealing washer	14	Pipe union
7	Bearing retaining plate	15	Vent hose
8	Screw	16	Spacer - 2 off

5 Note that when using the special tool in the manner described above, extreme force will not be required at any time. If the crankcase right-hand half does start to lift at an angle, release the tool centre bolt slightly, tap the casting down until it is level, and start again, using the soft-faced mallet to assist the tool. Apply the mallet only to the gearbox shaft ends and around the joint area. It is permissible to tap upwards on the large lugs which form the engine rear mounting bosses, but be careful never to hit any of the machined sealing faces of the casting. This method is much the safest way of separating the crankcase halves and is to be recommended if the special tool can be acquired.

6 There is, however, an alternative method which does not require the use of a special tool, but requires instead some care and patience. Support the engine/gearbox unit on wooden blocks so that it is lying horizontally with the crankcase right-hand half uppermost. The object is to draw the crankcase right-hand half off the left-hand half, leaving the crankshaft, the balancer shaft, and the gearbox components in the left-hand casting.

7 When using this method a soft-faced mallet of hide or rubber must be employed, although a metal hammer may be used in conjunction with a soft wood drift if such a mallet is not available. Do not at any time strike the machined sealing

surfaces of either of the crankcase castings and never attempt to lever the castings apart with any kind of tool. The risk of severe and expensive damage is too great. Fit the primary drive gear retaining nut on the crankshaft right-hand end, threading the nut down until it is flush with the end of the crankshaft. Similarly fit the clutch centre retaining nut to the input shaft right-hand end. These nuts will protect the threaded ends of their respective shafts.

8   Tap gently but firmly on the right-hand end of all the shafts which protrude from the crankcase right-hand half. These are the crankshaft, balancer shaft, input shaft, output shaft, selector drum, and the rear selector fork shaft. Tap also on the area around the joint face of the two crankcase halves and repeat the process until the joint separates. Then carry on tapping on the shaft ends while lifting the crankcase right-hand half with the free hand. Ensure by careful tapping that the right-hand half lifts squarely and evenly at all times. Remove the shaft-protecting nuts when they begin to foul the casting as it is lifted away. When this latter method of crankcase separation was employed on the machine featured in this Manual, it was found that the crankcase right-hand half lifted easily away, with the bare minimum of force being required.

9   Whichever of the above methods is employed to separate the crankcase halves, care must be taken to ensure that all components remain in the crankcase left-hand half. The only component which is likely to prove troublesome is the front selector fork shaft which may stick in the right-hand casing. Pay attention to this point when lifting the casing away. Similarly check that the two locating dowels set in the gasket surface of the two crankcase halves do not fall clear.

13.1 Rotate selector drum so that it can pass through cut-outs in crankcase

14 Dismantling the engine/gearbox unit: removing the crankshaft and gearbox components from the crankcase

1   When the crankcase halves have been separated, the crankshaft, the balancer shaft, and the gearbox components may be lifted out.

2   Start by pulling out the front, tubular, selector fork shaft and the rear, solid, selector fork shaft. Pivot all three selector forks about their respective pinions so that the guide pins come away from the selector drum and provide sufficient clearance for the drum to be withdrawn. This is done by pulling the drum out of its locating hole in the crankcase.

3   Withdraw the selector forks, taking great care to note exactly where each one is fitted and in what way it is fitted. It is useful to degrease each fork as it is withdrawn and to mark it with a spirit-based felt pen as a guide to correct assembly but

note that the selector forks fitted to the machine featured in the photographs were identified by numbers and model codes cast in each side, and if such is found to be the case on the machine being worked on, these marks will be sufficient provided notes are made of their position. As each fork is identified and withdrawn, replace it on its shaft to prevent any confusion on reassembly.

4   The gear clusters should be removed now. Carefully pull them out of their bearings, treating both as a single assembly. It may be necessary to tap on the output shaft left-hand end to free the shaft from its bearing.

5   The balancer shaft can be lifted out easily by hand, leaving only the crankshaft to be removed from the crankcase left-hand side. The manufacturer recommends the use of the same special tool used for separating the crankcase halves. The two mounting bolts of the tool should be threaded into the drilled and tapped holes of the two bosses cast on either side of the left-hand main bearing housing. The tool centre bolt is then tightened down on the crankshaft left-hand end and is used to press the crankshaft out. If, as is likely, the tool is not available, an alternative method is given which requires only a soft-faced mallet and some care.

6   To use this method, replace the rotor nut on the crankshaft end to avoid damaging the thread and place the crankcase on two wooden blocks situated as close around the crankshaft as possible to give maximum support. The blocks should be of a size to hold the crankcase half far enough from the workbench to allow the crankshaft to be removed. Using a soft-faced mallet with one hand, and supporting the crankshaft with the other, carefully tap the crankshaft out of its housing. Do not use excessive force and do not allow the crankshaft to drop away.

7   Put all the removed components to one side to await examination and reassembly.

15 Dismantling the engine/gearbox unit: removal of ancillary components from the crankcase halves

1   If any work is to be done on the crankcase halves themselves, removal of additional items will be required. Using an impact driver to release the fixing screws, remove any oil seal or bearing retaining plates from the crankcase halves, and pull the spacer out of the right-hand main bearing oil seal.

2   As a general rule, all the oil seals should be discarded and renewed whenever the crankcases are separated. This is especially true of the crankshaft main bearing oil seals, which are essential to the continued good performance and reliability of the engine. To remove them, a tool must be fabricated from an old broad-bladed screwdriver. Heat the tip with a blowlamp to a point where it can be bent into a slightly curved shape by hammering. Remove any sharp edges with a file. When cooled, insert the tip under one lip of the seal to be removed and lever that side of the seal out, pivoting the screwdriver if necessary on a piece of wood to prevent damage to the casting. Be careful that the screwdriver tip does not scratch the seal housing. Once one side of the seal is levered from its position, it can be easily pulled away by hand. If there is no bearing behind the seal, and if the seal is not retained by a lip in the casting, it can be driven out using a hammer and a socket or tubular drift of suitable size.

3   Unscrew the neutral indicator switch if necessary. This should be done using the bare minimum of force possible as the switch is only a light plastic moulding.

4   If any of the bearings are to be removed, the casting must first be heated. This is to take advantage of the fact that an aluminium alloy crankcase casting has a higher coefficient of expansion than a steel bearing outer race and that therefore the casting will, when heated, expand faster than the bearings. This will have the effect of loosening the bearings in their housings to the point where they can relatively easily be drifted or pulled away. To achieve this, the complete crankcase half must be heated to approximately 100°C by a gradual application of heat over its entire surface to prevent the distortion which would

result from a fierce local application of heat. An oven is the best method of heating the casting, but an alternative is to place the casting in a suitable container and then pour boiling water over it. Do not use a welding torch or blowlamp for this operation as the even application of heat cannot be guaranteed by the inexperienced. It will be evident that, whichever method is used, great care must be taken both with the method of heating and in subsequently handling the heated casting, if serious personal injury is to be avoided.

5   Bearings which are to be used again must be removed by tapping them out of their housings using a hammer and a socket spanner or tubular drift which bears only on the outer race of the bearing concerned. If any other part of the bearing is used, unacceptably high side loadings will be placed on the balls or rollers and their cages, causing premature failure. If the entire outer race is not accessible, for example where a bearing is fitted in a blind housing, some other means of bearing removal will have to be found. The most widely used method is to prepare a clean, flat wooden surface, heat the casting as previously described and tap the casting firmly and squarely on to the wooden surface with enough force to jar the bearings free. Great care must be taken that the casting is tapped squarely on to a clean surface to avoid damaging the gasket surface. An alternative method is to heat the casting as described, rest it on a clean wooden surface to support it evenly, and to tap the casting with a soft-faced mallet directly behind the bearing, again with the object of jarring the bearing free. It is stressed that in both cases, tapping must be as gentle as possible to avoid damaging the casting. If either method fails, take the casting to an authorised Yamaha dealer for the bearings to be removed using a slide hammer with an internal puller attachment.

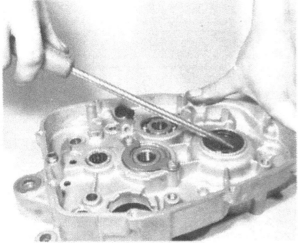

15.2 Using screwdriver to lever out oil seals

16   Examination and renovation: general

1   Before examining the parts of the dismantled engine unit for wear it is essential that they should be cleaned thoroughly. Use a petrol/paraffin mix or a high flash-point solvent to remove all traces of old oil and sludge which may have accumulated within the engine. Where petrol is included in the cleaning agent normal fire precautions should be taken and cleaning should be carried out in a well-ventilated place.
2   Examine the crankcase castings for cracks or other signs of damage. If a crack is discovered it will require a specialist repair.
3   Examine carefully each part to determine the extent of wear, checking with the tolerance figures listed in the Specifications section of this Chapter or in the main text. If there is any doubt about the condition of a particular component, play safe and renew.

4   Use a clean lint free rag for cleaning and drying the various components.
5   Various instruments for measuring wear are required, including a vernier gauge or external micrometer and a set of standard feeler gauges. Also an internal and an external micrometer will be required to check wear limits. Additionally, although not absolutely necessary, a dial gauge and mounting bracket is invaluable for accurate measurement of end float, and play between components of very low diameter bores, where a micrometer cannot reach. After some experience has been gained the state of wear of many components can be determined visually or by feel and thus a decision on their suitability for continued service can be made without resorting to direct measurement.

17   Cylinder head: examination and renovation

1   Remove all traces of carbon from the cylinder head, using a blunt-ended scraper. Finish by polishing with metal polish, to give a smooth, shiny surface. This will aid gas flow and will also prevent carbon from adhering so firmly in the future.
2   Check the condition of the threads in the spark plug holes. If the threads are worn or stretched as the result of overtightening the plugs, they can be reclaimed by a 'Helicoil' thread insert. Most dealers have the means of providing this cheap but effective repair.
3   Inspect the water passages cast into the cylinder head, and where necessary remove any accumulated corrosion or scale. As mentioned previously, this can result from failure to use the recommended coolant mixture. Be sure to remove any debris from the passages by flushing them through with clean water.
4   Lay the cylinder head on a sheet of plate glass to check for distortion. Aluminium alloy cylinder heads will distort very easily, especially if the cylinder head bolts are tightened down unevenly. If the amount of distortion is only slight, it is permissible to run the head down until it is flat once again by wrapping a sheet of very fine emery cloth around the plate glass sheet and rubbing with a rotary motion.
5   If the cylinder head is distorted badly, it is advisable to fit a new replacement. Although the head joint can be restored by skimming, this will raise the compression ratio of the engine and may adversely affect performance.

18   Cylinder barrel: examination and renovation

1   The usual indication of a badly worn cylinder barrel and piston is piston slap, a metallic rattle that occurs when there is little or no load on the engine. If the top of the bore of the cylinder barrel is examined carefully, it will be found that there is a ridge on the thrust side, the depth of which will vary according to the amount of wear that has taken place. This marks the limit of travel of the uppermost piston ring.
2   Measure the bore diameter just below the ridge, using an internal micrometer. Compare this reading with the diameter at the bottom of the cylinder bore, which has not been subjected to wear. If the difference in readings exceeds 0.08 mm (0.003 in) the cylinder should be rebored and fitted with an oversize piston and rings.
3   Bore ovality should also be checked, the maximum allowable being 0.05 mm (0.002 in). Given that the bore is within the above limits and that the piston is in serviceable condition (see Section 19) the parts may be re-used. Ovality may be corrected to some extent by honing, provided that this does not cause the maximum piston to bore clearance to be exceeded. A Yamaha dealer or a reputable engineering company will be able to assist with honing work should this prove necessary.
4   If scoring of the cylinder walls is evident it will normally prove necessary to have it re-bored to the next oversize, though

light scratching may sometimes be removed by careful honing or by judicious use of abrasive paper. If the latter approach is adopted be careful to avoid removing more than the absolute minimum of material. The paper should be applied with a rotary motion **never** up and down the bore, which would cause more problems than it solves. One of the proprietary 'glaze busting' attachments for use in electric drills can be used to good effect for this operation. Even where the bore is in good condition, the glaze busting operation should be undertaken prior to reassembly. The light scratch marks around the bore surface assist in bedding in the rings and help initial lubrication by holding a certain amount of oil.

5   If reboring is necessary, obtain the piston first, then have the boring done to suit the new piston. Most Yamaha dealers have an arrangement with a local engineering company and will be able to get the reboring work carried out promptly. Note that oversizes of 0.25, 0.50, 0.75, and 1.0 mm (0.010, 0.020, 0.030 and 0.040 in) are available from the manufacturer.

6   Carefully remove any accumulated carbon deposits from the cylinder bore and ports, taking care not to damage the bore surface. It is recommended that the ports are cleaned completely but carefully, taking great care to avoid burning the edges of the ports where they enter the bore. To prevent the rings from becoming chipped or broken dress any burrs with fine emery paper.

7   It is inadvisable to attempt modification of the port sizes or profiles to obtain more power from the engine. Such modifications are feasible but should only be considered for racing purposes. Generally speaking, the changed characteristics of the engine would make it unwieldy for road use, and it should be noted that the machine's warranty would be invalidated.

8   Check the water passages for rust and scale. These may have built up, especially where the correct coolant has not been used. If necessary, scrape the passages clean using wire or an old screwdriver, taking care to flush out any debris. Bear in mind that any residual debris may clog the radiator pump if it is not removed.

18.1 Check cylinder bore carefully for wear or damage (see text)

19 Piston and piston rings: examination and renovation

1   If a rebore is necessary, the existing piston and piston rings can be disregarded because they will have to be replaced with their new oversize equivalents as a matter of course.

2   Remove all traces of carbon from the piston crown, using a

blunt-edged scraper to avoid scratching the surface. Finish off by polishing the crown with metal polish, so that carbon will not adhere so readily in the future. Never use emery cloth on the soft aluminium.

3   Piston wear usually occurs at the skirt or lower end of the piston and takes the form of vertical streaks or score marks on the thrust face. There may also be some variation in the thickness of the skirt, in an extreme case. The overall diameter of the piston is measured 10 mm (0.4 in) above the bottom of the piston skirt, at right angles to the gudgeon pin axis. This figure is subtracted from the minimum cylinder bore diameter measurement taken as described in the previous Section to obtain the piston/cylinder bore clearance figure.

4   The piston ring grooves may have become enlarged in use, allowing the rings to have greater side float. If the clearances exceed those given, the rings, and possibly the piston, must be renewed. Piston ring/groove clearance is measured with feeler gauges, noting that in the case of the keystone-type top ring, the gauge must be inserted between the piston ring lower face and the piston, as the tapered top surface of the piston ring might give an inaccurate reading. If the clearance is found to be beyond the limits given in the Specifications Section, measure the thickness of the piston ring. Should the rings be found to be worn they must be renewed, but if they are found to be of the correct thickness, the piston grooves are at fault and the piston must be renewed.

5   Piston ring wear is measured by removing the rings from the piston and inserting them in the cylinder, using the crown of the piston to locate them about 20 mm from the bottom of the bore. Make sure they rest squarely in the bore. Measure the end gap with a feeler gauge; if the gap exceeds that given, the rings must be renewed. Note that if new piston rings are to be fitted, their end gaps must be measured in the same way, and if too small, the gaps must be opened up to the correct limits by filing the ring ends. If the gaps are too large, and the bore is known to be within the set wear limits, another set of rings should be tried. Do not assume that new components will automatically fit your machine perfectly as even the most carefully made parts will have some variation due to manufacturing tolerances.

6   Whenever new piston rings or a new piston are fitted in a worn bore, the cylinder bore surface must be prepared by honing. This will remove the glazed finish present in any used cylinder bore and will provide a slightly roughened surface which will enable the new components to bed in properly; if it is not carried out the components will not be able to bed in as well or as fast as they should, giving a poor seal and subsequent poor performance. Additionally any wear ridge which has been formed at the top of the bore by the top ring **must** be removed if new rings are fitted. The honing operation can be carried out by the owner if he has the necessary equipment. As previously mentioned, glaze-busting tools are readily available in motor accessory shops, usually being sold as attachments for an electric drill. If such is the case, ensure that the equipment is small enough for use in a motorcycle cylinder barrel and take great care to follow the manufacturer's instructions. Use the glaze-busting equipment as little as possible so that the required finish is achieved with the bare minimum of metal being removed. If glaze-busting equipment is not available, a local motorcycle dealer should be able to carry out the work for a small charge. If the cylinder barrel is being rebored, the bore surface should be honed as a matter of course to ensure that the correct finish is achieved and that the edges of the ports are suitably chamfered.

7   When examining the piston, finish by checking that the piston ring pegs are securely fixed in the piston grooves and by checking the gudgeon pin/piston clearance. It is imperative that the piston ring pegs are securely fixed in the grooves, as otherwise the rings would be free to rotate, with the danger that one of the ring ends might be trapped in the ports. The gudgeon pin should be a tight push fit in the piston and have no discernible free play in the installed position. If this is not the case, the gudgeon pin, or possibly even the piston, must be renewed.

19.3 Measuring piston overall diameter

19.4 Measuring piston ring/groove clearance

19.5 Measuring piston ring fitted end gap

20 Crankshaft and main bearings: examination and renovation

1   Check the crankshaft assembly visually for damage, paying particular attention to the slot for the Woodruff Key and the threads at each end of the mainshaft. Should these have become damaged specialist help will be needed to reclaim them.

2   The connecting rod should be checked for big-end bearing play. A small amount of endfloat is normal, but any up and down movement will necessitate renewal. Grasp the connecting rod and pull it firmly up and down. Any movement will soon become evident, but be careful that axial clearance (endfloat), a set amount of which is permissible, is not mistaken for wear. Assuming that the big-end bearing is in good order, attention should be turned to the rest of the connecting rod. Visually check the rod for straightness, particularly if the engine is being rebuilt after a seizure or other catastrophe. Look also for signs of cracking. This is extremely unlikely, but worthwhile checking. Spotting a hairline crack at this stage may save the engine from an untimely end. Should any wear or damage be found, it will

be necessary to consult an expert for advice as to whether repair is possible or if the purchase of a new crankshaft assembly is advisable.

3   If measuring facilities are available, set the crankshaft in V-blocks and check the big-end radial clearance using a dial gauge mounted on a suitable stand. Big-end radial clearance is measured as the amount of lateral deflection at the connecting rod small-end eye, thus magnifying any wear present in the bearing. Refer to the accompanying illustration if clarification of this point is required. Crankshaft run-out can also be checked with the V-blocks and dial gauge; the gauge pointer being applied to the straight, non-tapered, part of each mainshaft. Big-end axial clearance (endfloat) is measured with feeler gauges of the correct thickness which should be a firm sliding fit between the thrust washer next to the big-end eye and the machined shoulder on the flywheel. If any clearance exceeds the wear limits given in the specifications Section of this Chapter the crankshaft assembly should be taken to an authorised Yamaha dealer or similar repair agent for repair or renewal.

4   It should be noted that crankshaft repair work is of a highly specialised nature and requires the use of equipment and skills not likely to be available to the average private owner. Such work should not be attempted by anyone without this equipment and the skill to use it.

5   A caged roller bearing is employed as the small-end bearing. This bearing can be removed quite easily for examination. Check the rollers for any imperfection, renewing the bearing if less than perfect. The gudgeon pin and small-end eye should be checked where the rollers bear upon them, and remedial action taken where the surface(s) are marked. Assemble the small-end bearing and gudgeon pin in the connecting rod eye, and check for radial play. If any movement is found, renew the bearing, particularly if the engine has produced a characteristic rattle in the past, indicating that all might not be well with this bearing.

6   The crankshaft main bearings are of the ball journal type and usually remain in place in their respective crankcase halves on removal of the crankshaft. In this case, if they are to be removed for examination, the removal procedure is as described in Section 15 of this Chapter. If, however, they stick on the crankshaft as it is removed, a conventional knife-edged bearing puller must be used to remove them for examination or renewal. They should be washed thoroughly in clean petrol (gasoline) and checked for radial and axial play. Spinning the bearing

when dry will highlight any rough spots, producing obviously excessive amounts of noise once the lubricating film has been removed. Any signs of pitting or scoring of the bearing tracks or balls indicates the need for renewal. If there is any doubt about the condition of the main bearings, they should be renewed.

7   Failure of both the big-end bearing and the main bearings may not necessarily occur as the result of high mileage covered, if the machine is used only infrequently. It is possible that condensation within the engine may cause premature bearing failure. The condition of the flywheels is usually the best guide. When condensation troubles have occurred, the flywheels will rust and become discoloured. Note too that lack of care when disturbing the cylinder head or barrel can allow coolant to find its way into the crankcase. This will soon corrode and destroy the bearings and should be avoided for obvious reasons.

20.3 Measuring big-end axial clearance

21 Oil seals: examination and renovation

1   The crankshaft oil seals form one of the most critical parts in any two-stroke engine because they perform the dual function of preventing oil from leaking along the crankshaft and preventing air from leaking into the crankcase when the incoming mixture is under crankcase vacuum during induction.
2   Oil-seal failure is difficult to define precisely, although in most cases the machine will become difficult to start, particularly when warm. The engine will also tend to run unevenly and there will be a marked fall-off in performance, especially in the higher gears. This is caused by the intake of air into the crankcases which dilutes the mixture whilst it is in the crankcase, giving an exceptionally weak mixture for ignition.
3   It is unusual for the crankshaft seals to become damaged during normal service, but instances have occurred when particles of broken piston ringss have fallen into the crankcases and lacerated the seals. A defect of this nature will immediately be obvious.
4   In view of the foregoing remarks it is recommended that the two crankshaft oil seals are renewed as a matter of course during engine overhaul.

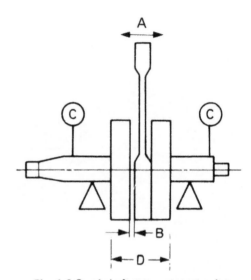

Fig. 1.6 Crankshaft measurement points

A   Big-end radial clearance
B   Big-end axial clearance
C   Crankshaft runout
D   Crankshaft width across flywheels

22 Crankcase castings: examination and renovation

1   The crankcase halves should be thoroughly degreased, using one of the proprietary water-soluble degreasing solutions such as Gunk. When clean and dry a careful examination should be made, looking for signs of cracks or other damage. Any such fault will probably require either professional repair or renewal of the crankcases as a pair. Note that any damage around the various bearing bosses will normally indicate that crankcase renewal is necessary, because a small discrepancy in these areas can result in serious mis-alignment of the shaft concerned. It is important to check crankcase condition at the earliest opportunity, because this will permit remedial action to be taken and any necessary machining or welding to be done whilst attention is turned to the remaining engine parts.
2   In most cases, badly worn or damaged threads can be reclaimed by fitting a thread insert. This is a simple and inexpensive task, but one which requires the correct taps and fitting tools. It follows that the various threads should be checked and the cases taken to a local engineering works or motorcycle dealer offering this service so that repair can take place while the remaining engine parts are checked.
3   Damaged cylinder head barrel studs may be removed by locking two nuts together on the stud and by unscrewing them with a spanner applied to the lower nut.

21.4 Renew main bearing oil seals as a matter of course

23 Primary drive: examination and renovation

1   The primary drive consists of a crankshaft pinion which engages a large gear mounted on the inner face of the clutch drum. Both components are relatively lightly loaded and will not normally wear until very high mileages have been covered.

2   If wear or damage is discovered it will be necessary to renew the component concerned. In the case of the large driven gear it will be necessary to purchase a complete clutch drum because the two items form an integral unit and cannot be obtained separately. Note that the large driven gear/clutch outer drum assembly has a shock absorber fitted. Check for play in this by holding the clutch outer drum and attempting to twist or rotate the primary driven gear backwards and forwards. Unfortunately no figures are given with which to assess the state of the shock absorber unit, and it will therefore be a matter of experience to decide whether renewal is necessary or not. Seek an expert opinion if any doubt exists about the amount of play discovered.

3   When obtaining new primary drive parts note that the two components are matched to give a prescribed amount of backlash. To this end, ensure that the match marks etched on the inner face of each are similar to avoid excessive or insufficient clearance. The marks, in the form of the letters 'B', 'C', or 'D', must be the same on both components, as shown in the accompanying photograph.

24 Clutch assembly: examination and renovation

1   After an extended period of service, the friction plates will have become worn sufficiently to warrant renewal, to avoid subsequent problems with clutch slip. The lining thickness is measured across the friction plate using a vernier caliper. When new, each plate measures 3.0 mm (0.118 in). If any plate is worn to 2.7 mm (0.106 in) or less, the friction plates must be renewed, preferably as a complete set. Note that if new friction plates are fitted, they must be coated with a light film of engine oil.

2   The plain plates should be free from any signs of blueing, which would indicate that the clutch had overheated in the past. Check each plate for distortion by laying it on a flat surface, such as a sheet of plate glass or similar, and measuring any detectable gap using feeler gauges. The plates must be less than 0.05 mm (0.002 in) out of true, and should preferably be renewed as a complete set if any are found to be distorted beyond the set limit.

3   The clutch springs may, after a considerable mileage, require renewal, and their free length should be checked as a precautionary measure. When new, each spring measures 34.5 mm (1.36 in) and the set should be renewed if they have compressed to 33.5 mm (1.32 in) or less.

4   Check the condition of the slots in the outer surface of the clutch centre and the inner surfaces of the outer drum. In an extreme case, clutch chatter may have caused the tongues of the inserted plates to make indentations in the slots of the outer drum, or the tongues of the plain plates to indent the slots of the clutch centre. These indentations will trap the clutch plates as they are freed and impair clutch action. If the damage is only slight the indentations can be removed by careful work with a file and the burrs removed from the tongues of the clutch plates in similar fashion. More extensive damage will necessitate renewal of the parts concerned.

5   The clutch lifting mechanism consists of the actuating lever assembly, a left-hand pushrod, a steel ball, and a right-hand pushrod which is threaded at its outer end to accept an adjuster and locknut. The mechanism is not likely to require any attention other than routine adjustments unless an extremely high mileage has been covered or if the transmission oil has not

been maintained at the correct level, causing a lack of lubrication. Check that the lever assembly is unworn, remove all traces of corrosion from the shaft bearing surfaces, and check that the return spring is in good order. Examine the pushrods for wear and distortion, rolling them on a sheet of plate glass or other flat surface and using feeler gauges to measure any discernible gap. If bent, the pushrods may be straightened, but if incorrect adjustment or lack of lubrication have caused such a build-up of friction that the hardening on the pushrod ends has broken down, the pushrods and the steel ball must be renewed. Such a case will be easy to see due to the heavy blueing and distorted shape which will result. The steel ball plays an important role in that it permits the right-hand pushrod to rotate with the clutch assembly, but does not transmit this rotary motion to the left-hand pushrod, thus protecting the lifting cam of the lever assembly. Check that the ball is perfectly round. Any flaws in the ball will mean that it must be renewed. All these clutch lifting components must be well greased on reassembly.

23.3 Note matching marks to set required backlash in primary drive gears

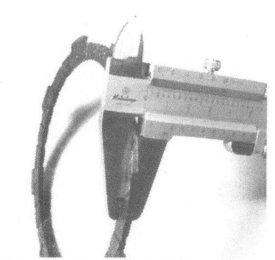

24.1 Measuring the thickness of the clutch friction plates

24.3 Measuring the clutch spring free length

24.4a Check the slots in the clutch centre and in ...

24.4b ... the clutch outer for indentations or burring

24.5 Check push lever assembly for corrosion or wear

25 Gearbox components: examination and renovation

1   Give the gearbox components a close visual inspection for signs of wear or damage such as broken or chipped teeth, worn dogs, damaged or worn splines and bent selectors. Replace any parts fround unserviceable because they cannot be reclaimed in a satisfactory manner.

2   The gearbox shafts are unlikely to sustain damage unless the lubricating oil has been run low or the engine has seized and placed an unusually high loading on the gearbox. Check the surfaces of the shaft, especially where a pinion turns on it, and renew the shaft if it is scored or has picked up. The shafts can be checked for trueness by setting them up in V-blocks and measuring any bending with a dial gauge. The procedure for dismantling and rebuilding the gearbox shaft assemblies is given in Section 27 of this Chapter.

3   Examine the gear selector claw assembly noting that worn or rounded ends on the claw can lead to imprecise gear selection. The springs in the selector mechanism and the detent or stopper arm should be unbroken and not distorted or bent in any way.

4   The gearbox bearings must be free from play and show no signs of roughness when they are rotated. If any sign of wear indicates the need for renewal of any of the bearings, the procedure for removal and refitting of those bearings is described in Section 15 of this Chapter.

5   It is advisable to renew the gearbox oil seals irrespective of their condition. Should a re-used oil seal fail at a later date, a considerable amount of dismantling is necessary to gain access and renew it.

6   Check the gear selector shafts for straightness by rolling them on a sheet of plate glass. A bent shaft will cause difficulty in selecting gears and will make the gear change action particularly heavy. Similarly, check the truth of the gearchange shaft.

7   The selector forks should be examined closely, to ensure that they are not bent or badly worn. Wear is unlikely to occur unless the gearbox has been run for a period with a particularly low oil content. Note that the selector fork guide pins are fitted into holes in the base of each fork and may be removed separately if worn. Check that they are a tight press fit in the base of their respective forks. A proprietary compound such as Loctite Bearing Fit may be used to take up any free play when the guide pins are fitted to the forks.

8   The tracks in the gear selector drum, with which the selector forks engage, should not show any undue signs of wear unless neglect has led to under lubrication of the gearbox. Do

not forget the bearing on the selector drum right-hand end. This should be checked as described for the gearbox shaft bearings and, if found to be worn, should be renewed. Clamping the selector drum in a vice, slacken and remove the single counter-sunk screw using an impact screwdriver if necessary, then pull off the selector cam and the Woodruff key. The bearing may then be levered off the drum end using two large flat-bladed screwdrivers or tyre levers. If the bearing is a particularly tight fit, it will be necessary to acquire a knife-edged bearing puller to carry out this operation safely.

9   To reassemble the selector drum, place the large plain washer over the drum right-hand end, followed by the new bearing, which is fitted with its sealed surface facing outwards. Using a hammer and a small socket or other tubular drift which bears only on the inner race of the bearing, tap the bearing firmly and fully into place. Replace the key in its keyway. Having checked the selector cam and found it free of damage or renewed it, refit the cam, aligning its cut-out with the key. Refit the plain washer and countersunk screw, applying a few drops of Loctite to the screw threads. Tighten the screw securely, using an impact driver, but note that if the necessary adaptors are available, the screw should be tightened with a torque wrench to a setting of 1.4 kgf m (10 lbf ft).

26 Kickstart mechanism: examination and renovation

1   The kickstart mechanism is a robust assembly and should not normally require attention. Apart from obvious defects such as a broken return spring, the friction clip is the only component likely to cause problems if it becomes worn or weakened. The clip is intended to apply a known amount of drag on the kickstart pinion, causing the latter to run up its quick thread and into engagement when the kickstarter lever is operated.

2   The clip can be checked using a spring balance. Hook one end of the balance onto the looped end of the friction clip. Pull on the free end of the balance and note the reading at the point where pressure overcomes the clip's resistance. This should normally be 1.0 kg (2.2 lb). If the reading is higher or lower than this and the mechanism has been malfunctioning, renew the clip as a precaution. Do not attempt to adjust a worn clip by bending it.

3   Examine the kickstart pinion for wear or damage, re-membering to check it in conjunction with the output shaft-mounted idler pinion. In view of the fact that these components are not subject to continuous use a significant amount of wear or damage is unlikely to be found.

25.3 Check gear selector claw ends carefully for signs of wear

25.7 Check that guide pins are a tight fit in the selector fork base

25.8 Remove countersunk screw to release selector cam and ball bearing behind

25.9 Do not forget to install Woodruff key on reassembly

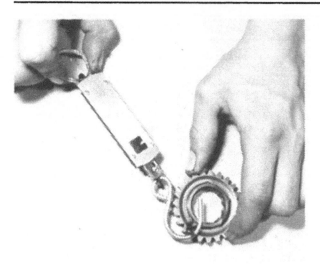

26.2 Checking tension of kickstart friction clip

27 Gearbox input and output shafts: dismantling and reassembly

1   The gearbox clusters should not be disturbed needlessly, and need only be stripped where careful examination of the whole assembly fails to resolve the source of a problem or where obvious damage, such as stripped or chipped teeth is discovered.

2   The input and output shaft components should be kept separate to avoid confusion during reassembly. Using circlip pliers, remove the circlip and plain washer which retain each part. As each item is removed, place it in order on a clean surface so that the reassembly sequence is self-evident and the risk of parts being fitted the wrong way round or in the wrong sequence is avoided. Care should be exercised when removing circlips to avoid straining or bending them excessively. The clips must be opened just sufficiently to allow them to be slid off the shaft. Note that a loose or distorted circlip might fail in service, and any dubious items must be renewed as a precautionary measure. The same applies to worn or distorted thrust washers.

3   Having checked and renewed the gearbox components as required (see Section 25) reassemble each shaft, referring to the accompanying line drawing and photographs for guidance. The correct assembly sequence is detailed below. Note that the manufacturer specifies a particular way in which the circlips are to be fitted. If a non-wire type circlip is examined closely, it can be seen that one surface has rounded edges and the other has sharply-cut square edges, this feature being a by-product of any stamped component such as a circlip or thrust washer. The manufacturer specifies that each circlip must be fitted with the sharp-edged surface facing away from any thrust received by that circlip. In the case of the gearbox shaft circlips this means that the rounded surface of any circlip must face towards the gear pinion that it secures. Furthermore, when a circlip is fitted to a splined shaft, the circlip ears must be positioned in the middle of one of the splines. These two simple precautions are specified to ensure that each circlip is as secure as is possible on its shaft and is best able, therefore, to carry out its task. The accompanying illustrations will clarify the point.

Input shaft

4   The input shaft is readily identified by its integral 1st gear pinion. Slide the 5th gear pinion into position with the selector dogs facing away from the 1st gear, followed by a plain thrust washer.

5   Ensuring that the circlip is positioned correctly as described above, slide the circlip down the length of the shaft to secure the 5th gear pinion. The double 3rd/4th gear pinion is fitted next

with the smaller diameter (18T on DT125 LC models, 19T on RD125 LC models) 3rd gear towards the 5th gear pinion. Secure the double gear with a circlip, followed by another plain thrust washer.

6   Slide the 6th gear pinion into place with its selector dogs facing the 4th gear, followed by the 2nd gear pinion. This last pinion is secured by a single circlip. Lubricate all bearing surfaces and gear teeth with a generous supply of transmission oil and put the completed input shaft assembly to one side.

Output shaft

7   The output shaft is readily identified because it has no integral gear pinion. It has a shoulder towards its left-hand end. Holding the shaft by its left-hand end, slide the 2nd gear pinion over the shaft right-hand end and down the length of the shaft to butt against the shoulder with its recessed surface facing away from the shoulder. Follow the pinion with a plain thrust washer and retain the two with a circlip.

8   The 6th gear pinion is then fitted with its selector fork groove facing the shaft right-hand end, away from the 2nd gear pinion, and is also retained by a circlip. Slide a splined thrust washer down to rest against the circlip.

9   The 4th gear pinion is then fitted with its recessed surface facing to the left so that it can engage the selector dogs on the 6th gear pinion and is retained in place by a circlip. The 3rd gear pinion is fitted next, with its flat surface facing to the left, ie towards the 4th gear pinion, and is followed by a splined thrust washer. Secure the 3rd gear pinion and thrust washer with a circlip.

10   The 5th gear pinion can now be slid into place with the selector fork groove facing to the left, towards the 3rd gear pinion, and should be followed by the large 1st gear pinion which is fitted with its recessed face inwards, to match up with the selector dogs on the 5th gear pinion. The 1st gear pinion is located by a large plain thrust washer and is retained by a circlip. Lubricate all bearing surface and gear teeth with a generous supply of transmission oil and put the completed output shaft assembly to one side.

General

11   It should be noted that if problems arise in identifying the various gear pinions which cannot be resolved by reference to the accompanying photographs and illustrations, the number of teeth on each pinion will identify it. Count the number of teeth on the pinion and compare this figure with that given in the Specifications Section of this Chapter, remembering that the output shaft pinions are listed first, followed by those on the input shaft. Note also that the internal gear ratios of the RD125 LC model are different from those on the DT125 LC model. The problem of identification of the various components should not arise, however, if the instructions given in paragraph 2 of this Section are followed carefully.

27.4a Take the bare input shaft ...

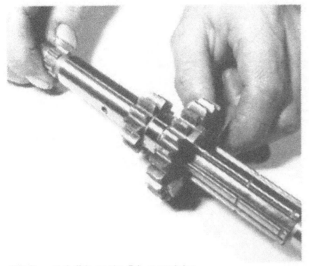

27.4b ... and slide on the 5th gear pinion ...

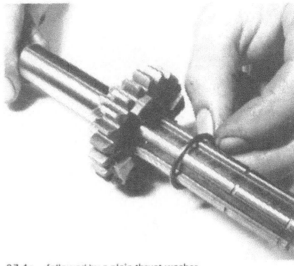

27.4c ... followed by a plain thrust washer

27.5a Secure 5th gear pinion and thrust washer with a circlip

27.5b Note correct direction of double 3rd/4th gear pinion

27.5c Secure 3rd/4th gear pinion with a circlip ...

27.5d ... then fit another plain thrust washer

27.6a Selector dogs on 6th gear pinion face inwards

27.6b Slide 2nd gear pinion into position ...

27.6c ... and secure with a circlip

27.7a Output shaft has plain shoulder at sprocket (left-hand) end

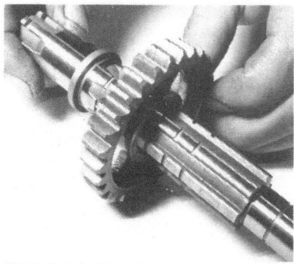

27.7b Fit 2nd gear pinion as shown ...

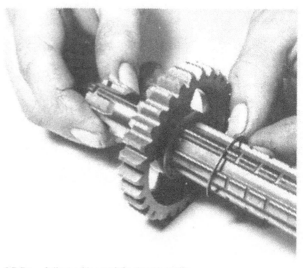

27.7c ... followed by a plain thrust washer ...

27.7d ... and secured with a circlip

27.8a Fit 6th gear pinion, noting position of selector groove

27.8b Secure with a circlip as shown

27.8c Slide a splined thrust washer against circlip

27.9a Note correct direction of 4th gear pinion

27.9b Retain 4th gear pinion with a circlip ...

27.9c ... then fit 3rd gear pinion as shown ...

27.9d ... secured by a splined thrust washer ...

27.9e ... and a circlip

27.10a Fit 5th gear pinion, noting position of selector groove

27.10b Flat surface of large 1st gear pinion faces outwards

27.10c Secure 1st gear pinion with large thrust washer and a circlip

Fig. 1.7 Gearbox components

27.11 The completed gearbox shaft assemblies

1	Input shaft	13	Output shaft 2nd gear pinion
2	Input shaft 5th gear pinion	14	Output shaft
3	Thrust washer	15	Output shaft left-hand bearing
4	Circlip	16	Oil seal
5	Input shaft 3rd and 4th gear pinion	17	Final drive sprocket
6	Input shaft 6th gear pinion	18	Lock washer
7	Input shaft 2nd gear pinion	19	Bolt - 2 off
8	Output shaft 1st gear pinion	20	Output shaft right-hand bearing
9	Output shaft 5th gear pinion	21	Circlip
10	Output shaft 3rd gear pinion	22	Thrust washer
11	Input shaft left-hand bearing	23	Splined washer
12	Output shaft 4th gear pinion	24	Output shaft 6th gear pinion
		25	Thrust washer
		26	Screw - 2 off
		27	Bearing retaining plate
		28	Input shaft right-hand bearing

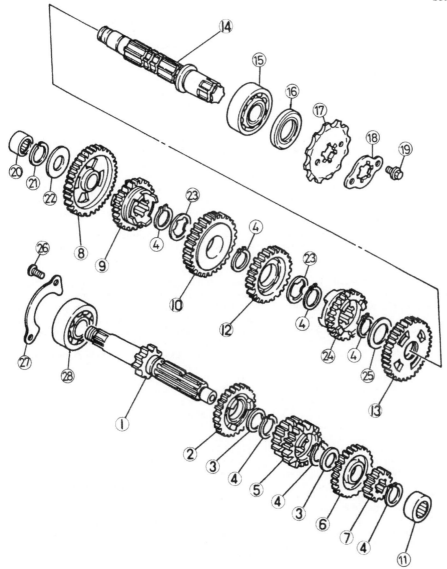

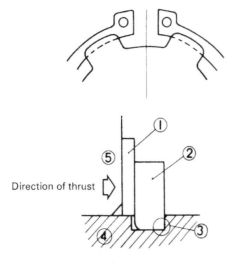

Direction of thrust

Fig. 1.8 Correct fitting of a circlip to a splined shaft

1	Thrust washer	4	Shaft
2	Circlip	5	Gear pinion
3	Square edge		

28 Engine reassembly: general

1   Before reassembly of the engine/gear unit is commenced, the various components parts should be cleaned thoroughly and placed on a sheet of clean paper, close to the working area.

2   Make sure all traces of old gaskets has been removed and that the mating surfaces are clean and undamaged. Great care should be taken when removing old gasket compound not to damage the mating surface. Most gasket compounds can be softened using a suitable solvent such as methylated sprits, acetone or cellulose thinner. The type of solvent required will depend on the type of compound used. Gasket compound of the non-hardening type can be removed using a soft brass-wire brush of the type used for cleaning suede shoes. A considerable amount of scrubbing can take place without fear of harming the mating surfaces. Some difficulty may be encountered when attempting to remove old gaskets of the self-vulcanising type, the use of which is becoming widespread, particularly as cylinder head and base gaskets. The gasket should be pared from the mating surface using a scalpel or a small chisel with a finely honed edge. Do not, however, resort to scraping with a sharp instrument unless necessary.

3   Gather together all the necessary tools and have available an oil can filled with clean engine oil. Make sure that all new gaskets and oil seals are to hand, also all replacement parts required. Nothing is more frustrating than having to stop in the middle of a reassembly sequence because a vital gasket or replacement has been overlooked. As a general rule each moving engine component should be lubricated thoroughly as it is fitted into position.

4   Make sure that the reassembly area is clean and that there is adequate working space. Refer to the torque and clearance settings wherever they are given. Many of the small bolts are easily sheared if overtightened. Always use the correct size screwdriver bit for the cross-head screws and never an ordinary screwdriver or punch. If the existing screws show evidence of maltreatment in the past, it is advisable to renew them as a complete set. It is strongly recommended that if renewal of the screws is required, a set of Allen screws is purchased instead. Allen screw sets are available through most good accessory retailers and are an inexpensive but thoroughly practical improvement to most Japanese machines.

29 Engine reassembly: preparing the crankcase halves

1   The crankcase castings should be absolutely clean and dry at this stage, and ready to accept any bearings or oil seals which have been renewed.

2   The manufacturer recommends that all the bearings are fitted with any marks or numbers facing outwards so that they can be seen when the bearing is installed. Refer to the photographs accompanying the text if in doubt.

3   When installing a bearing, the crankcase casting must be heated as described in Section 15 of this Chapter. Apply a thin smear of grease to the bearing outside diameter and place the bearing squarely in its housing. Using a hammer and a socket or tubular drift which bears only on the outer race of the bearing, tap the bearing fully into place, ensuring that it is kept absolutely square in its housing at all times.

4   When all the bearings have been installed, lubricate them generously. Use transmission oil for the gearbox bearings and two-stroke oil (from the engine oil tank) for the crankshaft main bearings.

5   Broadly speaking, oil seals are refitted in the same way as described above. Again any marks or numbers must face to the outside so that they can be seen when in the installed position, but examine each seal carefully for any further information which may be moulded into its surface. For example when refitting the crankshaft left-hand main bearing oil seal it was noticed that the seal had the words 'outside' on one surface and 'crankside' on the other, thus indicating quite clearly which way round the seal was to be fitted.

6   It is not necessary to heat the crankcase casting when fitting an oil seal, but it will be necessary to ensure that the bore of the seal housing is quite clean and free from dirt or oil. Apply a very thin smear of grease to the seal outside diameter and place the seal squarely over its housing. Being very careful to ensure that the seal remains absolutely square, press it into its housing as far as possible by hand. Using a hammer and a socket or a tubular drift which bears only on the hard outer diameter of the seal, tap the seal fully into place. Some oil seals locate against a shoulder machined in the casting, but where this is not the case, tap the seal down until its outer diameter is flush with the surrounding casting. Refer to the accompanying photographs if in doubt. Be careful at all times to ensure that the seal is kept absolutely square in its housing and do not risk distorting the seal by tapping it too far.

7   When the seals are refitted, liberally grease their sealing lips to protect them when the various shafts are installed. Insert the spacer carefully into the crankshaft right-hand main bearing oil seal. Whilst the spacer can be fitted after installation of the crankshaft it is easier to do so now. When refitting the retaining plates to the right-hand main bearing oil seal and to the input shaft right-hand bearing, apply a few drops of thread locking compound to the threads of each of the plate fixing screws and tighten the screws securely using an impact driver. Note, however, that if the necessary adaptors are available, a torque wrench may be used to tighten each of the screws to the torque setting given in the Specifications Section of this Chapter. It is essential that these screws are securely fastened as there are no locking washers provided to prevent them from slackening.

8   If any of the cylinder barrel studs are to be refitted, apply a few drops of thread locking compound to that part of the thread which will fit into the crankcase and, as far as possible, screw the stud in by hand. Lock two nuts together on the exposed thread and, using a torque wrench on the upper nut, tighten the stud to a torque setting of 1.0 kgf m (7 lbf ft). Release and remove the two nuts.

9   Install the neutral indicator switch but be very careful not to overtighten the switch, which is only a light plastic moulding and is easily cracked. Replace the crankcase breather tube and any grommets which may have been removed. If these are a slack fit, apply a thin smear of jointing compound to stick them in place, thus preventing any oil leaks or the entry of dirt or water when the machine is in use.

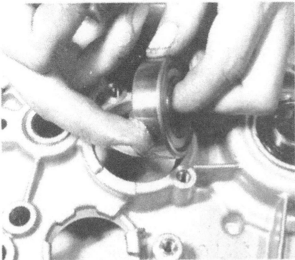

29.3a Apply a thin smear of grease to bearings and oil seals to assist refitting

29.3b Socket must bear on bearing outer race only

29.6 Be very careful to fit oil seals correctly, without damage

29.7a Do not omit spacer from crankshaft right-hand oil seal

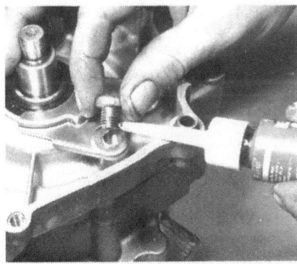

29.7b Use thread locking compound on threads of screws for oil seal ...

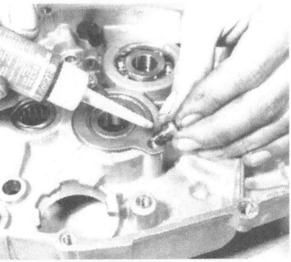

29.7c ... and bearing retaining plates

29.7d Tighten screws securely (see text)

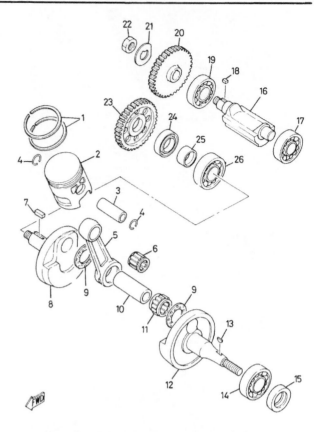

Fig. 1.9 Crankshaft and balancer shaft

1	Piston rings	15	Oil seal
2	Piston	16	Balancer shaft
3	Gudgeon pin	17	Balancer shaft left-hand
4	Circlip - 2 off		bearing
5	Connecting rod	18	Key
6	Small-end bearing	19	Balancer shaft right-hand
7	Key		bearing
8	Right-hand flywheel	20	Balancer shaft pinion
9	Thrust bearing - 2 off	21	Lock washer
10	Crank pin	22	Nut
11	Big-end bearing	23	Balancer shaft drive gear
12	Left-hand flywheel	24	Oil seal
13	Woodruff key	25	Spacer
14	Left-hand main bearing	26	Right-hand main bearing

30 Engine reassembly: refitting the crankshaft, balancer shaft and gearbox components

1   The method of crankshaft installation recommended by the manufacturer is to use the special tool which consists of the tool body, Part number 90890-01274, a large stud, Part number 90890-01275, and an adaptor, Part number 90890-01278. The tool is used as follows.
2   Check that the left-hand main bearing is lubricated with two-stroke engine oil and that the sealing lips of the oil seal are greased. Pass the crankshaft through the main bearing as far as possible and screw the adaptor on to the thread on the crankshaft left-hand end. Place the tool body over the crankshaft and adaptor, aligning the key on the adaptor with the slot in the tool body. Pass the long stud down through the tool body and screw it as far as possible into the adaptor. Screw the nut down the length of the stud until it comes into contact with the tool body and retains the whole assembly in place on the crankcase left-hand half. Check that the crankshaft is aligned correctly, ie absolutely square to the bearing and crankcase. If necessary, slacken the nut and re-position the tool body until this is achieved. When the crankshaft is correctly aligned, hold the connecting rod in the top dead centre position with one hand and, using a suitable spanner, tighten the nut down on the tool body. This will draw the crankshaft through the main bearing and oil seal with the minimum of effort and risk. Be careful that the connecting rod does not foul the crankcase casting and carry on tightening the nut until resistance indicates that the shoulder on the crankshaft left-hand flywheel is in contact with the left-hand main bearing. When this point has been reached, dismantle the tool and check that the crankshaft rotates freely and easily.
3   If as is likely, the special tool is not available, the following method can be used, providing that great care is exercised. Support the crankcase left-hand half on two blocks of wood so that its right-hand face is upwards. Place the two wooden blocks closely around the main bearing, but leave enough room for the crankshaft left-hand end to pass between them. Lightly oil the crankshaft left-hand end and the internal diameter of the left-hand main bearing. Check that the main bearing is lubricated with two-stroke oil and that the sealing lips of its oil seal are greased. Screw the primary drive gear retaining nut down on to the thread on the crankshaft right-hand end so that the nut is flush with the end of the crankshaft. Insert the crankshaft carefully into the left-hand main bearing ensuring that it is held absolutely square to the bearing and the

crankcase. Using a soft-faced mallet, tap firmly on the crankshaft right-hand end to drive the crankshaft into place. Be careful at all times that the connecting rod does not foul the crankcase casting and that the crankshaft is held square in the bearing. Carry on tapping gently but firmly on the crankshaft right-hand end until the shoulder on the left-hand flywheel comes into contact with the left-hand main bearing.
4   When using this alternative method of crankshaft installation, great care must be taken to use only the barest minimum of force. If resistance is encountered before the crankshaft is in place, stop tapping and find out why. Never resort to heavy blows to force the crankshaft into position, as the risk of distorting the assembly is too great. If the task appears to be too difficult, take the crankshaft and crankcase to a Yamaha dealer for the work to be carried out. When the crankshaft is in place, remove the nut from the shaft right-hand end and check that the crankshaft revolves easily and freely with no distortion apparent.
5   The remaining items can now be fitted, but work must be

carried out in a precise order. Check that the balancer shaft left-hand bearing is lubricated and insert the balancer shaft into it. Check that the shaft revolves easily and freely. Check that the gearbox input and output shaft left-hand bearings are lubricated and that the oil seal next to the output shaft bearing has plenty of grease on its sealing lips. Place both shafts together, ensuring that all matching gear pinions are correctly mated, and insert the shafts into their bearings treating the two as a single assembly. Take great care not to damage the oil seal as the output shaft splines pass through it, and tap gently on the shafts' right-hand ends, if necessary, to drive them into place.

6 Fit the selector forks. The single front fork fits on the input shaft double 3rd/4th gear pinion's selector fork groove, and of the two rear forks, the left-hand one fits on the 6th gear pinion fork groove, and the right-hand one on the 5th gear pinion fork groove. Lubricate the claw ends of each fork as it is fitted and use the marks made on dismantling to identify each one. When the forks are fitted, pivot each one about its respective pinion so that all three are well clear of the selector drum. Make a final check that the guide pins are firmly inserted into the base of each fork.

7 Take the selector drum, identify the protrusion on its left-hand end which forms the contact for the neutral indicator switch, lightly oil the boss on the drum left-hand end, and insert the drum into its housing in the crankcase left-hand half, ensuring that the switch contact is aligned with the switch itself so that the drum is in the neutral postion. Lubricate the selector fork guide pins and the tracks of the selector drum, then pivot the selector forks back so that the guide pin fitted to each engages with its respective track in the drum.

8 Oil the selector fork shafts and the bores of the selector forks and insert the fork shafts through the forks and into their respective housings in the crankcase. Remember that the short tubular shaft fits in front of the selector drum, retaining the single selector fork, and that the longer, solid, shaft fits to the rear of the selector drum and retains the other two forks. Check that the circlip fitted to this latter shaft is securely fixed.

9 When all the gearbox components are in place, check that all shafts are free to rotate easily and freely, then check that all gears can be selected by rotating the selector drum. This latter check is difficult to carry out at this stage but will save a lot of work if a fault remains undiscovered until the engine/gearbox unit is refitted in the frame. In neutral, it should be possible to hold either shaft still while rotating the other. Gear selection while conducting this test is made much easier if one of the shafts is rotated while turning the selector drum.

30.3a Prepare crankcase left-hand half as described in the text

30.3b Ensure crankshaft is absolutely square on refitting

30.5a Insert balancer shaft and check it is free to rotate

30.5b Fit gearbox shaft assemblies as a single unit

30.6a Fit selector forks using marks or notes made on dismantling ...

30.6b ... to identify each one and to position it correctly

30.6c Swing forks away to provide clearance for selector drum

30.7a Protrusion on selector drum end must be aligned with neutral indicator switch

30.7b Fit selector drum and swing selector forks into correct position

30.8a Tubular selector fork shaft is fitted in front of selector drum

30.8b Check circlip is secure on long selector fork shaft

30.9 Check all shafts are free to rotate and are well oiled

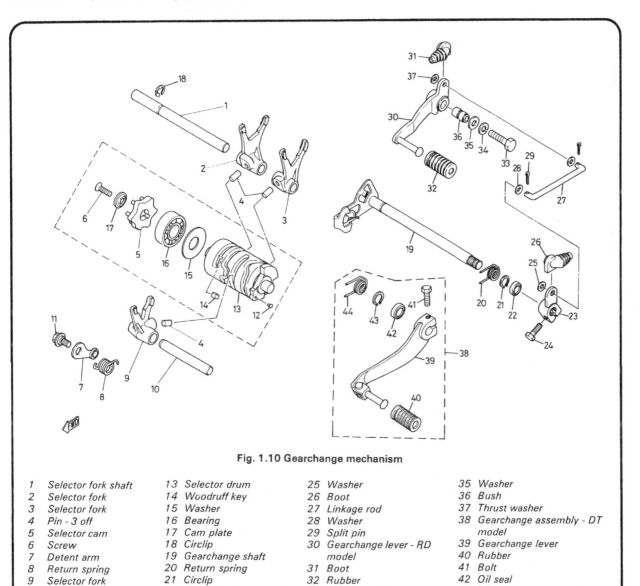

Fig. 1.10 Gearchange mechanism

1	Selector fork shaft	13	Selector drum	25	Washer	35	Washer
2	Selector fork	14	Woodruff key	26	Boot	36	Bush
3	Selector fork	15	Washer	27	Linkage rod	37	Thrust washer
4	Pin - 3 off	16	Bearing	28	Washer	38	Gearchange assembly - DT
5	Selector cam	17	Cam plate	29	Split pin		model
6	Screw	18	Circlip	30	Gearchange lever - RD	39	Gearchange lever
7	Detent arm	19	Gearchange shaft		model	40	Rubber
8	Return spring	20	Return spring	31	Boot	41	Bolt
9	Selector fork	21	Circlip	32	Rubber	42	Oil seal
10	Selector fork shaft	22	Oil seal	33	Bolt	43	Circlip
11	Shouldered bolt	23	Splined cam	34	Spring washer	44	Return spring
12	Neutral contact pin	24	Bolt				

31 Engine reassembly: refitting the crankcase right-hand half

1 Make sure that the crankcase halves are clean and completely free of grease. To this end it is sound practice to give the jointing faces a final wipe with a clean rag moistened with methylated spirit or clean petrol. Allow solvent to evaporate completely, then apply a thin film of jointing compound to the gasket face of one half. One of the RTV (room temperature vulcanising) silicone compounds, often sold as 'Instant Gasket' is recommended. Allow the compound to cure for a few minutes and in the meantime fit the two locating dowels to their recesses in the crankcase left-hand half.
2 Place the crankcase right-hand half over the ends of the shafts and lower it as far as possible by hand only, taking care to ensure that the casting stays square on the shafts. Position the change drum so that the cam 'ears' pass through the profiled hole in the casing. Using a soft-faced mallet, tap the right-hand half gently but firmly into place, checking frequently that the various shafts and dowel pins are correctly aligned. Do not forget that the selector drum must be rotated so that the protruding ears on the selector cam are aligned with the cut-outs in the crankcase. Be careful to use only the bare minimum of force and to stop tapping immediately any severe resistance is encountered so that the problem can be found and rectified. If necessary, remove the crankcase right-hand half and start again.
3 Once the crankcase right-hand half is firmly in position, check that all shafts are free to rotate easily and all the gears can be selected. If any doubts arise from these checks the problem should be cured before any more assembly work is done; if necessary, by separating the crankcases again to investigate. Note that it is sometimes possible to ease a shaft that is stiff to rotate by giving it a smart tap on both ends with a soft-faced mallet. This will centralise the shaft in its bearings, enabling it to rotate more easily. Do not tap too hard as excessive force is neither necessary nor desirable, and may force the crankcase halves apart.
4 When it is established that the crankshaft and gearbox components are properly installed, invert the crankcase assembly so that the left-hand side is now uppermost. Refit the crankcase securing screws in their correct positions, remembering to include the breather tube clamp, and tighten them down. Start in the middle, working outwards in a diagonal sequence and tighten the bolts in two or three stages to ensure an even and progressive application of pressure. Finally, tighten the screws securely; if the necessary adaptors are available, use a

torque wrench to tighten the screws to a torque setting of 0.8 kgf m (6 lbf ft).
5 Place the crankcase assembly upright on the workbench and check again that all the shafts are free to rotate. If not, find out why, even if it is necessary to separate the crankcases again to do so. If all is well, wipe away all surplus grease, oil, and jointing compound, tuck the crankcase breather tube through its clamp above the output shaft left-hand end, raise the connecting rod to the top dead centre position and stuff the crankcase mouth with clean rag to prevent the entry of dirt or debris. If the gearbox drain plug was disturbed for any reason, check that its sealing washer is in good condition, renewing it if necessary, and refit the drain plug, tightening it to a torque setting of 2.0 kgf m (14.5 lbf ft). Rotate the selector drum to the neutral position.

32 Engine reassembly: refitting the external gear selector components

1 Assemble the selector detent arm and its return spring together using the accompanying photographs and the notes made on dismantling. Apply a few drops of thread locking compound to the threads of the shouldered pivot bolt and fit all three parts as a single assembly. Screw the bolt in as far as possible, then pull or lever the detent arm so that the roller on its tip comes into correct contact with the selector cam plate. Check that the spring is correctly positioned, then tighten the pivot bolt to a torque setting of 1.5 kgf m (11 lbf ft). Check that the detent arm is free to move easily.
2 If the gearchange shaft return spring was removed for any reason it must now be refitted. Ensure that the spring is fitted the correct way round and that its ends engage correctly with the protruding tang of the selector claw plate. Refit the retaining circlip using a suitable pair of pliers.
3 Wrap the splines at the end of the gearchange shaft in a strip of PVC insulating tape or use a heavy application of grease to protect the sealing lips of the oil seal in the crankcase left-hand half. Carefully insert the gearchange shaft assembly and push it into place, taking great care to use a slight twisting action as the shaft splines pass through the oil seal. This will minimise the risk of damaging the delicate sealing lips of the seal.
4 When the shaft is in place, move it upwards and downwards, releasing it each time, to check the operation of the selector mechanism. The shaft should return quickly and positively to the centre position with no trace of free play or weakness.

31.1 Note two locating dowels (arrowed) and correct use of jointing compound

31.2 Remember to rotate selector drum during crankcase refitting

31.4a Tighten securely crankcase screws (see text)

31.4b Do not forget breather tube clamp

32.1a Note correct position of detent arm return spring and selector drum position in neutral

32.1b Apply thread locking compound to detent arm pivot bolt ...

32.1c ... and tighten pivot bolt to correct torque setting

32.2 Gearchange shaft return spring and retaining circlip in correct position

32.3 Note use in this case of grease to protect oil seal lips

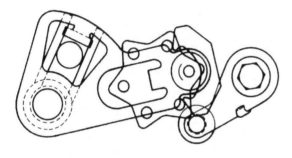

Fig. 1.11 Correct fitting of selector mechanism detent arm

with the raised boss cast in the crankcase wall. This is shown in the illustration and photographs accompanying the text. Holding the crankshaft in the TDC position, refit the balancer shaft drive gear over the crankshaft end. Ensure that the gear is located correctly on the Woodruff key and push it into position while rotating the balancer shaft gear pinion so that the two sets of teeth mesh with the two punch marks aligned exactly. Rotate the crankshaft to check that the two punch marks remain in exact alignment when the crankshaft is in the TDC position.

4   When the balancer shaft timing has been set and checked refit the primary drive gear with its recessed face outwards, followed by the large spring washer and the nut. Tighten the nut by hand only at first, then lock the crankshaft and tighten both the primary drive gear retaining nut and the balancer shaft gear retaining nut to the same torque setting of 6.5 kgf m (47 lbf ft). Using a hammer and a suitable punch, lock the balancer shaft gear retaining nut by bending an unused portion of its tab washer up against one of the flats of the nut.

5   When refitting the tachometer drive assembly of the RD125 LC model, lightly oil the vertical driven gear and insert it down through the opening in the crankcase top surface into its housing, then check the condition of the sealing O-ring fitted around the spigot of the black plastic housing. Renew the O-ring if necessary, apply a thin smear of grease to the housing spigot and fit the housing with a twisting motion as if screwing it in, and using the grease to assist the task. When the housing is fully in place, fit and tighten securely the single retaining bolt. Do not overtighten the bolt, or the housing will crack. If the clutch is not in place the white nylon drive gear may be assembled on its shaft with a thrust washer on each side and offered up as a complete assembly. If, however, the clutch is in place, a slightly different procedure must be used. Place one of the thrust washers on the crankcase wall around the housing machined to accept the drive gear shaft, and use a smear of grease to stick it in place. Insert the drive gear, ensuring that it meshes correctly with both the clutch outer drum and the vertical driven gear, and hold it in place while pushing the drive gear into place. The second thrust washer may then be fitted over the shaft end to rest against the drive gear. Take great care to check that the assembly is still in place before the crankcase right-hand cover is refitted, as it is only the latter which retains the drive gear assembly.

33 Engine reassembly: refitting the primary drive gear, balancer drive gear and tachometer drive gear

1   When refitting the primary drive and balancer shaft gear pinions it will be necessary to lock the crankshaft to permit the retaining nuts to be tightened. This is best achieved by placing a close-fitting round bar through the connecting rod small-end eye and resting the bar on two wooden blocks placed on the crankcase mouth, although if the work is being carried out with the engine/gearbox unit in the frame, the method used on dismantling will be sufficient.

2   Insert the short, square-section Woodruff key into the balancer shaft keyway and refit the balancer shaft gear pinion noting that the recess machined in one surface must face outwards. Place the tab washer over the shaft end and refit the retaining nut noting that the nut is a non-reversible type and must be fitted with its recessed face inwards, towards the balancer shaft gear pinion. Tighten the nut as far as possible by hand and rotate the pinion so that the alignment punch mark on its perphery is at the 3 o'clock position.

3   Insert the long, square-section Woodruff key into the crankshaft keyway, then turn the crankshaft to the TDC position, which is identified precisely by aligning the keyway

33.2a Short square section Woodruff key fits in balancer shaft keyway

33.2b Note correct position of recess in balancer shaft pinion

33.2c Fit lock washer and retaining nut as shown

33.3a Long square section Woodruff key fits in crankshaft keyway. Note raised boss (arrowed)

33.3b Align balancer shaft timing marks (see text)

33.4a Fit primary drive gear as shown

33.4b Do not omit spring washer

33.4c Lock crankshaft to tighten primary drive gear and balancer shaft gear retaining nuts

33.4d Lock balancer shaft retaining nut as shown

33.5a Fit a plain thrust washer to the tachometer drive location – RD125 ...

33.5b ... then insert drive gear and shaft ...

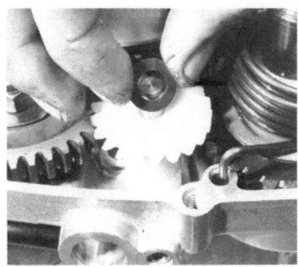

33.5c ... followed by second thrust washer

33.5d Oil tachometer driven gear before fitting

33.5e Check condition of sealing O-ring and renew if necessary

33.5f Do not overtighten housing retaining bolt

34 Engine reassembly: refitting the kickstart assembly and idler gear

1   Using a suitable pair of circlip pliers, refit the circlip on the innermost of the two grooves machined in the output shaft right-hand end. Place a thrust washer over the shaft end, followed by the kickstart idler gear; note that one side of the idler gear has a protruding boss while the other side is heavily chamfered around its periphery. It is the chamfered side which must face outwards. Secure the idler gear with the second thrust washer and a circlip.

2   If the kickstart assembly was dismantled for repair work, it must now be rebuilt and offered up to the crankcase as a single unit. Place the kickstart pinion on the workbench with its flat surface downwards and install the friction clip so that its looped end points upwards. Place the kickstart pinion over the end of the kickstart shaft with the friction clip pointing inwards, towards the shaft left-hand end, and slide the pinion down the shaft to engage with the large diameter splines, then refit the large plain washer which locates against the shaft shoulder. Place the kickstart return spring over the shaft with both spring ends facing inwards, towards the large diameter splines, and insert the spring short inner end into its locating hole in the kickstart shaft. Refit the white nylon spring guide onto the shaft, ensuring that the cut-out on one end is facing inwards and engages with the spring inner end.

3   Lubricate all bearing surfaces of the kickstart assembly and insert the assembly into the crankcase. As the end of the shaft fits into its housing in the crankcase wall, ensure that the large shaft stopper arm locates correctly against the stop cast in the crankcase wall, and that the friction clip looped end fits into the recess provided for it. Make a quick check to ensure that the kickstart assembly is operating correctly by rotating the shaft as far as possible anti-clockwise. The shaft should rotate smoothly and easily until the stopper arm comes into contact with its second stop, and the kickstart pinion should move easily up its shaft splines to engage with the teeth of the idler gear. Return the shaft to its normal position by rotating it clockwise and use a suitable pair of pliers to bring the return spring long hooked end around so that it can be engaged in its locating hole in the crankcase wall. Take care at this point to ensure that the spring end is hooked firmly into the locating hole, as the spring will be under quite heavy pressure.

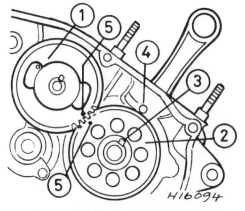

Fig. 1.12 Balancer shaft timing marks

1	Balancer shaft gear pinion	4	Crankshaft index mark
2	Balancer shaft drive gear		(raised boss)
3	Crankshaft keyway	5	Pinion alignment punch marks

34.1a Fit a circlip to output shaft end as shown ...

34.1b ... followed by first plain thrust washer

34.1c Note correct position of kickstart idler gear

34.1d Fit second plain thrust washer ...

34.1e ... and secure with a second circlip

34.2a Note direction of kickstart friction clip

34.2b Slide large plain washer into position

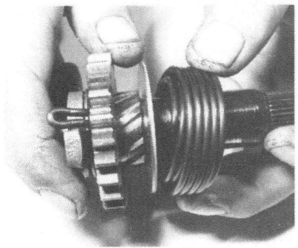

34.2c Fit kickstart return spring as shown

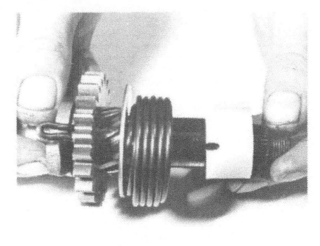

34.2d Note notch in spring guide which fits around inner end of return spring

34.3a Install kickstart assembly as shown

34.3b Ensure return spring is correctly engaged in locating hole

35 Engine reassembly: refitting the clutch assembly

1   Commence refitting the clutch assembly by fitting the dished thrust washer over the input shaft right-hand end, noting that the washer is fitted with its convex side towards the crankcase wall. Using a pair of needle-nose pliers, refit the key in its locating hole in the output shaft and slide the spacer down the length of the shaft to engage correctly with the key as shown in the accompanying photographs and illustration.

2   Lubricate the clutch outer drum centre bush, the impact shaft and the spacer. Place the clutch outer drum over the input shaft end and slide it into place, ensuring that the kickstart idler gear and primary drive gear mesh correctly with their respective sets of teeth on the outer drum. Refit the second, splined, spacer with its slightly chamfered surface facing outwards, then fit the clutch centre over the shaft end. The tab washer is fitted with its small protruding tang located in the cut-out in the clutch centre boss, whereupon the retaining nut may be replaced on the shaft end and tightened down as far as possible by hand.

3   Lock the clutch centre using the method employed on dismantling and tighten the retaining nut to a torque setting of 6.5 kgf m (47 lbf ft). Secure the nut by bending an unused portion of the tab washer up against one of the flats of the nut. Check that the clutch outer drum and the clutch centre can rotate freely, easily, and independently of one another.

4   Refit the return spring and the plain washer to the clutch actuating lever, and grease liberally the vertical part of the actuating assembly. Insert the assembly, ensuring that the pointed end of the lever is aligned with the index mark cast in the crankcase top. Check that the return spring is correctly engaged on both the lever and the crankcase top and check that the lever operates freely. Thoroughly grease the left-hand pushrod and insert it into the input shaft, followed by the steel ball. Similarly grease the right-hand pushrod and insert this as far as possible.

5   The clutch plain and friction plates should be coated with engine oil prior to installation. It will be noted that each of the plain plates has a part of its outer edge machined off. This effectively makes the plate become slightly out of balance,

causing each plate to be thrown outwards under the centrifugal force and thus reducing clutch noise. To prevent the whole clutch from getting out of balance it is necessary to arrange the plates so that the machined areas are spaced evenly around its circumference. This can be achieved by arranging each cutaway area to be approximately 70° from the previous one. Starting with a friction plate, and then a plain plate, insert the plates in succession until all are fitted.

6   When fitting the clutch pressure plate, note that the arrow mark cast on it must align with the stamped circular mark on the periphery of the clutch centre. Lift out the right-hand pushrod and fit the pressure plate over the rod end, ensuring that it is located correctly on the machined hexagon of the lifting plate which is threaded on to the pushrod right-hand end. Hold the pushrod thread with one hand, retaining the pressure plate in position, and guide the pressure plate into plate with the other, ensuring that it fits over the clutch spring pillars and aligns with the clutch centre mark at the same time.

7   Refit the clutch springs and their retainig bolts, and secure the assembly by tightening the bolts evenly and in a diagonal sequence to a final torque setting of 1.0 kgf m (7 lbf ft). Refit the plain washer and the adjuster locknut to the exposed threaded end of the pushrod.

8   The clutch lifting mechanism must now be set. Slacken fully the adjuster locknut and, using a suitable screwdriver, turn the pushrod as necessary to bring the pointed end of the clutch actuating lever into exact alignment with the index mark cast in the crankcase top, noting that it will be necessary at the same time to press the lever forwards against the pressure of its return spring to ensure that all free play is removed from the mechanism and that the correct setting is achieved.

9   When the correct initial setting has been achieved, the lever should return to a point behind the index mark when released, so that if it is pushed forward by hand, all free play in the mechanism will be eliminated just as the lever pointed end aligns precisely with the index mark. It will now be necessary to introduce a set amount of free play to allow for expansion of the various components when the engine is warmed up to normal operating temperature, thereby preventing clutch slip. To achieve this free play, turn the pushrod anti-clockwise by $\frac{1}{4}$ turn only and secure the mechanism by tightening the locknut to a torque setting of 0.8 kgf m (6 lbf ft).

35.1a Dished thrust washer is fitted first ...

35.1b ... followed by the key ...

35.1c ... and the spacer

35.2a Refit the clutch outer drum and then fit the spacer as shown

35.2b Note tab washer tang fitted into cut-out in clutch centre

35.3 Tighten clutch centre nut to recommended torque setting

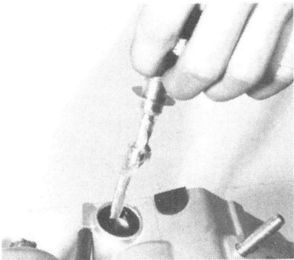
35.4a Liberally grease push lever assembly before fitting

35.4b Check that push lever return spring is fitted as shown

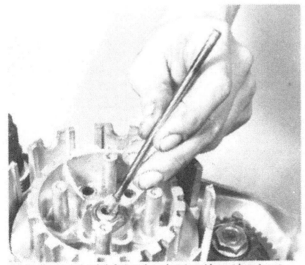
35.4c Grease clutch left-hand pushrod and insert into input shaft

35.4d Do not forget to fit steel ball ...

35.4e ... before inserting clutch right-hand pushrod

35.5a Friction plate is fitted first, followed by ...

35.5b .. plain plate. Note cutaway area (arrowed). Arrange plain plates as described in the text

35.6a Align arrow mark on pressure plate with circular mark on clutch centre

35.6b Fit pressure plate as shown

35.8 Screw and locknut are used to adjust clutch lifting mechanism

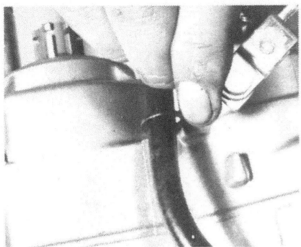

35.9 Pointed end of push lever must align with raised mark on crankcase

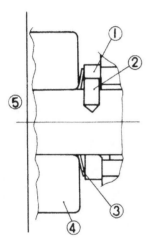

Fig. 1.13 Clutch component locations

1 Spacer   4 Bearing
2 Key   5 Crankcase
3 Dished thrust washer

36 Engine reassembly: refitting the crankcase right-hand cover

1 Make a final check that all components are correctly in place, securely fastened, and lubricated. Ensure that the two locating dowels are securely in place in the gasket face of the crankcase right-hand half and, on RD125 LC models only, check that the tachometer drive pinion outer thrust washer is still in position.

2 Thoroughly clean both gasket surfaces and place a new gasket on the crankcase sealing face, using a smear of grease to stick it in place. Smear a small amount of grease on the sealing lips of the kickstart shaft oil seal, and liberally grease the splines of the kickstart shaft itself. Offer up the cover, taking great care not to damage the oil seal lips as the kickstart shaft

passes through them. Ensure also that the water pump and oil pump drive pinions engage correctly with the crankshaft pinions, rotating the crankshaft if necessary to achieve this.

3 Once the cover is fully home, refit the retaining screws in their correct locations, not forgetting the oil feed pipe clamp and, on DT125 LC models only, the stop lamp switch rod bracket. Tighten the retaining screws securely using an impact driver in progressive stages and working from the centre outwards in a diagonal sequence. Note, however, that if the necessary equipment is available, the screws should be tightened to a torque setting of 1.0 kfg m (7 lbf ft).

4 If the work is being carried out with the engine/gearbox unit in the frame, it will now be necessary to reassemble all those components dismantled as described in Section 8 of this Chapter, the necessary work being described in Sections 39 and 40 of this Chapter.

37 Engine reassembly: refitting the flywheel generator

1 When refitting the generator stator, take great care to ensure that the wires running behind it are not trapped or damaged in any way. Offer up the stator, sliding the sealing grommets of the electrical leads into their locations in the crankcase. Push the stator home, ensuring that the temporary timing marks made on dismantling are aligned exactly, and tighten securely the two retaining screws. If a new stator is being fitted to an RD125 LC model, ensure that the stator is refitted so that the retaining screws are in the exact centre of their respective holes in the stator plate. Due to variations caused by manufacturing tolerances, the screws are not a tight fit in the stator plate casting and cannot be relied upon to position the plate accurately. In such a case, the ignition timing must be checked very carefully. The stator plates fitted to DT125 LC models will have an index mark stamped on them when delivered, and this mark must be used until the ignition timing can be checked and if necessary re-set.

2 Refit the Woodruff key to the crankshaft keyway, ensuring that it is located correctly. Wipe the crankshaft taper clean of any oil or grease and then refit the rotor. Replace the plain washer, spring washer, and the retaining nut, lock the crankshaft by the method employed on dismantling and tighten the nut to a torque setting of 5.0 kgf m (36 lbf ft).

3 Place the neutral switch lead grommet into its slot in the crankcase wall and connect the lead to the switch terminal. Do not refit the crankcase left-hand cover until the ignition timing has been checked.

36.1 Gasket surface must be clean and locating dowels (arrowed) in position

36.3 Tighten securely all cover screws

36.4 Kickstart may be refitted at this stage if desired

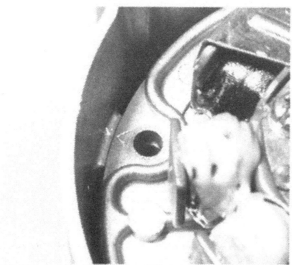

37.1 Use marks made on dismantling to position stator accurately

37.2a Refitting the rotor – note sealing grommets correctly installed

37.2b Do not forget to refit the plain washer ...

37.2c ... and the spring washer

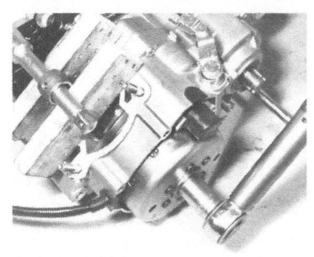

37.2d Lock crankshaft to tighten rotor nut to correct torque setting

38 Engine reassembly: refitting the piston, cylinder barrel, and cylinder head

1   If any of the cylinder head locating studs have been renewed, they should now be refitted as described in Section 29 of this Chapter. Refit the piston rings to the piston, noting that the thin expander ring must be fitted first, to the lower ring groove. Remember that the two compression rings are different, the lower one having a square section and the top one having a tapered top surface; the top surface of the lower ring can be identified only by the fact that the end gap is chamfered and is wider at the top to accommodate the ring locating peg. The top surface of each ring may be marked with a letter, as shown in photo 38.1b, if not the rings should be fitted as described above. It is essential that the rings are fitted the right way up in their respective grooves and that their end gaps are located correctly by the retaining pegs.

2   Lubricate the crankshaft main bearings and big-end bearing with two-stroke oil, then lubricate also the small-end bearing and insert this into the connecting-rod small-end eye. Check that the sealing surfaces of the crankcase mouth and the cylinder barrel are completely clean and free from grease, oil, or dirt, then lightly grease a new cylinder base gasket. Check that the two locating dowels are in place around the cylinder right-hand mounting studs and lower the gasket into place over the studs and dowels, ensuring that it aligns correctly with the ports. Pack the crankcase mouth with a piece of clean rag.

3   Insert a new gudgeon pin circlip in its groove, ensuring that it is seated fully, then push the gudgeon pin into the piston from the other side. If the gudgeon pin is a tight fit, the task will be made much easier if the piston is warmed first by wrapping it in a rag soaked in boiling water, although care must be taken to prevent personal injury due to burnt or scalded hands if this method proves necessary. Position the piston over the connecting rod small-end eye so that the arrow cast in the piston crown points to the front, or towards the exhaust port, and secure it by pushing the gudgeon pin through the small-end bearing and the piston itself. Fit the second new circlip, ensuring that it is also correctly seated.

4   Liberally lubricate, using two-stroke oil, the cylinder bore and the outer surface of both the piston and piston rings. Raise the piston to TDC and refit the barrel, which has a tapered lead-in machined at the bottom of its skirt to facilitate this operation. Hold the piston and compress one ring at a time with one hand while pushing the barrel carefully down with the other. Be careful to check that the rings are correctly located by their retaining pegs and that the piston is entering the bore absolutely squarely. Due to the light weight and small size of the components, the task is easily carried out by one person working alone, but requires some care if components are not to be broken. Once both rings are fully engaged in the bore, remove the rag from the crankcase mouth and push the barrel carefully down to rest on the crankcase mouth. Refit the clutch cable bracket on the cylinder left-hand rear mounting stud, then fit and tighten the four cylinder retaining nuts. The nuts should be tightened in a diagonal sequence and in stages to a final torque setting of 2.8 kgf m (20 lbf ft).

5   Renew if necessary the rubber O-rings which are fitted at each end of the crankcase cover/cylinder barrel coolant pipe, grease lightly both O-rings and insert the pipe into its bore in the crankcase right-hand cover. Take great care not to damage the sealing O-rings. When the coolant pipe is in place, refit its two retaining screws, tightening them to a torque setting of 1.2 kgf m (9 lbf ft).

6   Refit the reed valve assembly, noting that it is advisable to apply a very thin coating of jointing compound to seal the joint between the reed valve case and the inlet stub, and to fit a new gasket between the reed valve case and the cylinder barrel. Tighten the reed valve/inlet stub assembly mounting bolts progressively and in a diagonal sequence to a torque setting of 0.8 kgf m (6 lbf ft).

7   Rotate the crankshaft so that the piston is at the top of its stroke and wipe away any surplus oil. Check that the mating surfaces of the cylinder head and barrel are clean and dry, then place the cylinder head gasket in position over the studs. The use of jointing compound is not recommended at this joint area. Refit the cylinder head, ensuring that the tapped hole for the temperature gauge sender unit faces forwards, then refit the plain washers and domed retaining nuts to each of the four studs.

8   Tighten the four nuts in stages, working in a diagonal sequence to a final torque setting of 2.2 kgf m (16 lbf ft). Check that the sender unit sealing washer is in good condition, renewing it if necessary, and refit the sender unit, which should be tightened to a torque setting of 1.0 kgf m (7 lbf ft). The radiator hose union fitted to the cylinder head has an O-ring fitted which must be checked and renewed if damaged or worn. Smear the O-ring with a small amount of grease and insert the union into the head casting. Tighten the single retaining screw to a torque setting of 1.2 kgf m (9 lbf ft).

9   If work is being carried out with the engine/gearbox unit out of the frame, the spark plug should not be fitted at this stage. Cover the head with a piece of rag to prevent the entry of dirt or debris while the engine/gearbox unit is refitted.

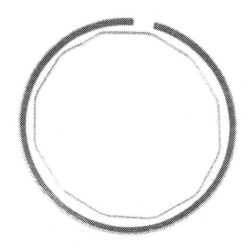

38.1a Expander ring is fitted inside bottom piston ring

38.1b Letter marked in top surface of top piston ring to show correct position

38.1c Ensure piston ring end gaps are correctly located at retaining pegs

38.2a Note two dowels which must be fitted around cylinder right-hand studs

38.2b Oil crankshaft bearings and fit new base gasket

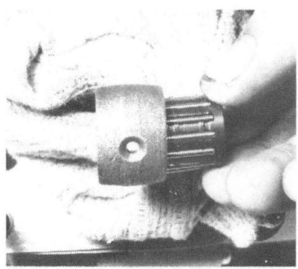

38.2c Oil small-end bearing before inserting into gudgeon pin

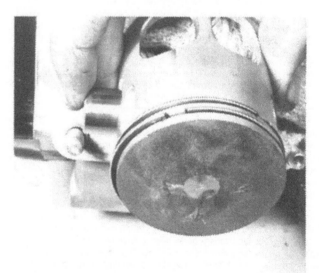

38.3a Retain piston with gudgeon pin. Note direction of arrow on piston crown

38.3b Check circlips are located securely in grooves

38.4 Note clutch cable bracket on cylinder left-hand rear stud

38.5 Check condition of sealing O-ring before refitting coolant pipe

38.6 Use new gasket and tighten inlet stub bolts securely to prevent leaks

38.7a Bring piston to TDC and wipe away surplus oil

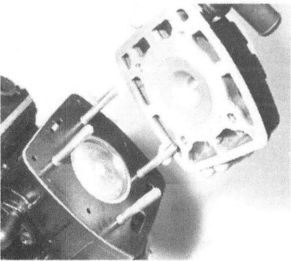

38.7b Refit head gasket and cylinder head as shown

38.8 Use torque wrench to tighten cylinder head nuts

39 Fitting the engine/gearbox unit in the frame

1   The engine/gearbox unit is refitted into the frame by reversing the removal procedure. Protect the paintwork of the frame by wrapping rag around the cradle tubes and insert the unit from the right-hand side. Push the engine front mounting bolt through from left to right and refit the plain washer and self-locking nut on the bolt right-hand end. Note that the self-locking nut should be renewed if it no longer grips the bolt securely. Once the engine/gearbox unit is aligned correctly on its front mounting, check that the plain washers are in place in their recesses on each side of the crankcase rear upper mounting lug, raise the unit at the rear so that it is aligned with the subframe pivot, and push the pivot bolt fully through from right to left on the RD125 LC model and from left to right on the DT125 LC model. The pivot bolt should pass through the crankcase rear upper mounting lug, through both rear sub-frame pivot bearings, and through both sides of the main frame.
2   Once the engine/gearbox unit is retained in position by the front and rear upper mounting bolts, refit the spring or plain washer and the self-locking retaining nut to the pivot bolt end, and refit the rear lower mounting bolt. Tighten the nuts and bolts by hand only, remove the protective rag from the frame tubes and check that the engine/gearbox unit is correctly aligned on its mountings and that nothing is trapped between the engine/gearbox unit and the frame. Tighten the two engine mounting bolts to a torque setting of 2.5 kgf m (18 lbf ft), and the rear sub-frame pivot bolt to a torque setting of 4.3 kgf m (31 lbf ft). Check that the spark plug is clean and correctly gapped, then refit it, tightening it to a torque setting of 2.0 kgf m (14.5 lbf ft). Connect the spark plug cap and the temperature gauge sender unit lead.
3   Remove the gearbox oil filler plug and pour in 600 cc (1.06 pint) of SAE 10W/30SE engine oil. Note that this amount is slightly more than that required during the course of routine maintenance, the difference being accounted for by the amount of residual oil trapped in the crankcase and components which cannot be removed by normal draining. Refit the filler plug and remember to check the oil level after the engine has been started and warmed up to normal operating temperature.
4   Engage the gearbox sprocket on the chain and slide it over the splines of the output shaft. Replace the sprocket retaining plate, rotating it to lock the sprocket on the shaft. Apply a few drops of thread locking compound to the threads of the two

bolts and refit these, tightening them to a torque setting of 1.0 kgf m (7 lbf ft). Check that the final drive chain is adjusted correctly, resetting it, if necessary, as described in the Routine Maintenance Section of this Manual.
5   When working on a DT125 LC model, refit the rear brake pedal assembly, and connect the stop lamp rear switch. Adjust the rear brake so that there is 20-30 mm (0.8-1.2 in) of free play at the pedal tip before the brake shoes are felt coming into contact with the drum, and adjust the stop lamp switch height so that the lamp lights just as the pedal has reached the end of its free play and the brake is starting to operate. When working on an RD125 LC model these components need not be disturbed during the removal and refitting of the engine/gearbox unit, but the opportunity should be taken to ensure that they are correctly adjusted as described above.
6   Connect the main generator stator lead to the main loom at its various connecting blocks or snap connectors, and clamp the lead to the frame tubes using the clamps or guides provided for this purpose. Refit the kickstart lever on its shaft. Apply a few drops of thread locking compound to the threads of the kickstart shaft and refit the lever retaining nut, tightening it to a torque setting of 6.5 kgf m (47 lbf ft). On DT125 LC models only, refit the gearchange pedal using the marks made on dismantling to align it correctly, then fit and tighten the retaining pinch bolt.
7   Connect the clutch cable to the end of the push lever assembly by reversing the removal sequence. Ensure that the thin metal tang is bent back across the slot in the trunnion to prevent the cable end nipple from jumping out. The clutch cable must now be adjusted but it should be noted that the procedure varies slightly between the two models described in this Manual. When the clutch is perfectly adjusted on either model there should be 2-3 mm (0.08-0.12 in) free play in the cable, measured between the handlebar lever and its clamp, and the pointed end of the clutch actuating lever should align exactly with the raised index mark on the crankcase top when the handlebar lever is released. When working on a DT125 LC model this setting is achieved by screwing the handlebar adjuster in as far as possible and by rotating the cable in-line adjuster as necessary to align the clutch push lever end with the index mark. On the RD125 LC model, the setting must be achieved by the use of the handlebar lever adjuster alone, as this is the only one fitted to the clutch cable of this model. If the mechanism free play has been correctly set as described in Section 35 of this Chapter, using the cable adjuster to align the push lever and index mark will produce approximately the correct amount of free play at the handlebar lever. Any difference can then be made up by rotating the handlebar lever adjuster. When the clutch cable is correctly adjusted, lubricate the exposed length of cable inner and the adjuster threads, tighten securely the adjuster locknuts and replace any rubber sleeves or grommets fitted to protect the cable or adjusters.
8   Refit the carburettor assembly, sliding it first into the inlet stub and then connecting the air filter hose. Refit the throttle valve assembly, ensuring that it is the right way round, and tighten securely the carburettor top. Tilt the carburettor upright and tighten securely its retaining clamps, then route all carburettor breather pipes through their respective guides and pass all lengths of pipe to the rear of the engine/gearbox unit.
9   Connect the oil tank/oil pump feed pipe to its union on the oil pump and secure it with its wire clip. Slacken and remove the bleed screw situated next to the union and allow oil to drip out. Watch carefully for air bubbles and replace the screw when no more bubbles can be seen. Tighten the bleed screw securely and connect the lower end of the oil pump/carburettor feed pipe to its oil pump union, then secure the feed pipe with its tubular clip. Unplug the upper end of this latter oil feed pipe and check to see if oil is present. If this is the case, then it will not be necessary to bleed the oil pump/carburettor feed pipe, and the pipe can be refitted on its carburettor union and secured with the tubular clip. If oil is not present it will be necessary to bleed the pipe to remove any traces of air; a task which can only be carried out when the engine is running, and the feed pipe should not be connected to the carburettor to act as a reminder of this.

Check that the oil feed pipes are not twisted or kinked and are retained correctly by the feed pipe clamp (where fitted) and the sealing grommet.

10 Route the oil pump cable down to the crankcase right-hand cover so that it is not twisted or kinked, insert it into the passage in the cover and connect it to the oil pump pulley. Note that the cable inner must be retained by the separate spring clip or by the extension of the pulley return spring and snap the throttle twistgrip open and closed several times to check that the cable is functioning properly. There should be 4-7 mm ($\frac{1}{8}$ - $\frac{1}{4}$ in approx) of free play in the throttle cable, measured around the circumference of the twistgrip rubber inner flange. If this is not the case, slacken the locknut of the twistgrip adjuster and rotate the adjuster to achieve the necessary free play. Snap the throttle open and closed a few times to settle the cable and recheck the setting.

11 Do not refit the oil/water pump cover as the oil pump settings must be checked and, if necessary, altered. This can only be done when the engine has been started and warmed up sufficiently to permit the tickover speed to be reset. As the oil pump is controlled by the throttle cable, any variation in the free play of the throttle cable caused by resetting the carburettor will alter the oil pump setting.

12 On RD125 LC models only, refit the small oil seal around the lower end of the tachometer drive inner cable and insert the cable into the tachometer drive assembly. Tighten securely the knurled retaining ring, but be careful not to overtighten it.

13 Refit the YEIS hose to the inlet stub, using a suitably heavy pair of pliers to secure the retaining clip, then refit the YEIS chamber to the frame, tightening securely the single mounting bolt. Fit a new gasket to the exhaust port, using a smear of grease to stick it in place and refit the exhaust system. Offer the exhaust system up to the frame, securing it in position by tightening the various mounting nuts and bolts by hand only at first. Ensure that the exhaust is aligned correctly on its mountings, then tighten the two front mounting nuts to a torque setting of 1.8 kgf m (13 lbf ft), and tighten securely the remaining mounting bolts. Ensure that the rubber joint seal fitted to DT125 LC models is correctly positioned and securely fastened.

14 Refit and connect the radiator hoses, ensuring that their securing clips are in good order and correctly fastened, then check that the coolant drain plugs have been refitted and are tightened to a torque setting of 1.0 kgf m (7 lbf ft). Fill the cooling system using a mixture of 50% distilled water and 50% ethylene glycol anti-freeze. Do not use ordinary tap water in this mixture as the impurities contained in it will promote corrosion and furring-up of the system, but it is permissible to use rainwater or water that has been boiled and allowed to stand until cool as alternatives to distilled water. The total coolant capacity of both of the machines described in this Manual is 800 cc (1.41 pint); a great amount should be prepared so that a surplus of mixture is left over for any subsequent topping up. Fill the radiator (and cooling system) to the base of the radiator neck, and top up the expansion tank to the 'Full' level mark, then refit the radiator cap. Note that it will be necessary to check the level after the engine has been run for the first time.

15 Refit the battery, if this was removed, and reconnect it, using the colour-coded wires to ensure correct polarity. Refit the petrol tank, the seat, the sidepanels, and either the belly fairing on RD125 LC models, or the bashplate on DT125 LC models. Make a final check that all the components fitted thus far are correctly in place and securely fastened, and that all controls are adjusted properly and functioning correctly. Switch on the ignition and check that all electrical components are operating correctly.

39.1a Check that engine/gearbox unit is aligned correctly on its mountings before ...

39.1b ... tightening the engine mounting bolts ...

39.1c ... to recommended torque settings. Check self-locking nuts, where fitted

39.4a Fit gearbox sprocket retaining plate

39.4b Use thread locking compound on sprocket mounting bolts

39.7a Bend tang up to secure clutch cable end nipple

39.7b Adjust clutch cable as described in text – DT125 shown

39.9 Connect oil feed pipes and bleed oil pump to remove air bubbles

39.13a Always fit a new exhaust gasket to prevent leaks

39.13b Tighten exhaust front mountings to correct torque setting ...

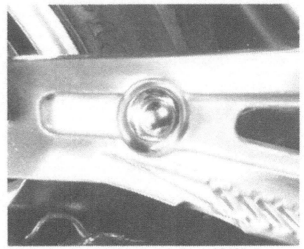

39.13c ... and then tighten rear mountings – RD125 shown

40 Starting and running the rebuilt engine

1   Attempt to start the engine using the usual procedure adopted for a cold engine. Do not be disillusioned if there is no sign of life initially. A certain amount of perseverance may prove necessary to coax the engine into activity even if new parts have not been fitted. Should the engine persist in not starting, check that the spark plug has not become fouled by the oil used during re-assembly. Failing this, go through the fault finding charts and work out what the problem is methodically.

2   When the engine does start, keep it running as slowly as possible to allow the oil to circulate. Open the choke as soon as the engine will run without it. During the initial running, a certain amount of smoke may be in evidence due to the oil used in the reassembly sequence being burnt away. The resulting smoke should gradually subside.

3   Check the engine for blowing gaskets and water or oil leaks. Before using the machine on the road, check that all the gears select properly, and that the controls function correctly.

4   As soon as the engine is warmed up to its normal operating temperature, allow it to idle and check the carburettor idle speed adjustment as described in Chapter 3 of this Manual. If any repair work has been carried out and new components have been fitted to the engine, carburettor adjustment will almost certainly be required. Once the engine is ticking over smoothly at its correct speed, check that the throttle cable free play has not altered, resetting it if necessary.

5   If it is necessary to bleed the delivery side of the oil pump and the oil pump/carburettor feed pipe to remove traces of air, this must now be done. With the engine running at tickover speed, pull the oil pump cable out of the crankcase right-hand cover, thus rotating the pump control pulley to set the pump stroke at its maximum. Keep the engine running, preferably at idle speed and at no more than 2000 rpm, until oil can be seen issuing from the carburettor end of the feed pipe. When no air bubbles can be seen in the oil, fit the feed pipe on to its carburettor union and secure it with its tubular clip. Restore the oil pump cable to its normal position and allow the engine to idle once more. Watch the pump adjustment plate, noting that it moves in and out as the pump rotates. When the pump adjustment plate has moved out as far as possible, stop the engine and check the pump minimum stroke setting as described in Section 19 of Chapter 3 of this Manual, starting the engine and rechecking the stroke setting as necessary. Once the oil pump stroke has been checked and reset, if necessary,

the oil pump cable adjustment must be checked to ensure that the oil pump and throttle are correctly synchronized. This is also described in Section 19 of Chapter 3. When the oil pump has been bled and its setting checked, apply a smear of grease to the control pulley cable groove and lubricate all exposed moving parts, including the oil pump cable inner. Refit the oil/water pump cover, ensuring that the oil feed pipe grommet is correctly located.

6   Before proceeding to the next step, stop the engine and very carefully remove the radiator filler cap. Place a thick rag over the cap and twist it slowly, allowing any surplus pressure to escape before fully removing it, to minimise the risk of personal injury due to escaping steam or coolant. Top the radiator up to the base of its filler neck, if necessary, and refit the cap. Note that while this task should be carried out when the engine is cold, it is unlikely that it will be convenient to wait for the engine to cool at this stage of the reassembly procedure, and great care must be taken therefore, to avoid personal injury.

7   The ignition timing can only be checked using a strobe timing lamp as described in Section 9 of Chapter 4 of this Manual. Connect the strobe as described in the manufacturer's instructions, start the engine and allow it to idle. Aim the strobe at the timing marks (DT125 LC) or at the ignition trigger (RD125 LC) and accelerate the engine to 3000 rpm. If the timing marks align correctly (DT125 LC) or if the ignition trigger appears to move clockwise underneath the pulser coil (RD125 LC), the ignition timing is correct. If such is not the case, while the ignition timing of the DT125 LC model can be reset, the ignition system of the RD125 LC must be tested and the faulty components renewed. In the case of both models, the necessary operations are described in Chapter 4 of this Manual and must be carried out before the machine is taken on the road.

8   Assuming that the timing is correct, refit the crankcase left-hand cover and its gasket, tightening securely the retaining screws. On RD125 LC models only, grease the bearing surfaces of the gear linkage components and offer up the linkage as shown in the photographs accompanying the text. Align the linkage with the footrest in the position required by rotating the linkage front arm around the gearchange shaft, remembering that the linkage will operate most efficiently when the angles formed between the link rod and the linkage front and rear arms are at 90°. When the linkage is aligned satisfactorily, push the front arm on the splines of the gearchange shaft, then fit the bolt at the linkage pivot point. Tighten the pivot bolt to a torque setting of 1.5 kgf m (11lbf ft) and tighten securely the linkage

front arm pinch bolt. Check that the linkage operates correctly.
9   With the machine standing upright on a level surface, check that the transmission oil level is correct. A sight glass is fitted in the crankcase right-hand cover, below the kickstart lever, and is marked with maximum and minimum levels. The oil level should be between these marks. Add oil through the gearbox filler plug if necessary.
10 Allow the engine to cool down, then check that the cylinder barrel and head retaining nuts are fastened to their correct torque settings. The cylinder barrel nuts are tightened to 2.8 kgf m (20 lbf ft), the cylinder head nuts to 2.2 kgf m (16 lbf ft), the coolant pipe screws to 1.2 kgf m (9 lbf ft), and the exhaust front mounting nuts to 1.8 kgf m (13 lbf ft). Check carefully for coolant or oil leaks and rectify any that may have occurred. Make a final check of the coolant level, both in the radiator itself and in the expansion tank. The radiator should be filled to the base of its filler neck and the expansion tank level should be between the level marks. Top up, if necessary, using the correct antifreeze/distilled water mixture. Refit securely the filler caps and the radiator cover.

40.5 Check oil feed pipes are routed correctly when refitting oil/water pump cover. Note position of white band (arrowed) – See Chapter 3

41 Taking the rebuilt machine on the road

1   Any rebuilt machine will need time to settle down, even if parts have been replaced in their original order. For this reason it is highly advisable to treat the machine gently for the first few miles to ensure oil has circulated throughout the lubrication system and that new parts fitted have begun to bed down.
2   Even greater care is necessary if the engine has been rebored or if a new crankshaft has been fitted. In the case of a rebore, the engine will have to be run in again, as if the machine were new. This means greater use of the gearbox and a restraining hand on the throttle until at least 500 miles have been covered. There is no point in keeping to any set speed limit; the main requirement is to keep a light loading on the engine and to gradually work up performance until the 500 mile mark is reached. These recommendations can be lessened to an extent when only a new crankshaft is fitted. Experience is the best guide since it is easy to tell when an engine is running freely.
3   Remember that a good seal between the piston and the cylinder barrel is essential for the correct functioning of the engine. A rebored two-stroke engine will require more careful running-in, over a long period, than its four-stroke counterpart. There is a far greater risk of engine seizure during the first hundred miles if the engine is permitted to work hard.
4   If at any time a lubrication failure is suspected, stop the engine immediately, and investigate the cause. If an engine is run without oil, even for a short period, irreparable engine damage is inevitable.
5   Do not on any account add oil to the petrol under the mistaken belief that a little extra oil will improve the engine lubrication. Apart from creating excess smoke, the addition of oil will make the mixture much weaker, with the consequent risk of overheating and engine seizure. The oil pump alone should provide full engine lubrication.
6   Do not tamper with the exhaust system. Unwarranted changes in the exhaust system will have a marked effect on engine performance, invariably for the worse. The same advice applies to dispensing with the air cleaner or the air cleaner element.
7   When the initial run has been completed allow the engine unit to cool and then check all the fittings and fasteners for security. Re-adjust any controls which may have settled down during initial use.

Chapter 2 Cooling system

For modifications, and information relating to later models, see Chapter 8

Contents

Specifications

Coolant

Mixture type	50% distilled water, 50% corrosion-inhibited ethylene glycol antifreeze
Total capacity	800 cc (1.41 Imp pint)

Water pump

Type	Centrifugal impeller
Drive	Gear, from crankshaft via idler pinion

Radiator

Core width	122.5 mm (4.82 in)
Core height	180.0 mm (7.09 in)
Core thickness	32.0 mm (1.26 in)
Testing pressure	1.0 kg/cm² (14 psi)
Cap valve opening pressure	0.9 kg/cm² (12.8 psi)
Tolerance	0.75 – 1.05 kg/cm² (10.7 – 14.9 psi)

Torque settings

Component	kgf m	lbf ft
Coolant drain plugs	1.0	7.0
Pump cover screws	1.0	7.0
Coolant pipe mounting screws	1.2	9.0
Temperature gauge sender unit	1.0	7.0

1 General description

The Yamaha LC models are provided with a liquid cooling system which utilises a water/antifreeze coolant to carry away excess energy produced in the form of heat. The cylinders are surrounded by a water jacket from which the heated coolant is circulated by thermo-syphonic action in conjunction with a water pump fitted in the engine right-hand cover and driven via a pinion and shaft from a crankshaft mounted pinion. The hot coolant passes upwards through a flexible pipe to the top of the radiator which is mounted on the frame downtubes to take advantage of maximum air flow. The coolant then passes downwards, through the radiator core, where it is cooled by the passing air and then to the water pump and engine where the cycle is repeated.

The complete system is partially sealed and is pressurised; the pressure being controlled by a valve contained in the spring-loaded radiator cap. By pressurising the coolant the boiling point is raised, preventing premature boiling in adverse conditions. The overflow pipe from the radiator is connected to an expansion tank into which excess coolant is discharged by pressure. The expelled coolant automatically returns to the radiator to provide the correct level when the engine cools again.

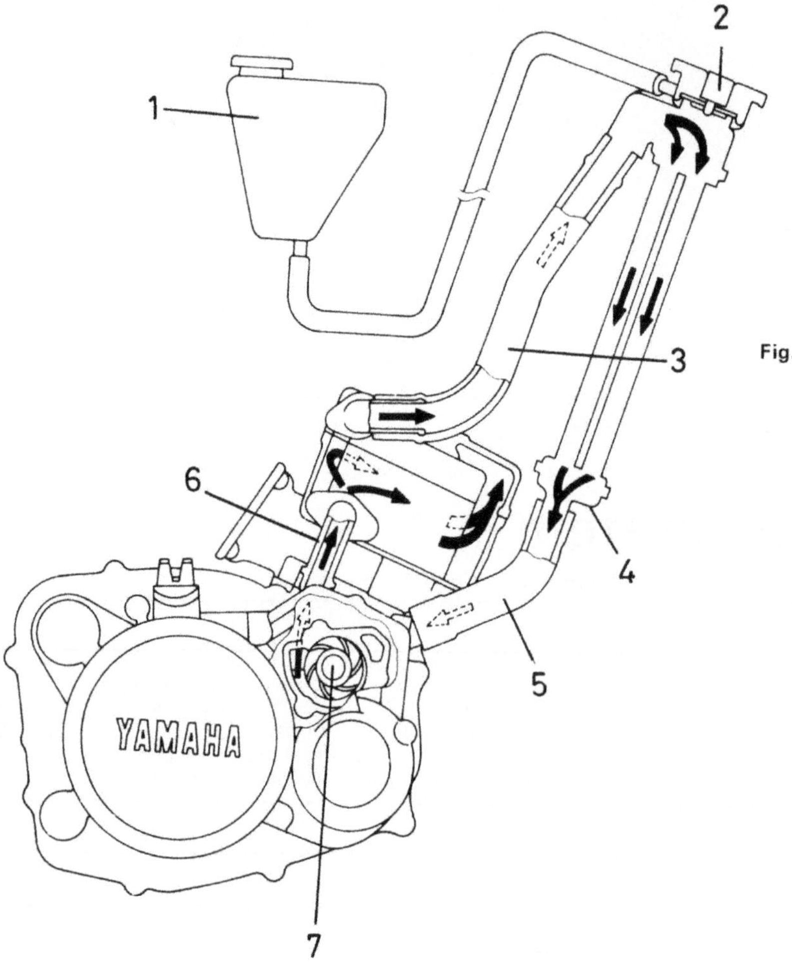

Fig. 2.1 Cooling system - DT model

1 Expansion tank
2 Radiator cap
3 Inlet top hose
4 Radiator
5 Outlet bottom hose
6 Transfer hose
7 Water pump

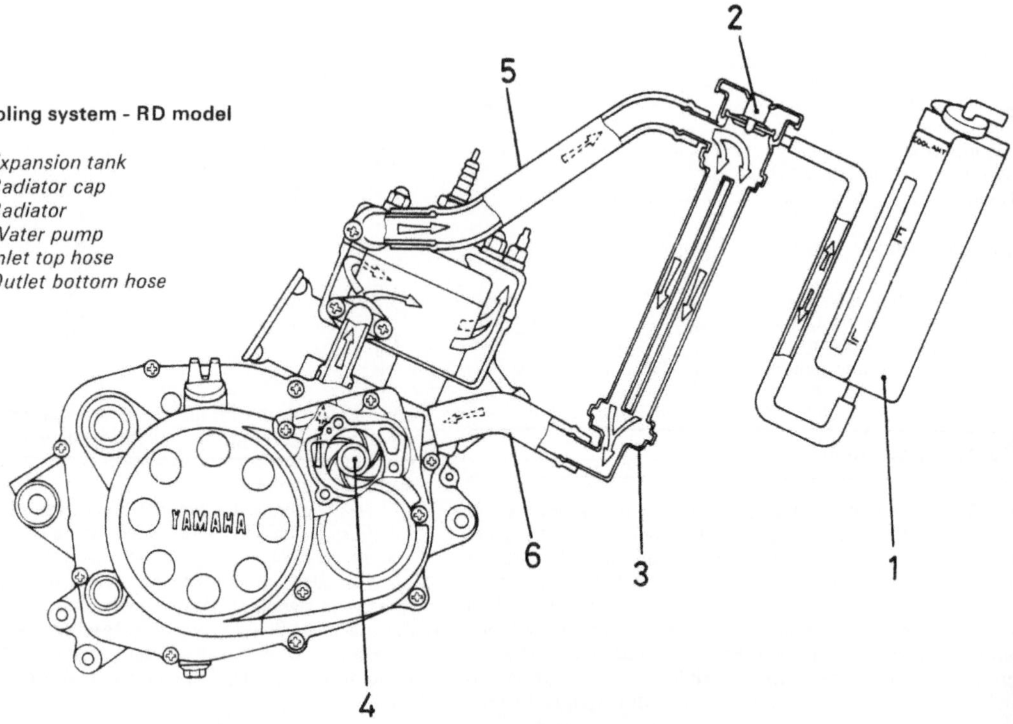

Fig. 2.2 Cooling system - RD model

1 Expansion tank
2 Radiator cap
3 Radiator
4 Water pump
5 Inlet top hose
6 Outlet bottom hose

2 Draining the cooling system

1 It will be necessary to drain the cooling system on infrequent occasions, either to change the coolant at two yearly intervals or to permit engine overhaul or removal. The operation is best undertaken with a cold engine to remove the risk of scalding from hot coolant escaping under pressure.

2 Place the machine upright on a suitable stand and gather together a drain tray or bowl of about 1 litre (2 pint) capacity, and something to guide the coolant from the drain plugs into the bowl. A small chute made from thick card will suffice for this purpose, but do not be tempted to allow the coolant to drain over the engine casings — the antifreeze content may discolour the painted surfaces.

3 Detach the radiator cover to permit the radiator cap to be removed. The cover fitted to RD125 LC models is retained by a single horizontally-mounted screw at its upper edge, and by two clips at the bottom which engage on rubber grommets placed over lugs on the frame, while that fitted to the DT125 LC model is secured to the radiator mounting bracket at three points, a single screw set horizontally in the cover right-hand side and two prongs which are held in rubber grommets mounted in the top and bottom rails of the mounting bracket. Slacken and remove the single screw and withdraw carefully the cover,

noting that on RD125 LC models only, the cover will also have to be disengaged from two clips on the belly fairing. Remove also the oil/water pump cover from the front of the crankcase right-hand cover to expose the drain plug situated in the water pump casing. Take great care when removing the cap from the radiator if the engine has been run recently, because there will be some residual pressure in the system. If the engine is hot, steam and boiling water may be ejected and can cause scalding. As a precaution, place some rag over the cap and remove it slowly to allow pressure to escape.

4 Slacken and remove each of the drain plugs in turn, using the chute to guide the coolant clear of the crankcase outer covers and into the bowl. If the system is to be drained fully pull off the pipe from the expansion tank and allow this to drain. Drain any residual coolant by detaching the bottom hose from its union at the front of the crankcase right-hand cover.

5 If the system is being drained as a precursor to engine overhaul little else need be done at this stage. If the coolant is reasonably new it can be re-used if it is kept clean and uncontaminated. If, however, the system is to be refilled with new coolant it is advisable to give it a thorough flushing with tap water, if possible using a hose which can be left running for a while. If the machine has done a fairly high mileage it may be advisable to carry out a more thorough flushing process as described below.

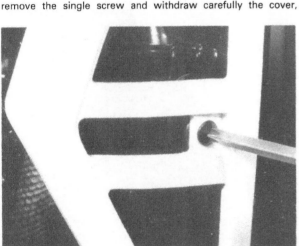

2.3 Radiator covers are retained by a single screw and two clips — DT125 shown

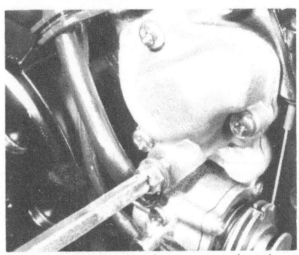

2.4a Coolant drain plugs are set in water pump casing and ...

2.4b ... in cylinder barrel. Note cardboard chute to assist in draining

2.4c Release bottom hose as shown to drain residual coolant

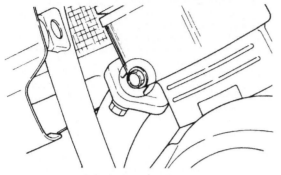

Cylinder barrel drain bolt

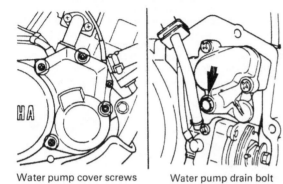

Water pump cover screws   Water pump drain bolt

Fig. 2.3 Cooling system drain plug locations

3 Cooling system: flushing

1   After extended service the cooling system will slowly lose efficiency, due to the build-up of scale, deposits from the water and other foreign matter which will adhere to the internal surfaces of the radiator and water channels. This will be particularly so if distilled water has not been used at all times. Removal of the deposits can be carried out easily, using a suitable flushing agent in the following manner.
2   After allowing the cooling system to drain, refit the drain plugs and refill the system with clean water and a quantity of flushing agent. Any proprietary flushing agent in either liquid or dry form may be used, providing that it is recommended for use with aluminium engines. NEVER use a compound suitable for iron engines as it will react violently with the aluminium alloy. The manufacturer of the flushing agent will give instructions as to the quantity to be used.
3   Run the engine for ten minutes at operating temperatures and drain the system. Repeat the procedure TWICE and then again using only clean cold water. Finally, refill the system as described in the following Section.

4 Cooling system: filling

1   Before filling the system, check that the sealing washers on the drain plugs are in good condition and renew if necessary. Fit and tighten the drain plugs and check all the hose clips. The torque setting for both drain plugs is 1.0 kgf m (7 lbf ft).
2   Fill the system slowly to reduce the amount of air which will be trapped in the water jacket. When the coolant level is up to

the lower edge of the radiator filler neck, run the engine for about 10 minutes at idle speed. Increase engine revolutions for the last 30 seconds to accelerate the rate at which any trapped air is expelled. Stop the engine and replenish the coolant level again to the bottom of the filler neck. Refill the expansion tank up to the 'Full' level mark. Refit the radiator cap, ensuring that it is turned clockwise as far as possible.
3   Ideally, distilled water should be used as a basis for the coolant. If this is not readily available, rain water, caught in a non-metallic receptacle, is an adequate substitute as it contains only limited amounts of mineral impurities. Similarly, it is permissible to use tap water which has been thoroughly boiled and allowed to cool down. In emergencies only, tap water can be used, especially if it is known to be of the soft type. Using non-distilled water will inevitably lead to early 'furring-up' of the system and the need for more frequent flushing. The correct water antifreeze mixture is 50/50; do not allow the antifreeze level to fall below 40% as the anti-corrosion properties of the coolant will be reduced to an unacceptable level. Antifreeze of the ethylene-glycol-based type should always be used. Never use alcohol-based antifreeze in the engine.

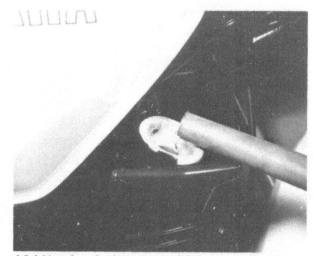

4.2 Add coolant slowly to prevent air being trapped in system

5 Radiator and radiator cap: removal, cleaning, examination and refitting

1   Drain the radiator as described in Section 2 of this Chapter. Disengage the clips at the radiator ends of both hoses, using a suitably heavy pair of pliers to compress the protruding ends of each clip, then carefully pull the hoses off their respective radiator unions. Disengage the wire clip, then pull the radiator/expansion tank tube off its union at the radiator filler neck.
2   The radiator fitted to the RD125 LC model is secured to the frame front downtubes by a single bolt on its right-hand side, and by two rubber grommets set in its left-hand mounting flange which engage on two lugs mounted on the frame left-hand downtube. To remove the radiator, disengage the expansion tank from its clip on the radiator left-hand side and put the expansion tank to one side, then slacken and remove the single mounting bolt, pull the radiator to the right to free the grommets from their lugs, and remove the radiator from the machine.
3   The radiator fitted to the DT125 LC model is mounted on a

separate bracket which is bolted to the frame downtube. This position is a compromise adopted by the designer in an attempt to place the radiator as high and as far to one side as possible so that it is out of the direct line of the mud and stones that are thrown up by the front wheel during the course of off-road riding, and yet to ensure that such a bulky and vulnerable component is as unobtrusive and as well protected as possible. To remove the radiator from the machine, slacken and remove the two bolts which secure the mounting bracket to the frame, and withdraw the radiator and mounting bracket as a single unit. If it is wished to detach the radiator from its mounting bracket, slacken and remove the three bolts, two of which are rubber-mounted, which fasten the two components together.

4   The exterior of the radiator core can be cleaned by blowing from the rear surface forwards with compressed air. Most garages have an air line which can be used for this purpose, but if none is available a garden hose can be substituted, using water instead of air. It is essential that all dirt and debris are removed from the radiator matrix as any obstructions to the constant flow of air over its surface will severely reduce the radiator's cooling efficiency. While this is not a particularly severe problem on a road-going machine, dead insects and leaves being the commonest obstructions to be found, if a DT125 LC model is used regularly on off-road excursions, a good proportion of the dust or mud, stones, assorted vegetation, and other debris which inevitably accompany such usage will find its way on to the radiator and must be removed at frequent intervals if overheating is to be avoided. In such cases the owner should avail himself of the first opportunity which presents itself to flush out the radiator matrix, preferably before the material has a chance to dry out, thus making its removal even more difficult.

5   The interior of the radiator can most easily be cleaned while the radiator is in situ on the motorcycle, using the flushing procedure described in Section 3 of this Chapter. Additional flushing can be carried out by placing the hose in the filler neck and allowing the water to flow through for about ten minutes. Under no circumstances should the hose be connected to the filler neck mechanically as any sudden blockage in the radiator outlet would subject the radiator to the full pressure of the mains supply (about 50 psi). The radiator should not be tested to greater than 14 psi (1.0 kg/cm^2).

6   If care is exercised, bent fins can be straightened by placing the flat of a screwdriver either side of the fin in question and carefully bending it into its original shape. Badly damaged fins cannot be repaired. If bent or damaged fins obstruct the air flow more than 20%, a new radiator will have to be fitted.

7   Generally, if the radiator is found to be leaking, repair is impracticable and a new component must be fitted. Very small leaks may sometimes be stopped by the addition of a special sealing agent in the coolant. If an agent of this type is used, follow the manufacturer's instructions very carefully. Soldering, using soft solder, may be efficacious for caulking large leaks but this is a specialised repair best left to experts.

8   Inspect the radiator mounting rubbers for perishing or compaction. Renew the rubbers if there is any doubt as to their condition. The radiator may suffer from the effect of vibration if the isolating characteristics of the rubber are reduced.

9   If the radiator cap is suspect, have it tested by a Yamaha dealer. This job requires specialist equipment and cannot be done at home. The only alternative is to try a new cap, but it should be noted that as the cap is very similar in size to those fitted to cars, a local car garage may have the necessary equipment; it is, therefore worthwhile to find out if this is the case before going to the possibly unnecessary expense of substituting a new component. If the equipment is available, the cap is fitted to one end, using an adaptor if necessary, and a pressure of 11 – 15 psi (0.75 – 1.05 kg/cm^2) applied to it, usually by means of a hand-operated plunger. The pressure must be held for a period of 10 seconds, during which time there should be no measurable loss. If the cap is found to be faulty, it must be renewed as repairs are not possible.

10   It should be noted that when tracing an elusive leak the entire cooling system can be pressurised to its normal operating pressure by connecting the test equipment described above to the radiator filler orifice. Remove the radiator cap, check the coolant level, topping it up if necessary, and apply a pressure of no more than 14 psi (1.0 kg/cm^2) by means of the hand-operated plunger. Any leaks should soon become apparent. Most leaks will, however, will be readily apparent due to the tell-tale traces of antifreeze left on the components in the immediate area of the leak.

11   Refitting the radiator and cap is a straightforward reversal of the dismantling procedure described above. Fill the cooling system as described in Section 4 of this Chapter and check very carefully for leaks.

5.2a RD125 expansion tank is clipped to radiator left-hand side

5.2b Radiator is retained by a single bolt into frame right-hand downtube

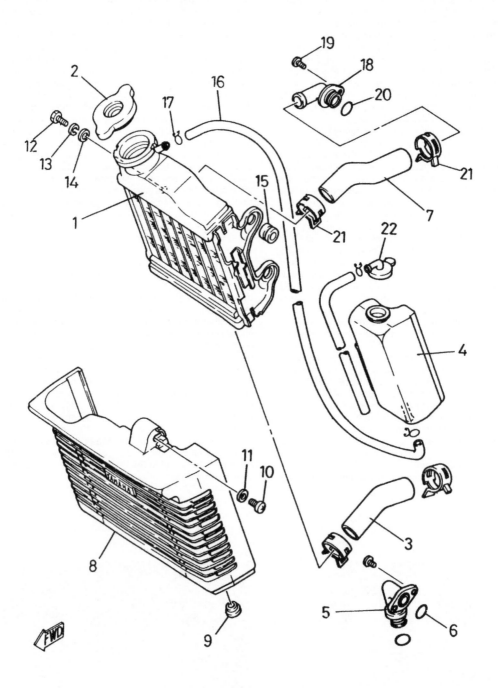

Fig. 2.4 Radiator - RD model

1	Radiator	7	Top hose	13	Spring washer	18	Top hose union
2	Radiator cap	8	Radiator cover	14	Washer	19	Screw - 3 off
3	Bottom hose	9	Grommet - 2 off	15	Grommet - 2 off	20	O-ring
4	Expansion tank	10	Screw	16	Radiator feed pipe	21	Hose clip - 4 off
5	Bottom hose union	11	Washer	17	Pipe clip	22	Expansion tank cap
6	O-ring - 2 off	12	Bolt				

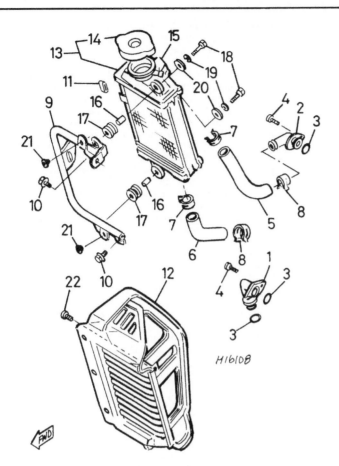

H16108

Fig. 2.5 Radiator - DT model

1   Bottom hose union
2   Top hose union
3   O-ring - 3 off
4   Screw - 3 off
5   Top hose
6   Bottom hose
7   Hose clip - 2 off
8   Hose clip - 2 off
9   Mounting bracket
10 Bolt - 2 off
11 Damping rubber
12 Radiator cover
13 Radiator
14 Radiator cap
15 Damping rubber
16 Spacer - 2 off
17 Grommet - 2 off
18 Bolt - 2 off
19 Spring washer - 2 off
20 Washer - 2 off
21 Grommet - 2 off
22 Screw

6 Hoses and connections: examination and renovation

1   The radiator is connected to the engine unit by two flexible hoses, there being an additional pipe between the water pump in the right-hand crankcase cover and the cylinder. The hoses should be inspected periodically and renewed if any sign of cracking or perishing is discovered. The most likely area for this is around the clips which secure each hose to its unions. Particular attention should be given if regular topping up has become necessary. The cooling system can be considered to be a semi-sealed arrangement, the only normal coolant loss being minute amounts through evaporation in the expansion tank. If significant quantities have vanished it must be leaking at some point and the source of the leak should be investigated promptly.

2   Serious leakage will be self-evident, though slight leakage can be more difficult to spot. It is likely that the leak will only be apparent when the engine is running and the system is under pressure, and even then the rate of escape may be such that the hot coolant evaporates as soon as it reaches the atmosphere, although traces of antifreeze should reveal the source of the leak in most cases. If not, it will be necessary to make use of testing equipment, as described in the previous Section, to pressurise the cooling system when cold, thereby enabling the source of the leak to be pinpointed. To this end it is best to entrust this work to an authorised Yamaha dealer who will have access to the necessary equipment if this is not available elsewhere, for example at a car garage or radiator repair agent.

3   In very rare cases the leak may be due to a broken head gasket in which case the coolant may be drawn into the engine and expelled as vapour in the exhaust gases. If this proves to be the case it will be necessary to remove the cylinder head for investigation. If the rate of leakage has been significant it may prove necessary to remove the cylinder barrel and piston so that the crankcase can be checked. Any coolant which finds its way that far into the engine can cause rapid corrosion of the main and big end bearings and must be removed completely.

4   Other possible sources of leakage are the water pump casing gasket, the water pump seal, and the O-rings sealing the joints in the crankcase right-hand cover, the cylinder barrel, and on the cylinder head. All these should be investigated and any leaks rectified by tightening the retaining screws, where applicable, or by renewing any seals or gaskets which are worn or damaged.

7 Water pump: removal, overhaul and refitting

1   The water pump will not normally require attention unless there is evidence of leakage of coolant into the transmission oil, as shown by excessive amounts of light grey-brown emulsified oil in the oil level sight glass and a persistent drop in the coolant level which cannot be accounted for by any other reason. If a very high mileage has been covered, noise from the water pump area might indicate a worn impeller shaft or crankcase right-hand cover.

2   In order to gain access to the impeller shaft assembly, the water pump seal, or the water pump drive gears, it will be necessary to remove the crankcase right-hand cover as described in Section 8 of Chapter 1. Note, however, that if it is wished merely to remove the water pump casing, to renew the gasket for example, it will suffice to drain the cooling system as described in Section 2 of this Chapter. Be careful at all times to mop up all residual coolant, and not to allow any coolant on to the oil pump.

3   If the water pump is to be dismantled fully, remove the crankcase right-hand cover and displace the circlip which retains the water pump driven gear on the impeller shaft. Withdraw first the thrust washer, then the driven gear, the gear

locating pin, and the second thrust washer. Invert the crankcase cover, placing it on a clean, flat surface and use an impact driver to slacken the pump casing screws. Remove the casing and its gasket, noting that the casing is located by two dowel pins which must not be lost, and lift out the impeller shaft assembly. The pump oil seal can be prised from its housing using a suitably-shaped screwdriver, but note beforehand the exact way in which the seal is fitted. The seal should be discarded, regardless of its condition, and a new component purchased and fitted on reassembly.

4   Inspect the impeller shaft assembly. Remove any scale deposits from it and from the two castings. Check carefully that the impeller shaft is clean and unworn, noting that any damage at the point where the shaft passes through the pump seal must be rectified, if necessary by removing the shaft assembly, although some minor repairs may be effected by the use of fine emery paper. Any damage at this point will result in rapid wear of the new pump seal. Insert the shaft into the crankcase cover and check for excessive free play which indicates a worn shaft or bearing surface. Compare the outside diameter of the shaft at two points, once where it is subject to wear as it passes through the crankcase cover bearing surface, and once at any point which protrudes beyond the bearing surface and can be considered unworn for all practical purposes. If any significant difference is found, the shaft assembly must be renewed.

5   If the impeller shaft is found to be unworn, or if a new component is purchased and fitted, and free play is still found, it must be assumed that the crankcase cover itself is worn. In such a case the cover must be renewed. The only alternative to this rather expensive solution is to find a local engineering company who is prepared to ream out the worn impeller shaft bore, and to fabricate and press in a bush to provide a new bearing surface. Fortunately, such wear is unlikely to be found until a very high mileage has been covered and, therefore, will not be encountered very often.

6   Although the water pump drive gears are well lubricated and under a very light load, wear therefore being negligible even after very high mileages, the opportunity should be taken to examine them, and to renew them if they should be damaged or worn. The gears rotate on a shaft which is pressed through the crankcase right-hand cover and are retained by a shoulder machined on the shaft left-hand end. To remove the gears, the shaft must be driven out of the cover. Remove the oil pump and any seals or other plastic components from the cover, then immerse the cover in water heated to boiling point (100°C), thus causing the alloy of the casting to expand enough to loosen the shaft. Remove the cover, taking care to prevent injury when handling the heated metal, and support it on two wooden blocks with its right-hand face upwards. Place the blocks to give maximum support to the casting while permitting the shaft and gears to be removed. Tap out the shaft using a hammer and a suitable metal drift on the shaft right-hand end which is visible projecting through the cover into the gasket surface of the water pump casing, immediately below the impeller shaft bore. Note that the water pump drive gears are a single unit, and must be renewed as such, even if only one is damaged. The gears are located on the shaft by two identical thrust washers, one on each side.

7   When refitting the water pump drive gears, reverse the dismantling procedure, but apply a locking compound such as Loctite Bearing Fit to ensure that the shaft remains in position. Ensure that the shaft right-hand end does not project beyond the gasket surface, or a coolant leak will develop.

8   Check the water pump casing for cracks or other damage, which may be repaired by welding, depending on the exact nature of the damage found. Such repairs should be carried out only by a person who has the necessary skill and access to the correct equipment. The usual solution to a damaged water pump casing would be to renew it. The casing gasket surface must be clean, undamaged and completely flat. If slight distortion or minor abrasions are found, lay a sheet of emery paper on a flat surface such as a sheet of plate glass, and rub the casing down with a rotary motion until the gasket surface

presents a uniform dull grey appearance, indicating that it is quite flat. Deeper cuts or scars may be filled with Araldite or a similar compound, and then rubbed down to restore the original surface.

9   On reassembly, clean carefully all the components to ensure that no dirt or debris finds its way into the cooling system to damage the pump seal. The pump seal is marked 'WATER SIDE' on one surface, which must face outwards, towards the impeller head. Apply a thin smear of grease to the outside diameter of the seal and insert it into the casing, ensuring that it is the correct way round, and that it is absolutely square in its housing. Using a hammer and a tubular drift or socket which bears only on the seal outer diameter, tap the seal into place until it is flush with the crankcase cover surface, taking great care that the seal remains square in its housing as it is driven home. When the seal is in place, apply a generous quantity of grease to its sealing lips and to the shaft of the impeller shaft assembly.

10   Insert the impeller assembly into the seal, taking great care not to damage the delicate sealing lips or to dislodge the spring behind them. Push the shaft in with a twisting motion to reduce further the risk of damage to the sealing lips. Invert the crankcase cover and refit the first thrust washer, the driven gear locating pin, the driven gear, the second thrust washer, and the retaining circlip. Ensure that the driven gear locates correctly on the locating pin. Place the crankcase cover on a flat surface with its right-hand side uppermost and fit the two dowel pins which locate the water pump casing.

11   Fit a new water pump gasket, noting that while it is permissible to apply a thin smear of grease to retain the gasket, no jointing compound should be used, and a new gasket must be fitted whenever the pump casing is disturbed. Tighten the casing screws securely; if the necessary equipment is available, a torque wrench should be used to tighten the screws to a torque setting of 1.0 kgf m (7 lbf ft).

12   Complete the reassembly procedure which is a straight-forward reversal of the dismantling procedure as described in Section 36 of Chapter 1. Refill the cooling system as described in Section 4 of this Chapter and check very carefully for leaks when the engine has completed its initial run.

8   Water temperature gauge and sender: testing

Water temperature is monitored by an electrically operated gauge in the instrument panel controlled by a sender unit which screws into the cylinder head water jacket. A description and test procedure for these components will be found in Chapter 7.

7.4 Impeller must be clean and unworn, renew if damaged

7.10a Grease lips of water pump seal before refitting impeller

7.10b Insert impeller shaft with a twisting motion, being careful not to damage oil seal

7.10c Position a thrust washer over impeller shaft end

7.10d Push driven gear locating pin through impeller shaft ...

7.10e ... and fit driven gear as shown

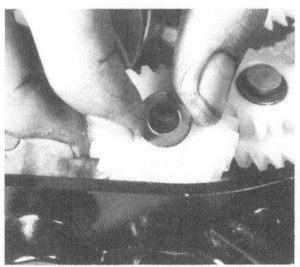

7.10f Then refit second thrust washer

7.10g Secure water pump driven gear with a circlip as shown

7.11a Always use a new water pump gasket

7.11b Note two locating dowel pins in water pump casing

7.11c Tighten securely the casing screws

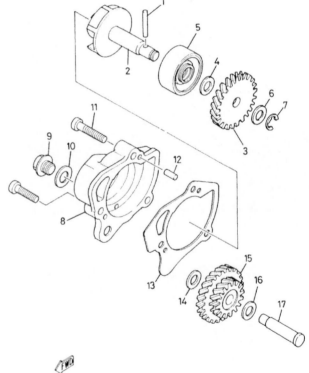

Fig. 2.6 Water pump assembly

1	Locating pin	10	Sealing washer
2	Impeller shaft	11	Screw - 3 off
3	Driven gear	12	Dowel pin - 2 off
4	Thrust washer	13	Gasket
5	Oil seal	14	Thrust washer
6	Thrust washer	15	Drive gears
7	Circlip	16	Thrust washer
8	Pump casing	17	Gear shaft
9	Drain plug		

Chapter 3 Fuel system and lubrication

For modifications, and information relating to later models, see Chapter 8

Contents

Specifications

Petrol tank capacity

	RD125 LC	DT125 LC
Overall	13.0 litre (2.86 gal)	9.0 litre (1.98 gal)
Reserve	1.9 litre (3.34 pint)	1.5 litre (2.64 pint)

Carburettor

	RD125 LC (10W)	RD125 LC (12A)	DT125 LC
Manufacturer	Mikuni	Mikuni	Mikuni
Type	VM24SS	VM24SS	VM24SS
ID Number	10W00	12A00	10V00
Main jet	155	195	80
Main air jet	1.0	2.5	0.5
Jet needle	5GN36	5GN36	4J13
Clip position	3	4	4
Needle jet	O-8	O-2	P-2
Throttle valve cutaway	1.5	1.5	2.0
Pilot jet	20	20	20
Pilot outlet	0.6	0.6	0.6
By-pass	1.4	1.4	1.4
Pilot mixture screw, turns out	$1\frac{1}{2}$	$1\frac{1}{2}$	$1\frac{1}{4}$
Float valve seat	2.5	2.5	2.0
Starter jet	25	25	20
Float height	21 ± 1.0 mm	21 ± 1.0 mm	21 ± 1.0 mm
	(0.83 ± 0.04 in)	(0.83 ± 0.04 in)	(0.83 ± 0.04 in)
Fuel level	26 mm (1.02 in)	26 mm (1.02 in)	26 mm (1.02 in)
Idle speed	1300 ± 50 rpm	1300 ± 50 rpm	1350 ± 50 rpm

Reed valve assembly

Stopper plate height	10.3 mm (0.406 in)
Tolerance	9.9 – 10.7 mm (0.390 – 0.421 in)
Reed thickness	0.2 mm (0.008 in)
Reed maximum warpage	0.5 mm (0.020 in)

Air filter
    Type .... Oiled polyurethane foam

Engine lubrication system
    Type .... Pump fed total loss system (Yamaha Autolube)
    Oil tank capacity:
        RD125 LC .... 1.1 litre (1.94 pint)
        DT125 LC .... 1.0 litre (1.76 pint)
    Oil pump delivery rate:
        Minimum (200 strokes) .... 0.38 – 0.48 cc (0.0134 – 0.0169 fl oz)
        Maximum (200 strokes) .... 3.56 – 3.94 cc (0.126 – 0.139 fl oz)
    Oil pump minimum stroke .... 0.20 – 0.25 mm (0.0079 – 0.0098 in)
    Oil pump maximum stroke .... 1.85 – 2.05 mm (0.073 – 0.081 in)

Gearbox lubrication
    Capacity:
        At oil change .... 550 cc (0.97 pint)
        At engine rebuild .... 600 cc (1.06 pint)

Torque settings

Component	kgf m	lbf ft
Oil pump mounting screws	0.5	3.5
Reed valve assembly mounting bolts	0.8	6.0
Reed fastening screws	0.1	0.7
Exhaust pipe front mounting nuts	1.8	13.0
Transmission oil drain plug	2.0	14.5

1 General description

1   The fuel system comprises a petrol tank, from which petrol is fed by gravity to the float chamber of the Mikuni carburettor, via a three position petrol tap. The tap has 'Off', 'On' and 'Reserve' positions, the latter providing a warning that the petrol level is low in time for the owner to find a garage.
2   The carburettor is of conventional concentric design, the float chamber being integral with the lower part of the carburettor. Cold starting is assisted by a separate starting circuit which supplies the correct fuel-rich mixture when the 'choke' control is operated.
3   Air entering the carburettor passes through a moulded plastic air cleaner casing which contains an oil-impregnated foam air filter. This effectively removes any airborne dust, which would otherwise enter the engine and cause premature wear. The air cleaner also helps silence induction noise, a common problem inherent with two-stroke engines.
4   The induction system fitted to the machines described in this Manual has two further features; the YEIS chamber, and a reed valve assembly. The principles by which each of these operate are discussed in detail later on in this Chapter.
5   Engine lubrication is catered for by the Yamaha Autolube system. Oil from a separate tank is fed by an oil pump to a small injection nozzle in the carburettor body. The pump is linked to the throttle twistgrip, and this controls the volume of oil fed to the engine.
6   The exhaust system fitted to the RD125 LC models is a single unit comprising the exhaust pipe and expansion chamber/silencer welded together. The unit is finished in a heat-resistant matt black coating and does not incorporate baffles that can be removed for cleaning purposes. The DT125 LC exhaust system is similar but incorporates a second, separate silencer assembly and is routed upwards over the engine/gearbox unit to pass down the left-hand side of the machine at a high level in keeping with the trail bike styling of the machine.
7   When working on any part of the induction and exhaust systems fitted to the machines described in this Manual, especially if alterations or modifications are being considered, bear in mind the fact that the system should be considered not as a collection of individual components to be modified or omitted at will, but as a single entity which has been very carefully designed, with each individual constituent being matched to the other and making its own contribution to the performance of the whole. In view of this, injudicious and unskilled modification should not be made to the system.

2 Petrol tank: removal, examination and refitting

1   The petrol tanks fitted to the machines described in this Manual are retained in a similar fashion; the nose of the tank is a push fit over two rubber buffers which are located on pegs which protrude from each side of the frame, behind the steering head, the tank being retained on these buffers by a single rear mounting. The rear mounting consists of a lug which projects upwards from a strut welded across the two frame top tubes. A moulded rubber 'mat' is laid on top of the strut, providing a cushion for the tank base and, with its raised moulded centre section, forms a positive method of tank location by engaging with the rectangular cut-out in the tank base. A further moulded rubber pad and a large metal 'washer' fit on top of the rear mounting and are held in place by a single bolt.
2   To remove the tank, unlock and lift away the seat, then remove both sidepanels to expose the tank rear mounting. Turn the petrol tap to the 'Off' position and use a suitable pair of pliers to release the wire petrol pipe retaining clip. Slide the clip down the petrol pipe until it is clear of the petrol tap spigot and carefully pull the pipe away from the tap. Slacken and remove the single tank rear mounting bolt, then lift away and put to one side the rear mounting top rubber, the large washer, and the bolt. Lift the tank up at the rear and ease it backwards off its front mountings, taking care not to damage the paintwork as the mountings are very tight when new and have a nasty habit of releasing their grip so that the tank comes away suddenly.
3   Inspect the tank mounting rubbers for signs of damage or deterioration and, if necessary, renew them before the tank is refitted. Store the tank in a safe place whilst it is removed from the machine, well away from any naked lights or flames. It will otherwise represent a considerable fire or explosion hazard. Check that the tap is not leaking and that it cannot be accidentally knocked into the 'On' position. It is well worth taking simple precautions to protect the paint finish of the tank whilst in storage. Placing the tank on a soft protected surface and covering it with a protective cloth or mat may well avoid damage being caused to the finish by dirt, grit, dropped tools, etc.

4   In the event that it is necessary to remove the petrol tank for repairs the following points should be noted. Petrol tank repair, whether necessitated by accident damage or by petrol leaks, is a task for the professional. Welding or brazing is not recommended unless the tank is purged of all petrol vapour; which is a difficult condition to achieve. Resin-based tank sealing compounds are a much more satisfactory method of curing leaks, and are now available through suppliers who advertise regularly in the motorcycle press. Accident damage repairs will inevitably involve re-painting the tank; matching of modern paint finishes, especially metallic ones, is a very difficult task not to be lightly undertaken by the average owner. It is therefore recommended that the tank be removed by the owner, and then taken to a motorcycle dealer or similar expert for professional attention.

5   Repeated contamination of the fuel tap filter and carburettor by water or rust and paint flakes indicates that the tank should be removed for flushing with clean fuel and internal inspection. Rust problems can be cured by using a resin tank sealant.

6   To refit the tank, reverse the procedure adopted for its removal. Move it from side to side before it is fully home, so that the rubber buffers engage with the guide channels correctly. If difficulty is encountered in engaging the front of the tank with the rubber buffers, apply a small amount of petrol to the buffers to ease location. Secure the tank with the single retaining bolt whilst ensuring that the mounting components are correctly located and that there is no metal to metal contact between the tank and frame.

7   Finally, always carry out a leak check on the fuel pipe connections after fitting the tank and turning the tap lever to the 'On' position. Any leaks found must be cured; as well as wasting fuel, any petrol dropping onto hot engine castings may well result in fire or explosion.

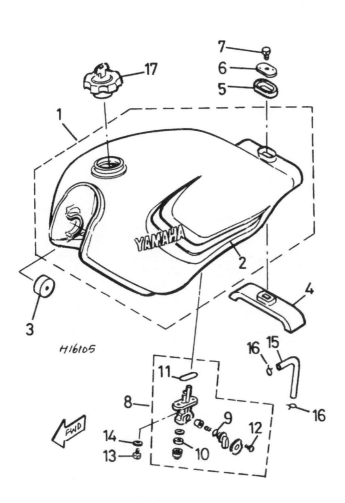

Fig. 3.1 Petrol tank – RD model

1 Petrol tank assembly	6 Washer	10 Sealing washer	14 Washer – 2 off
2 Petrol tank	7 Bolt	11 O-ring	15 Petrol feed pipe
3 Rubber buffer – 2 off	8 Petrol tap	12 Screw – 2 off	16 Pipe clip – 2 off
4 Rubber mat	9 Rubber seal	13 Screw – 2 off	17 Filler cap
5 Damping rubber			

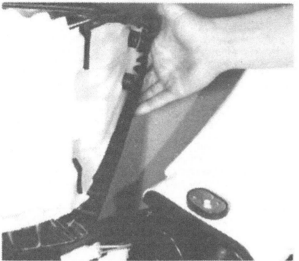

2.2a Unlock and lift away the seat

2.2b Switch petrol off before disconnecting petrol pipe

2.3 Inspect mounting rubbers and renew if damaged – do not lose or forget

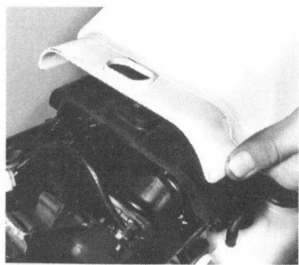

2.5a Settle tank on front and rear mounting rubbers

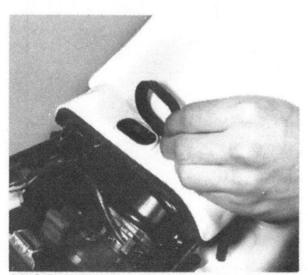

2.5b Refit moulded rubber pad around cut-out in tank base ...

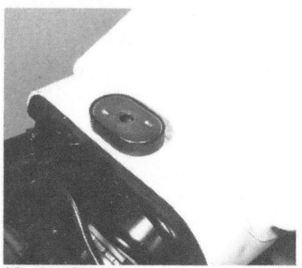

2.5c ... then refit metal plate ...

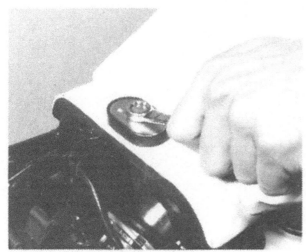

2.5d ... and tighten securely the mounting bolt

3 Petrol tap: removal, dismantling and reassembly

1 The petrol tap is secured to the underside of the tank by two screws. There is seldom need to disturb the main body of the petrol tap. In the event of a leak at the operating lever, the complete lever assembly can be dismantled (provided the petrol tank is drained first) with the main body undisturbed.

2 The tap lever assembly can be withdrawn for inspection, after releasing the single crosshead screw which locates it. If leakage has been evident, the most likely culprit will be the O-ring which seals the tap valve against the body. If this fails to effect a cure, it will be necessary to renew the complete tap assembly.

3 The tap is provided with a sediment bowl, in which any fine debris from the tank which has managed to get through the filter gauze, will be trapped, along with any water. The bowl should be periodically removed for cleaning. When refitting the sediment bowl, ensure that the O-ring is in good condition.

4 Before reassembling the petrol tap, check that all the parts are clean, especially the tube which forms the filter and main and reserve intakes.

5 Do not overtighten any of the petrol tap components during reassembly. The castings are in a zinc-based alloy, which will fracture if over-stressed. Most leakages occur as the result of defective seals.

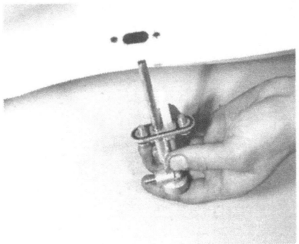

3.1 Tap is secured to tank by two screws and sealed by an O-ring

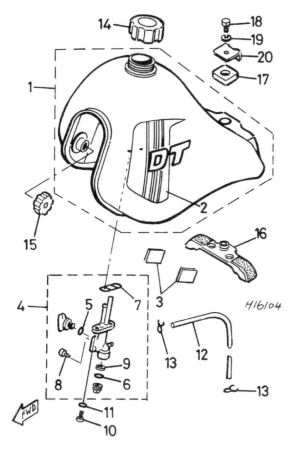

H16104

Fig. 3.2 Petrol tank – DT model

1	Petrol tank	11	Washer – 2 off
2	Tank emblem	12	Petrol feed pipe
3	Damping rubber – 2 off	13	Pipe clip – 2 off
4	Petrol tap assembly	14	Filler cap
5	Rubber seal	15	Rubber buffer – 2 off
6	Sealing washer	16	Rubber mat
7	Gasket	17	Damping rubber
8	Screw – 2 off	18	Bolt
9	Filter gauze	19	Spring washer
10	Screw – 2 off	20	Washer

3.2 Tap can be dismantled fully for cleaning

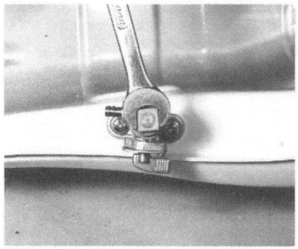

3.3a Use only close-fitting spanner to unscrew filter bowl

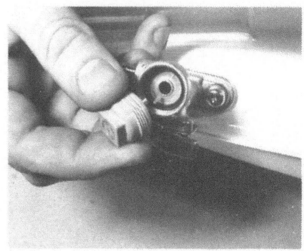

3.3b Bowl and filter gauze must be cleaned regularly

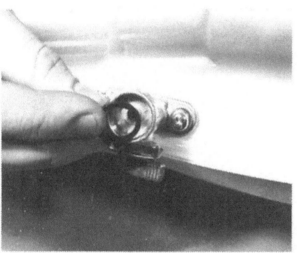

3.3c Check condition of O-ring and renew if necessary

4   Petrol feed pipe: examination

1   The petrol feed pipe is made from thin walled synthetic rubber and is of the push-on type. It is necessary to replace the pipe only if it becomes hard or splits. It is unlikely that the retaining clips will need replacing due to fatigue as the main seal between the pipe and union is effected by an interference fit.

2   If the petrol pipe has been replaced with the transparent plastic type for any reason, look for signs of yellowing which indicate that the pipe is becoming brittle due to the plasticiser being leached out by the petrol. It is a sound precaution to renew a pipe when this occurs, as any subsequent breakage whilst in use will be almost impossible to repair. **Note**: On no account should natural rubber tubing be used to carry petrol, even as a temporary measure. The petrol will dissolve the inner wall, causing blockages in the carburettor jets which will prove very difficult to remove.

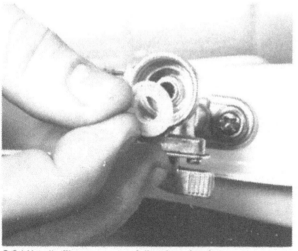

3.3d Handle filter gauze carefully when cleaning

5   Carburettors: removal and refitting

1   As a general rule, the carburettor should be left alone unless it is in obvious need of overhaul. Before a decision is made to remove and dismantle, ensure that all other possible sources of trouble have been eliminated. This includes the more obvious candidates such as a fouled spark plug, a dirty air filter element or choked exhaust system. If a fault has been traced back to the carburettor, proceed as follows.

2   Make sure that the petrol tap is turned off, then prise off the petrol feed pipe at the carburettor union. The oil delivery pipe is removed in a similar manner, noting that the small tubular clip should be displaced first. The pipe can then be eased away from its union with the aid of an electrical screwdriver.

3   Slacken the screws of the clips which secure the carburettor to its inlet stub and air filter hose so that the carburettor can be twisted free of them and partially removed. This affords access to the threaded carburettor top, which should be unscrewed to allow the throttle valve assembly to be withdrawn. It is not normally necessary to remove this from the cable and it can be left attached and taped clear of the engine. If removal is necessary, however, proceed as follows.

4   Holding the carburettor top, compress the throttle return spring against it and hold it in position against the cap. Invert the throttle valve and shake out the pressed steel spring seat. This component serves to prevent the cable from becoming detached when in position and once out of the way the cable can be pushed down and slid out of its locating groove. The various parts can now be removed and should be placed with the carburettor.

5   The carburettor is refitted by reversing the removal sequence. Note that it is important that the instrument is mounted vertically to ensure that the fuel level in the float bowl is correct. A locating tab is fitted to provide a good guide to alignment but it is worthwhile checking this for accuracy. Once refitted, check the carburettor adjustments as described later in this Chapter. Note too that the oil pump delivery pipe should be bled and the pump adjustments checked after overhaul.

6   **Note:** If the carburettor is to be set up from scratch it is important to check jet and float level settings prior to installation. To this end, refer to Sections 7 and 8 before the carburettor is refitted.

6   Carburettor: dismantling, examination and reassembly

1   Invert the carburettor and remove the float chamber by withdrawing the four retaining screws. The float chamber bowl will lift away, exposing the float assembly, hinge and float needle. There is a gasket between the float chamber bowl and the carburettor body which need not be disturbed unless it is leaking.

2   With a pair of thin-nose pliers, withdraw the pin that acts as the hinge for the twin floats. This will free the floats and the float needle. Check that neither of the floats has punctured and that the float needle and seating are both clean and in good condition. If the needle has a ridge, it should be renewed in conjunction with its seating.

3   The two floats are made of plastic, connected by a brass bridge and pivot piece. If either float is leaking, it will produce the wrong petrol level in the float chamber, leading to flooding and an over-rich mixture. The floats cannot be repaired successfully, and renewal will be required.

4   With the float assembly withdrawn, the various jets can be seen screwed into the underside of the carburettor. The first is the main jet which is a small hexagon-shaped brass component screwed into a brass holder in the base of the central pillar. Unscrew the main jet and then the hexagon-headed jet holder. Pick out the O-ring which fits around the bottom of the needle jet, invert the carburettor and push out the needle jet from above, using a soft wooden drift. Be careful to use only a soft drift, or the jet will be damaged, and note that moderate pressure will be quite enough to dislodge the jet from its seating. Turn the carburettor upside down again and use a small electrical screwdriver to unscrew the pilot jet from its seating adjacent to the main/needle jet holder.

5   Slacken and remove the hexagon-headed float needle valve seat and inspect both the float needle and the valve seat for signs of wear, which normally manifests itself in the form of a slight ridge or groove around the seating taper of the float needle. As previously mentioned, any such wear will allow leakage and raise the float level, thus richening the air/petrol mixture, and must be rectified by renewing both components together, although it should be noted that they are in fact obtainable only as a matched pair from Yamaha dealers. Inspect both components very carefully for signs of dirt or foreign matter, and remove any that is found. Lastly check the condition of the valve seat washer, and renew it if it is at all worn or damaged.

6   Remove the two screws which secure the needle retaining plate to the inside of the throttle valve and withdraw the plate and the needle. Check that the needle is straight by rolling it on a flat surface such as a sheet of plate glass and then examine both the needle and the needle jet for signs of wear or damage. Any damage to either component will mean that the two must be renewed together. Do not attempt to straighten a bent needle as they are easily broken; also, if the machine has been running for any length of time with a bent needle, the needle and needle jet must be renewed anyway to rectify the uneven wear which will have occurred.

7   The needle is suspended from the valve, where it is retained by a circlip. The needle is normally suspended from the groove specified at the front of this Chapter, but other grooves are provided as a means of adjustment so that the mixture strength can be either increased or decreased by raising or lowering the needle. Care is necessary when replacing the carburettor top because the needle is easily bent if it does not fit inside the needle jet.

8   After an extended period of service the throttle valve will wear and may produce a clicking sound within the carburettor body. Wear will be evident from inspection, usually at the base of the slide and in the locating groove. Worn slides should be replaced as soon as possible because they will give rise to air leaks which will upset the carburation.

9   The manually operated choke is unlikely to require attention during the normal service life of the machine. When the plunger is pulled out, fuel is drawn through a special starter jet by a partial vacuum that is created in the crankcase. Air from the float chamber passes through holes in the starter emulsion tube to aerate the fuel. The fuel then mixes with air drawn in via the starter air inlet to the plunger chamber. The resultant mixture, richened for a cold start, is drawn into the engine through the starter outlet, behind the throttle valve. The plunger assembly is removed by unscrewing the large brass securing nut. Check that the brass plunger itself is unworn, that the plunger seating is clean and undamaged and that the plunger is operating smoothly. Any fault will mean that the complete assembly must be renewed as repairs are not possible and the assembly is available only as a single unit. Check that the plunger housing and the various passages in the carburettor body are clean and free from any particles of foreign matter.

10   If removal of the throttle stop and pilot mixture adjustment screw is required, screw each one carefully in until it seals lightly, counting and recording the number of turns required, and then unscrew each one, complete with its spring and O-ring. Remove any dirt or corrosion and check for signs of wear or damage, renewing any component which needs it. When refitting the adjustment screws, ensure that the sealing O-rings are fitted first, then the screw retaining springs, and the screw itself. Tighten the screws carefully until they seat lightly, then unscrew each one by the number of turns counted on removal to return it to its original position; this will serve as a basis for subsequent adjustments.

11   Before the carburettor is reassembled, using the reversed dismantling procedure, it should be cleaned out thoroughly, preferably by the use of compressed air. Avoid using a rag because there is always risk of fine particles of lint obstructing the internal air passages or the jet orifices. Check carefully the condition of the carburettor body and float chamber, looking for distorted or damaged mating surfaces or any other signs of wear. If severe damage or wear is found, the carburettor assembly will have to be renewed. Check the condition of all O-rings and gaskets, renewing any that are worn or distorted.

12   Never use a piece of wire or sharp metal object to clear a blocked jet. It is only too easy to enlarge the jet under these circumstances and increase the rate of petrol consumption. Always use compressed air to clear a blockage; a tyre pump makes an admirable substitute when a compressed air line is not available.

13   Do not use excessive force when reassembling the carburettor because it is quite easy to shear the small jets or some of the smaller screws. Before attaching the air cleaner hose, check that the throttle slide rises smoothly when the throttle is opened.

6.6 Throttle needle is retained by needle plate inside throttle valve

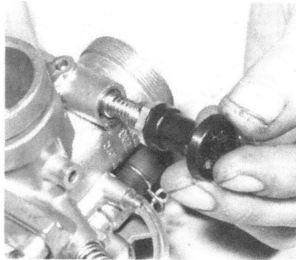

6.9 Choke plunger assembly screws into carburettor body

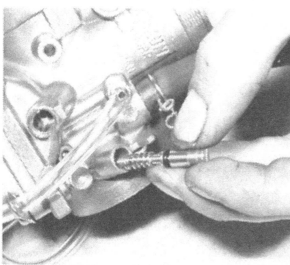

6.10a Check condition of mixture screw retaining spring and O-ring

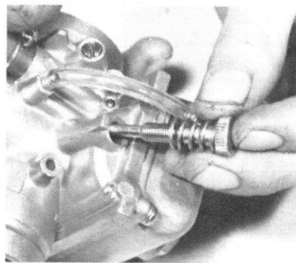

6.10b Throttle stop screw location

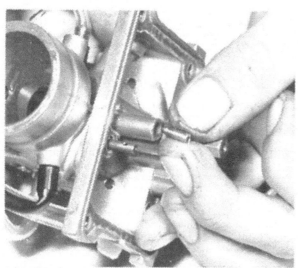

6.11a Pilot jet is screwed into position

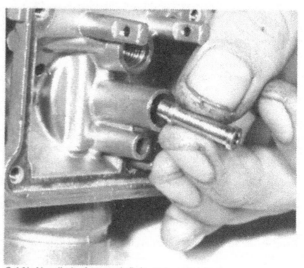

6.11b Needle jet is a push fit in carburettor body

6.11c Check condition of O-ring and renew if necessary

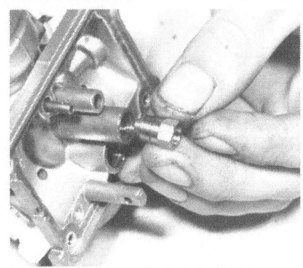

6.11d Main jet holder screws into base of carburettor

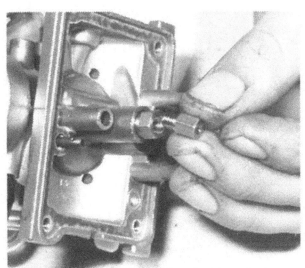

6.11e Main jet is screwed into holder – do not overtighten

6.11f Float needle seat is screwed into position – washer must be in good order

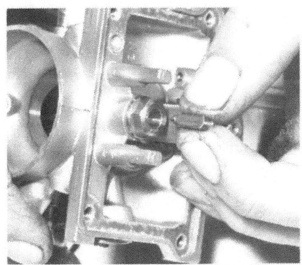

6.11g Ensure float needle taper is clean and unworn before refitting

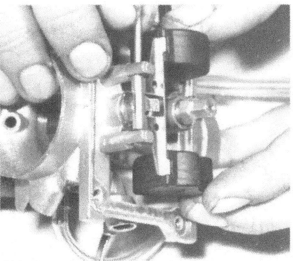

6.11h Float assembly is secured by pivot pin

Fig. 3.3 Carburettor

6.11i Note position of breather tube guides

1	Jet needle	20	O-ring
2	Throttle valve	21	Main jet holder
3	Pilot jet	22	O-ring
4	Needle jet	23	Spring
5	O-ring	24	O-ring
6	Main jet	25	Spring
7	Drain plug	26	Vent hose
8	Throttle stop screw	27	Circlip
9	Float	28	Needle retaining plate
10	Float needle and housing	29	Screw – 2 off
11	Choke knob	30	Spring seat
12	Pilot air screw	31	Return spring
13	Sealing washer	32	Circlip
14	Float pivot pin	33	Carburettor top
15	Float chamber gasket	34	Sealing washer
16	Float chamber	35	Rubber cap
17	Screw – 4 off	36	Cable sleeve
18	Spring washer – 4 off	37	Lock nut
19	Overflow pipe		

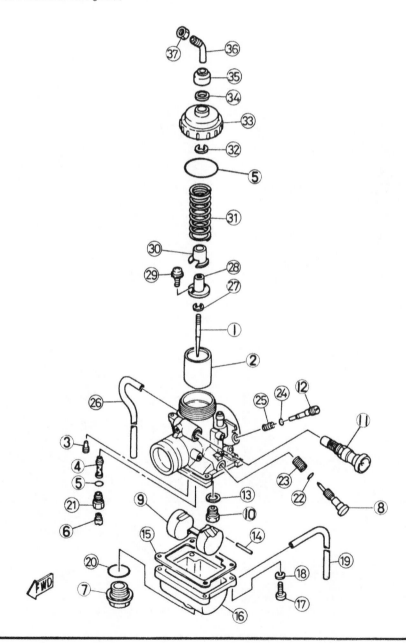

7 Carburettor: checking the settings

1   The various jet sizes, throttle valve cutaway and needle position are predetermined by the manufacturer and should not require modification. Check with the specifications list at the beginning of this Chapter if there is any doubt about the types fitted. If a change appears necessary it can often be attributed to a developing engine fault unconnected with the carburettor. Although carburettors do wear in service, this process occurs slowly over an extended length of time and hence wear of the carburettor is unlikely to cause sudden or extreme malfunction. If a fault does occur check first other main systems, in which a fault may give similar symptoms, before proceeding with carburettor examination or modification.

2   Where non-standard items, such as exhaust systems or air filters, have been fitted to a machine, some alterations to carburation may be required. Arriving at the correct settings often requires trial and error, a method which demands skill borne of previous experience. In many cases the manufacturer of the non-standard equipment will be able to advise on correct carburation changes.

3   As a rough guide, up to $\frac{1}{8}$ throttle is controlled by the pilot jet, $\frac{1}{8}$ to $\frac{1}{4}$ by the throttle valve cutaway, $\frac{1}{4}$ to $\frac{3}{4}$ throttle by the needle position and from $\frac{3}{4}$ to full by the size of the main jet. These are only approximate divisions, which are by no means clear cut. There is a certain amount of overlap between the various stages.

4   If alterations to the carburation must be made, always err on the side of a slightly rich mixture. A weak mixture will cause the engine to overheat which, particularly on two-stroke engines, may cause engine seizure. Reference to the chapter on the ignition system will show how, after some experience has been gained, the condition of the spark plug electrodes can be interpreted as a reliable guide to mixture strength.

7.1 Check carefully before altering carburettor settings from specification

8 Carburettor: adjustment

1   Before any dismantling or adjustment is undertaken, eliminate all other possible causes of running problems, checking in particular the spark plug, ignition timing, air cleaner and the exhaust. Checking and cleaning these items as appropriate will often resolve a mysterious flat spot or misfire.

2   The first step in carburettor adjustment is to ensure that the jet sizes, needle position and float height are correct, which will require the removal and dismantling of the carburettors as described in Sections 5 and 6 of this Chapter.

3   If the carburettor has been removed for the purpose of checking jet sizes, the float level should be measured at the same time. It is unlikely that once this is set up correctly there will be a significant amount of variation, unless the float needle or seat have worn. These should be checked and renewed, if necessary, as described in Section 6.

4   Remove the float bowl from the carburettor body, if this has not already been done, and very carefully peel away the float chamber gasket. Check that the gasket surface of the carburettor body is clean and smooth once the gasket is removed. Hold the carburettor body so that the venturi is now vertical with the air filter side upwards and the floats are hanging from their pivot pin. Carefully tilt the carburettor to an angle of about 60 – 70° from the vertical so that the tang of the float pivot is resting firmly on the float needle and the float valve is therefore closed, but also so that the spring-loaded pin set in the float needle itself is not compressed. Measure the distance between the gasket face and the bottom of one float with an accurate ruler or a vernier caliper; the distance should be 21 mm (0.83 in). A tolerance of 1 mm (0.04 in) above or below the set figure is allowed, but the more accurate the setting is, the better the engine's performance, reliability and economy will be.

5   If adjustment is required, remove the float assembly and bend by a very small amount the small tang which acts on the float needle pin. Reassemble the float and measure the height again. Repeat the process until the measurement is correct, then check that the other float is exactly the same height as the first. Bend the pivot very carefully and gently if any difference is found between the heights of the two floats.

6   When the jet sizes have been checked and reset as necessary, reassemble the carburettor and refit it to the machine as described in Sections 5 and 6 of this Chapter.

7   Start the engine and allow it to warm up to normal operating temperature, preferably by taking the machine on a short journey. Stop the engine and screw the pilot mixture screw in until it seats lightly, then unscrew it by the number of turns shown in the Specifications Section for the particular model. Start the engine and set the machine to its specified idle speed by rotating the throttle stop screw as necessary. Note that on DT125 LC models the idle speed should be regarded as the slowest speed at which the engine will tick over smoothly and reliably. Try turning the pilot mixture screw inwards by about $\frac{1}{4}$ turn at a time, noting its effect on the idling speed, then repeat the process, this time turning the screw outwards.

8   The pilot mixture screw should be set in the position which gives the fastest consistent tickover. The tickover speed may be reduced further, if necessary, by unscrewing the throttle stop screw the required amount. Check that the engine does not falter and stop after the throttle twistgrip has been opened and closed a few times.

9   Throttle cable adjustment should be checked at regular intervals and after any work is done to the carburettor, oil pump or to the cable itself. The amount of free play specified for the throttle cable is 4 – 7 mm ($\frac{1}{8} - \frac{1}{4}$ approx) measured at the inner flange of the twistgrip rubber. To measure this, use a piece of chalk or some paint to mark both the twistgrip rubber at its inner flange, and the twistgrip drum. These two marks will provide a convenient reference point for future adjustment. Carefully open the throttle by rotating the twistgrip rubber in the usual way until all the free play in the cable has been taken up. Measure the distance around the circumference of the drum between the static mark on the drum and the mark that has moved with the twistgrip rubber. If this distance is more or less than the specified amount the cable must be adjusted.

10 On RD125 LC models, check first that excessive play is not present in the lower throttle cable, between the junction box and the carburettor top. This is difficult to check as the adjusting nut is crimped on to the end of the cable at the carburettor top and the check must be made, therefore, at the junction box. Turn the adjusting nut as necessary so that all but the slightest trace of free play is eliminated from the lower throttle cable, and

then use the adjuster below the twistgrip to provide the correct amount of free play at the twistgrip itself. Fully open and close the throttle several times to settle the cables, then check that the adjustment has remained the same, resetting it if necessary. If correct, tighten the adjuster locknuts and slide the rubber sleeve back over the adjuster. Remember that on this model, any alteration made to the throttle cable will affect the adjustment of the oil pump cable. The oil pump cable setting must be checked, therefore, as described in the relevant part of Section 19 of this Chapter before the machine is used on the road.

11 On DT125 LC models the procedure for adjusting the throttle cable is slightly different. If the self-adjusting junction box is operating correctly, all free play will be eliminated automatically from the lower cables. Throttle cable adjustment is made, therefore, at the adjuster below the twistgrip. Rotate the adjuster as necessary to provide the correct amount of free play at the twistgrip itself, then fully open and close the throttle several times to settle the cables. Check that the adjustment has remained the same, resetting it if necessary. If correct, tighten the adjuster locknut and slide the rubber sleeve back over the adjuster. Note that while the adjustment of the throttle cable does not affect the oil pump cable on DT125 LC models, it is recommended that the opportunity be taken to check that the oil pump cable is operating correctly as described in the relevant part of Section 19 of this Chapter. If any fault is found the self-adjusting junction box must be checked as described in Section 20 of this Chapter.

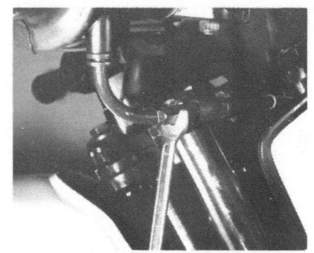

8.8 Use twistgrip adjuster to set throttle cable free play

9 Reed valve induction system: mode of operation

1 Of the various systems of controlling the induction cycle of a two-stroke engine, Yamaha has chosen to adopt the reed valve, a device which permits precise control of the incoming mixture, allowing more favourable port timing to give improved torque and power outputs. The reed valve assembly comprises a wedge-shaped die-cast aluminium alloy valve case mounted in the inlet tract. The valve case has rectangular ports which are closed off by flexible stainless steel reeds. The reeds seal against a heat and oil resistant synthetic rubber gasket which is bonded to the valve case. A special shaped valve stopped, made from cold rolled stainless steel plate, controls the extent of movement of the valve reeds.

2 As the piston ascends in the cylinder, a partial vacuum is formed beneath the cylinder in the crankcase. This allows atmospheric pressure to force the valve open, and a fresh charge of petrol/air mixture flows past the valve and into the crankcase. As the pressure differential becomes equalised, the valves close, and the incoming charge is then trapped. The charge of mixture in the cylinder is by this time fully compressed, and ignition takes place driving the piston downwards. The descending piston eventually uncovers the exhaust port and the hot exhaust gases, still under a certain amount of pressure, are discharged into the exhaust system. At this stage the reed valve, in conjunction with the 7th, or auxiliary scavenging port, performs a secondary function; as the hot exhaust gases rush out of the exhaust port a momentary depression is created in the cylinder, this allows the valve to open once more, but this time the incoming mixture enters directly into the cylinder via the 7th port and completes the expulsion of the now inert burnt gasses. This ensures that the cylinder is filled with the maximum possible combustion mixture. The charge of combustion mixture which has been compressed in the crankcase is released into the cylinder via the transfer ports, and the piston again ascends to close the various ports and begin compression. The reed valves open once more as another partial vacuum is created in the crankcase, and the cycle of induction thus repeats.It will be noted that no direct mechanical operation of the valve takes place, the pressure differential being the sole controlling factor.

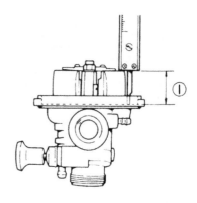

Fig. 3.4 Measuring the float height

1   21 ± 1.0 mm (0.83 ± 0.04 in)

10 Reed valves: removal, examination and renovation

1 The reed valve assembly is a precision component, and as such should not be dismantled unnecessarily. The valves are located in the inlet tract, covered by the carburettor flange.

2 Remove the carburettor as described in Section 5 of this Chapter thus exposing the four bolts retaining the inlet stub and the reed valve assembly to the cylinder. After removing these bolts, the assembly can be carefully lifted away.

3 The valves can now be washed in clean petrol to facilitate further examination. They should be handled with great care, and on no account dropped. The stainless steel reeds should be inspected for signs of cracking or fatigue, and if suspect, should be renewed. Remember that any part of the assembly which breaks off in service will almost certainly be drawn into the engine, causing extensive damage. Make a quick check of the state of the assembly by putting the carburettor side to the lips

and sucking hard. The reed petals should seal effectively against their seats, providing considerable resistance to the suction applied, although the manufacturer allows that some slight leakage is inevitable and permissible. The reeds should rest flush against their seats but a slight gap is quite normal.

4   Complete the visual inspection by checking the inlet stub for signs of cracking, perishing, or other deterioration, renewing it if necessary. The distance between the inner edge of the valve stopper and the top edge of the valve case is important as it controls the movement of the reed. If smaller than specified, performance will be impaired. More seriously, if larger than specified, the reed may fracture. The nominal setting is 10.3 mm (0.406 in). If the measurement is more than 0.4 mm (0.016 in) either side of the nominal figure, the stopper plate should be renewed.

5   To dismantle the reed valve assembly, first degrease it thoroughly and have a spirit-based felt marker pen ready. Mark each face of the valve case in such a way that they can be identified (eg face 'A' and face 'B'). Working on one face at a time, slacken and remove the two screws retaining the reed stopper plate and the reed and withdraw the two components, marking the outer surface of each one with the identifying mark for that face, so that each component of the reed valve assembly can be refitted in exactly the same position. Note that the manufacturer has assisted this by providing a small cutout in the lower right-hand cover of each stopper plate. Put the components to one side, but store them separately to minimise further the risk of incorrect refitting.

6   Check the condition of the various components. The small screws must be in excellent condition and should be renewed if there is any doubt at all about their condition. The stopper plates should be renewed, as previously mentioned, if the stopper plate/valve case heights are above or below the specified tolerances. Never attempt to bend the stopper plate to alter this distance. While it is recognised that this is accepted practice in certain tuning circles to gain more performance, it should be noted that it is not recommended by the manufacturer as the curve of the stopper plates is carefully designed to give maximum performance consistent with maximum reed life. Any attempt to alter the stopper plate heights by bending the plates will alter this curve and may overstress the reeds to the point where fatigue cracks develop and the reeds break-off, falling into the engine with disastrous consequences. If there is any evidence of the stopper plates having been bent, perhaps by a previous owner, they should be renewed.

7   Carefully examine the valve case itself, looking for cracks, distorted mating surfaces, or excessive wear or damage to the neoprene seating face. Remedial action will depend on the nature of the damage found; some resurfacing work may be possible, using a sheet of fine emery paper placed on a flat surface such as s sheet of plate glass, but due to the design and construction of the valve case, such work will be limited in its effect. Any serious damage or wear, or any damage to the neoprene reed seats will mean that the valve case must be renewed, both in the interests of safety and reliability, and in the interest of maximum performance.

8   The reeds themselves, once removed from the valve case, must be checked very carefully, with a magnifying glass if necessary, for signs of fatigue cracks. If undamaged, they must be measured to ensure that they are not worn or distorted. Each reed should be 0.2 mm (0.008 in) thick; there should be no measurable reduction in this thickness at the reed tips. Lay each reed on a completely flat surface such as a sheet of plate glass with its marked, or outer surface facing upwards and hold it by pressing down at the retaining screw holes. In this position any distortion which has taken place will be revealed immediately. The reed should not be distorted at any point more than 0.5 mm (0.020 in), a measurement which can be checked using feeler gauges. Any signs of damage, wear, or distortion revealed by the above tests will mean that the reeds must be renewed. Repairs are not possible and should never be attempted. Do not attempt to bend the reeds straight, if they are found to be distorted, in case fatigue failure is induced.

9   Reassemble the reed valve assembly using the marks made on dismantling and the cutout provided by the manufacturer to ensure that the reeds and stopper plates are refitted in their original positions. A thread locking compound, such as Loctite, must be applied to the four cross-headed screws, which should be tightened progressively to avoid warping the reed or stopper. Do not omit the locking compound, as the screws retain a component which vibrates many times each second and consequently are prone to loosening if assembled incorrectly. Note that a torque setting of 0.1 kgf m (0.7 lbf ft) is specified for these screws; this should be adhered to if the necessary equipment is available. If not, ensure that all four are tightened securely.

10 Refit the reed valve assembly to the cylinder barrel, using a new gasket at the valve case/cylinder barrel joint and applying a thin coat of jointing compound to seal the valve case/inlet stub joint. Tighten the four mounting bolts progressively and in a diagonal sequence to a torque setting of 0.8 kgf m (6 lbf ft). Refit the carburettor as described in Section 5.

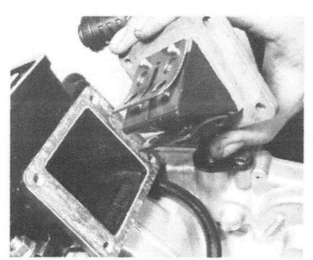
10.1 Reed valve assembly is located in the inlet tract

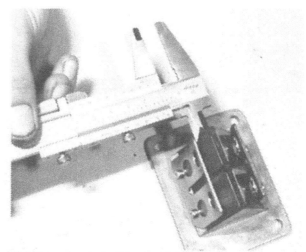
10.4 Measuring reed valve stopper plate height

10.9 Note cut-out in stopper plate corner to assist location on reassembly

10.10 Always use a new gasket to prevent induction leaks

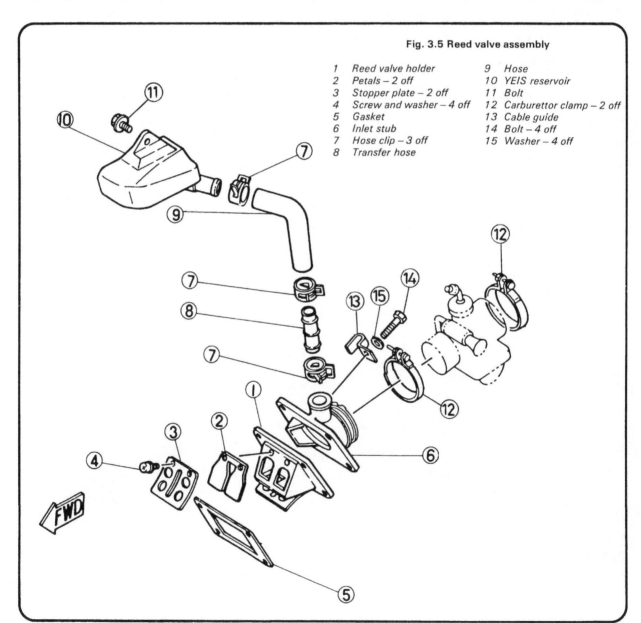

Fig. 3.5 Reed valve assembly

1	Reed valve holder	9	Hose
2	Petals – 2 off	10	YEIS reservoir
3	Stopper plate – 2 off	11	Bolt
4	Screw and washer – 4 off	12	Carburettor clamp – 2 off
5	Gasket	13	Cable guide
6	Inlet stub	14	Bolt – 4 off
7	Hose clip – 3 off	15	Washer – 4 off
8	Transfer hose		

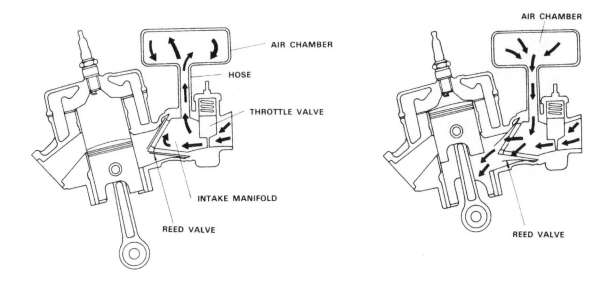

Reed valve open

Reed valve closed

Fig. 3.6 YEIS system operation

11 YEIS: mode of operation

1 The Yamaha Energy Induction System (YEIS), more popularly known as the boost bottle, is a feature developed by Yamaha's design staff to increase intake efficiency in the low to medium speed ranges, thus making the machine easier to ride and increasing fuel economy.

2 In a conventional induction system, whether two-stroke or four-stroke, considerable variations in pressure are induced in the inlet tract as the intake valve opens and closes. These pressure variations occur because as the fuel/air mixture is forced into the engine by atmospheric pressure it begins to acquire a momentum of its own which assists in filling the crankcase or combustion chamber; however, if the mixture flow is suddenly shut off by the closing of the inlet valve, this momentum carries on forcing the mixture along the inlet tract, causing a considerable build-up of pressure which can only be relieved by travelling backwards into the carburettor until the valve opens again and the cycle is repeated. This pressure build-up can interfere with the smooth flow of the fuel/air mixture to a significant extent, more particularly at the low to medium engine speeds.

3 Yamaha's solution to this problem is to connect a carefully designed reservoir to the inlet stub by means of a short length of hose. The system functions by making use of the pressure differentials in its various components. When the inlet reed valve is open, all components are at atmospheric pressure, but when the valve closes under the increasing pressure in the crankcase caused by the descent of the piston, the pressure build-up causes the surplus fuel/air mixture to be diverted upwards into the reservoir, which then becomes pressurised itself.

4 The reservoir remains at this higher pressure for long enough to allow the piston to complete its cycle and for the inlet valve to open again. The pressure in the inlet tract is then reduced to the level of atmospheric pressure, causing the mixture in the reservoir to flow out and join with the new charge of mixture.

5 The effect of this is to reduce greatly the pressure variations in the inlet tract, thus smoothing out the flow of fuel/air mixture. This increases the engine's efficiency at low to medium engine speeds, enabling it to develop more power at these speeds than

would otherwise be the case. The system has the added benefit of reducing fuel consumption thanks to this greater efficiency.

6 It will be evident from the above that while such a system is very simple in principle, it requires very careful design to ensure that the hose and reservoir are of the correct capacity and shape. It will also be evident that YEIS components designed to match one particular engine in a certain state of tune cannot be expected to function correctly if fitted to another model, or if the state of tune of the original engine is altered.

12 YEIS: removal, examination and refitting

1 The YEIS components are very simple in design and construction having no moving parts at all, and will require no maintenance at all in the life of the machine. The hose and reservoir should be examined whenever the petrol tank is removed during the course of routine maintenance or for repair work. Check carefully for any signs of splits or cracks, and that neither of the components is chafing against anything else. Check very carefully the inlet stub and hose connections for deterioration of any sort, as these are the only two areas likely to suffer. Ensure that the hose clips are securely fastened at all times.

2 If at any time a component is found to be defective, it must be renewed. Use only the genuine replacement parts supplied by a Yamaha dealer, as it is essential that the correct internal dimensions and total capacity of the hose and reservoir are preserved if the system is to function correctly.

3 The petrol tank must be removed first to gain access to the YEIS components, the necessary work being described in Section 2 of this Chapter. Using a suitable heavy pair of pliers, disengage the hose securing clips by compressing together the protruding ends of each clip and then sliding it up the length of the hose. Twist the hose off its union, then slacken and remove the single bolt which retains the reservoir to the frame top tubes.

4 Refitting is a straightforward reversal of the removal procedure but care must be taken to ensure that the components are mounted correctly and cannot chafe against anything. Fasten securely the hose clips to prevent leaks.

12.3a Location of YEIS chamber – RD125

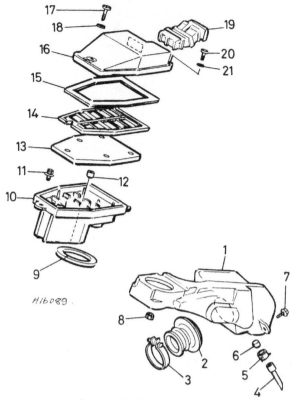

Fig. 3.7 Air filter – RD model

1	Air filter casing	12	Seal
2	Air outlet hose	13	Element
3	Hose clamp	14	Element frame
4	Drain pipe	15	Rubber sealing ring
5	Pipe clip	16	Filter cover
6	Seal	17	Screw
7	Bolt – 2 off	18	Washer
8	Nut	19	Air inlet hose
9	Sealing ring	20	Screw – 2 off
10	Air filter housing	21	Washer – 2 off
11	Bolt		

12.3b Use heavy pliers to release YEIS hose clamps

13 Air cleaner: removal and cleaning

RD125 LC models

1   The air cleaner casing is mounted on the frame beneath the fuel tank. It is connected to a second moulded plastic chamber immediately behind the carburettor. This functions as an intake silencer and conveys the cleaned air to the carburettor intake via a short hose.
2   Access to the air cleaner will require the prior removal of the sidepanels, the seat, and the petrol tank, as described in Section 2 of this Chapter.
3   The air cleaner cover is retained by three screws. Remove the screws and lift the cover clear to expose the plastic frame and flat foam element. These can be removed in turn by pulling them out of the casing together or separately. Wash the element in clean petrol to remove the old oil and any dust which has been trapped by it. When it is clean, wrap it in some clean rag and gently squeeze out the remaining petrol. The element should now be left for a while to allow any residual petrol to evaporate. Soak the cleaned element in engine oil and then squeeze out any excess to leave the foam damp but not

dripping. Refit the element, ensuring that the cover seat can now be refitted.
4   Note that a damaged element must be renewed immediately. Apart from the risk of damage from ingested dust, the holed filter will allow a much weaker mixture and may lead to overheating or seizure. It follows that the machine must never be used without the filter in position.
5   The rest of the air cleaner system requires little attention, other than checking that the connecting rubbers are undamaged. When checking these do not omit the stub which connects the carburettor to the cylinder barrel. This is prone to perishing and cracking around the YEIS hose union and should be renewed if leakage is suspected.

DT125 LC models

6   The air cleaner system fitted to the DT125 LC models is essentially the same as that described above, but is fitted behind the left-hand sidepanel and incorporates a two piece element to provide greater filtering efficiency. The outer element is constructed of a fibrous synthetic material which presents a closely-packed hairy surface to the incoming air, thus

trapping the larger particles of dirt. This is backed up by a conventional polyurethane foam element which removes any remaining particles of dust.

7   Remove the left-hand sidepanel to gain access to the filter assembly, taking care not to damage the panel moulding as it is withdrawn. The filter cover is retained by a single screw at its centre which must be removed to permit the withdrawal of the cover and the plastic element supporting frame which is situated immediately underneath. The elements can then be picked out and cleaned, then re-oiled, as described above for the RD125 LC model. Inspect the filter elements very carefully, renewing them as a pair at the first sign of damage or deterioration.

8   When refitting the filter assembly, the grey foam element is fitted first, then the orange fibrous element, which is fitted with its hairy surface facing outwards. The plastic supporting frame is then refitted, followed by the filter cover. Do not overtighten the cover retaining screw, or the cover will be distorted. A light application of grease to the cover seal will help prevent the entry of dirt. Check carefully, as described above, for any signs of damage or deterioration in any of the plastic or rubber

components of the induction system. Any such damage must be rectified immediately to prevent the risk of induction leaks or unfiltered air entering the engine.

General

9   When cleaning the air filter element of either of the machines described in this Manual, the following points must be borne in mind. While petrol is usually considered the most convenient cleaning medium, due to its availability, there is a very high risk of fire involved in its use. If petrol is beng used, take great care not to smoke or to have any naked lights or hot components in the immediate vicinity. A better solution is to employ a high flash point solvent such as white spirit for this task.

10   If, for any reason, the air filter casings are removed from the machine and separated, great care must be taken to ensure that induction leaks cannot occur. Check that the mating surfaces are clean and undamaged and that the seal is in good order, using a non-setting silicone RTV jointing compound to assist in sealing the casing joint. Do not overtighten the retaining screws or bolts to avoid distorting the casing halves.

13.3a Slacken and remove filter cover screws ...

13.3b ... then withdraw the cover and supporting frame

13.3c Do not forget to fit supporting frame on reassembly

13.8a When reassembling the air filter, refit the orange fibrous element, not omitting the supporting frame – DT125

13.8b Do not overtighten filter cover screw

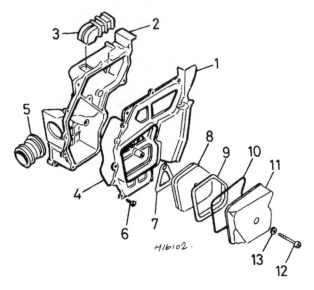

Fig. 3.8 Air filter – DT model

1	Air filter casing	8	Element
2	Casing cover	9	Element supporting frame
3	Air inlet hose	10	Sealing ring
4	Sealing ring	11	Filter cover
5	Air outlet hose	12	Screw
6	Screw – 13 off	13	Washer
7	Plate		

14 Exhaust system: removal and refitting

1 The exhaust systems fitted to the machines described in this Manual are essentially similar in design and in the fact that neither is fitted with baffles that can be removed for cleaning purposes. It will be necessary, therefore, to remove the complete system before any work can be undertaken.

RD125 LC models

2 The exhaust system fitted to these machines is a one-piece assembly retained at only two points: by two nuts at the cylinder barrel, and by a single bolt which passes through the centre of the right-hand footrest mounting plate. Remove first the belly fairing to gain sufficient access for the system to be withdrawn, then slacken and remove the two nuts and the single bolt. Withdraw the system, taking care not to damage the painted finish either of the system or of the machine itself.

3 Refitting is a straightforward reversal of the removal procedure. Always use a new gasket to prevent exhaust leaks, sticking it to the exhaust port with a thin smear of grease. Offer up the system to the machine and secure it in position by tightening the mounting nuts and bolts by hand only at first. Ensure that the system is aligned correctly on its mountings, tighten the two front mounting nuts to a torque setting of 1.8 kgf m (13 lbf ft) and then tighten securely the single rear mounting bolt. Check that all breather pipes are secured clear of the system and refit the belly fairing.

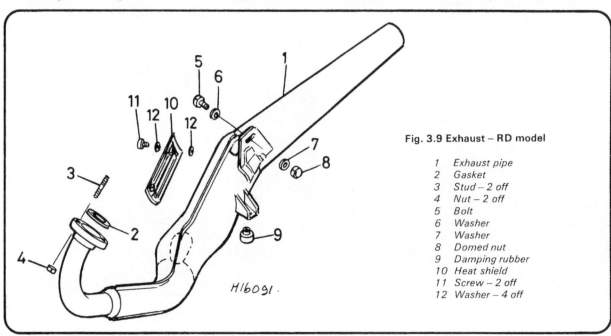

Fig. 3.9 Exhaust – RD model

1   Exhaust pipe
2   Gasket
3   Stud – 2 off
4   Nut – 2 off
5   Bolt
6   Washer
7   Washer
8   Domed nut
9   Damping rubber
10 Heat shield
11 Screw – 2 off
12 Washer – 4 off

DT125 LC models

4   The exhaust system fitted to these machines consists of two pieces which are retained to the machine at a total of five mounting points, two of which are obscured by the petrol tank; it is necessary, therefore, to remove the seat, sidepanels, and the petrol tank before the exhaust is withdrawn, as decribed in Section 2 of this Chapter. Slacken fully the two clips which secure the rubber sleeve sealing the joint between the two parts of the system, slacken and remove the two bolts which fasten the rear part to the frame, and carefully pull it backwards clear of the main exhaust section. Then slacken and remove the two front mounting nuts and the two mounting bolts to release this front part of the system. Carefully manoeuvre the front section forwards clear of the frame and withdraw it from the machine taking great care not to damage the paintwork of either the exhaust or the machine itself. Inspect all the rubber mounting components and the rubber sleeve, renewing any part that is worn or damaged.

5   Refitting is a straightforward reversal of the removal procedure. Always use a new gasket to prevent exhaust leaks, sticking it to the exhaust port with a thin smear of grease. The front section of the assembly must be offered up first and secured in position by refitting loosely the mounting nuts and bolts. Fit the rear section, ensuring that the rubber sleeve and its two clips are correctly fitted and aligned, then replace the two bolts which secure the rear section to the frame. Check that the system is correctly aligned on its mountings, then tighten the mounting nuts and bolts securely, working from the front back down the length of the system, tightening each fastener in succession but noting that the two front mounting nuts must be tightened to a torque setting of 1.8 kgf m (13 lbf ft). Finally, tighten securely the two clips which fasten the rubber joint sleeve and refit the petrol tank, the seat, and the sidepanels.

6   A final note on the exhaust fitted to the DT125 LC model: it will be noted that there is a rubber sleeve fitted around the small bracket protruding from the side of the exhaust rear section. This is fitted to hold the left-hand sidepanel away from the exhaust so that it cannot be melted by the heat of the exhaust. Check that the bracket has not been bent and that the rubber sleeve is in good order whenever the sidepanel is removed.

14.4a DT125 front exhaust mounting point

14.4b The rear mounting of the forward part of the system

14.4c The silencer front mounting

14.4d The silencer rear mounting

Fig. 3.10 Exhaust – DT model

1	Exhaust pipe	11	Joining piece
2	Gasket	12	Clamp – 2 off
3	Stud – 2 off	13	Bolt – 2 off
4	Nut – 2 off	14	Heat shield
5	Grommet	15	Screw – 3 off
6	Collar	16	Washer – 6 off
7	Mounting bracket	17	Heat shield
8	Bolt – 2 off	18	Screw – 2 off
9	Bolt	19	Washer – 4 off
10	Silencer	20	Damping rubber

14.5 Tighten securely the sleeve clips to prevent leaks

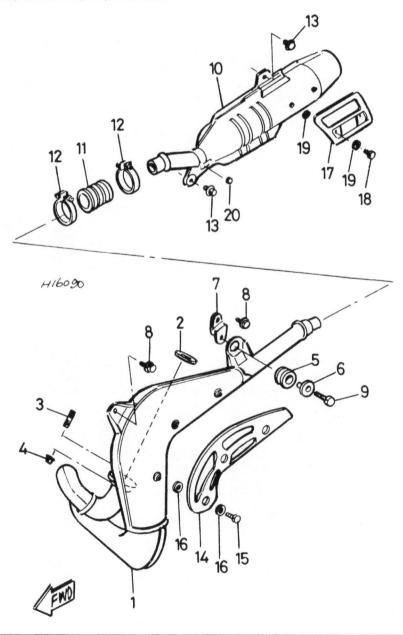

H16090

15 Exhaust system: cleaning

1   Due to the fact that the exhaust systems fitted to the machines described in this Manual have no removable baffles, exhaust cleaning will be rather more difficult than in most machines of this type.

2   The exhaust gases of any two stroke are particularly oily in nature and will produce a marked build-up of carbon deposits in the front length of the exhaust pipe and of an oily sludge in the rear part of the system. If not removed at regular intervals, these deposits will cause an undesirable increase in back pressure in the system, restricting the engine's ability to breathe and producing a marked loss in power output.

3   Three possible methods of cleaning the system exist, the first, which will only be really effective if the deposits are very oily, consists of flushing the system with a petrol/paraffin mixture. It will be evident that great care must be taken to prevent the risk of fire when using this method, and that the system must be hung overnight in such a way that all the mixture will drain away. It will be necessary to use a suitable scraping tool to remove any hardened deposits from the front length of the exhaust pipe and from the exhaust port of the cylinder barrel.

4   The second method will involve the use of a welding torch or powerful blow lamp to burn off the deposits. This usually results in the production of a great deal of smoke and fumes and so must be carried out in a well-ventilated area. It also requires some considerable skill in the use of welding equipment if it is to be fully effective, and if personal injury or damage to the exhaust system is to be avoided. Another drawback is that the excessive heat created will destroy the painted finish of the system, and that repainting will be necessary. With severely blocked systems, it may even be necessary to cut the exhaust open so that the flame can be applied to the blocked section, and the exhaust welded up afterwards. In short this method, while quite effective, should only be employed by a person with welding equipment and the necessary degree of skill in its use. The best solution would be for the owner to remove the exhaust from the machine and to take it to a local dealer or similar expert for the work to be carried out.

5   The third possible method of exhaust cleaning is to use a solution such as caustic soda to dissolve the deposits. While this is a lengthy and time-consuming operation, it is the simplest and most effective method that can be used by the average owner, and is therefore described in detail in the following paragraphs. To clean the silencer casing and exhaust pipe assembly, remove the system from the machine and suspend it from its silencer end. Block up the end of the exhaust pipe with a cork or wooden bung. If wood is used, allow an outside projection of three or four inches with which to grasp the bung for removal.

6   The method used to remove heavy carbon deposits from within the system is to dissolve them by using a chemical solution. Caustic soda dissolved in water is the solution most usually utilised as it is highly effective and relatively cheap. The mixture used is a ratio of 3 lbs caustic soda to a gallon of fresh water. This is the strongest solution ever likely to be required. Obviously the weaker the mixture the longer the time required for the carbon to be dissolved. Note, whilst mixing the solution, that the caustic soda should be added to the water gradually, whilst stirring. Never pour water into a container of caustic soda powder or crystals; this will cause a violent reaction to take place which will result in great danger to one's person.

7   Commence the cleaning operation by pouring the solution into the system until it is quite full. Do not plug the open end of the system. The solution should now be left overnight for its dissolving action to take place. Note that the solution will continue to give off noxious fumes throughout its dissolving process; the system must therefore be placed in a well ventilated area. After the required time has passed, carefully pour out the solution and flush the system through with clean,

fresh water. The cleaning operation is now complete.

8   Bear in mind that it is very important to take great case when using caustic soda as it is a very dangerous chemical. Always wear protective clothing, this must include proper eye protection. If the solution does come into contact with the eyes or skin it must be washed clear immediately with clean, fresh, running water. In the case of an eye becoming contaminated, seek expert medical advice immediately. Also, the solution must not be allowed to come into contact with aluminium alloy – especially at the above recommended strength – caustic soda reacts violently with aluminium and will cause severe damage to the component.

9   When cleaning the exhaust system, do not forget to remove any carbon deposits from the exhaust port itself. Use a blunt edged scraping tool of suitable shape and size, and take great care not to damage the piston or to scratch the surface of the port. If the port is severely blocked, the cylinder barrel should be removed for a major decarbonising operation.

10   Do not at any time attempt to modify the exhaust system in any way. The exhaust system is designed to give the maximum power possible consistent with legal requirements and yet to produce the minimum noise level possible. Quite apart from the legal aspects of attempting to modify the exhaust of one of the restricted machines described in this Manual, it is very unlikely that an unskilled person could improve the performance of any of the machines by working on the exhaust. If an aftermarket accessory system is being considered, check very carefully that it will maintain or increase performance when compared with the standard system. Very few 'performance' exhaust system live up to the claims made by the manufacturers, and even fewer offer any performance increase at all over the standard component.

11   The final point to be borne in mind when considering the exhaust system is the finish. The matt-black painted finish employed in pursuit of current styling trends is cheaper to renovate but less durable than a conventional chome-plated system. It is inevitable that the original finish will deteriorate to the point where the system must be removed from the machine and repainted, therefore some thought must be given to the type of paint to be used. As it is such a common problem, especially amongst the owners of modern trail bikes, several alternative finishes are offered. Reference to the advertisements in the national motorcycle press, or to a local Yamaha dealer and to the owners of machines with similarly-finished exhausts will help in selecting the most effective finish. The best are those which require the paint to be baked on, although some aerosol sprays are almost as effective. Whichever finish is decided upon, ensure that the surface is properly prepared according to the paint manufacturer's instructions and that the paint itself is correctly applied.

16 The engine lubrication system

1   In line with current two-stroke practice the Yamaha LC models utilise a pump-fed engine lubrication system and do not require the mixture of a measured quantity of oil to the petrol content of the fuel tank in order to utilise the so-called 'petroil' method. Oil of the correct viscosity is contained in a separate oil tank mounted on the right-hand side of the machine and is fed to a mechanical oil pump on the right-hand side of the engine which is driven from the crankshaft by reduction gear. The pump delivers oil at a predetermined rate, via a flexible plastic tube, to a union on the inlet side of the carburettor venturi. In consequence, the oil is carried into the engine by the incoming charge of petrol vapour, when the inlet port opens.

2   The oil pump is also interconnected by the twist grip throttle, so that when the throttle is opened, the oil pump setting is increased a similar amount. This technique ensures that the lubrication requirements of the engine are always directly related to the degree of throttle opening. This facility is

arranged by means of a control cable attached to a pulley on the end of the pump; the cable is joined to the throttle cable junction box.

17 Oil pump: removal and refitting

1   It is rarely necessary to remove the oil pump unless specific attention to it is required. It should be noted that the pump should be considered a sealed unit – parts are not available and thus it is not practicable to repair it. The pump itself can be removed quite easily leaving the drive shaft and pinion in place in the crankcase right-hand cover. If these latter components require atttention it will be necessary to remove the crankcase right-hand cover as described in Section 8 of Chapter 1.

2   The oil pump itself is a very reliable unit which does not give trouble as a rule. If a fault is suspected in the lubrication system, check first the following points:

(a) Sufficient oil in the oil tank, oil tank breather pipes clear

(b) Oil pump correctly adjusted

(c) Oil tank/oil pump and oil/pump carburettor feed pipes free from dirt, kinks, or other obstructions which might be causing a blockage and are correctly secured so that air is not entering the system

(d) The check valve ball and spring correctly fitted and functioning properly

3   While the first three of the above items are easy to check and rectify, the last will require some care. The check valve consists of a spring-loaded steel ball which is fitted behind the union of the oil pump/carburettor feed pipe. The valve assembly is secured by the union itself which is a tight press-fit in the pump body and can be removed, therefore, with the careful use of a suitable pair of pliers. Pull the union out, taking care not to damage it, and withdraw the check valve spring and ball. Remove all oil from the components and any particles of dirt, then examine the components, renewing any that are damaged or worn. If necessary, check their condition by comparing them with new components at a Yamaha dealer. Check that the oil pump passage is free from dirt and refit the check valve, taking great care to ensure that the union is a secure, tight fit in the pump body.

4   if the above items have been checked and the oil feed is still suspect, the pump must be removed as follows. Slacken and remove the four screws which fasten the oil/water pump cover to the front of the crankcase right-hand cover, noting that while an impact driver should be used to slacken any tightly fastened screws, great care must be taken in this case to avoid cracking the plastic moulding. Withdraw the cover, freeing it from the oil feed pipe retaining grommet. Disengage the clips of the oil feed pipes and pull each pipe carefully off its union, plugging it immediately with a screw or bolt of suitable size.

5   Disconnect the oil pump cable inner wire from the pump pulley, using a small screwdriver to push back the extension of the pulley return spring or withdrawing the spring clip which retains the inner wire in the pulley track and then disengaging the cable end nipple from the pulley.

6   The pump is secured to the cover by two screws which pass through its mounting flange. Once these have been removed the pump can be removed, noting that it may prove necessary to turn the pump slightly to free it from its drive shaft. Note also that a thrust washer is fitted over the end of the drive shaft. This must be picked out of the pump body and replaced on the shaft end so that it is not lost.

7   Further dismantling is not practicable, and it will be necessary to renew the pump if it is obviously damaged. Maintenance must be confined to keeping the pump clear of air and correctly adjusted, as described in the following sections.

8   Refit the oil pump to the crankcase cover, using a new gasket at the oil pump/crankcase cover joint. Replace and tighten the two crosshead mounting screws to a torque setting of 0.5 kgf m (3.5 lbf ft). The remainder of the reassembly is accomplished by reversing the dismantling procedure but note

that it will be necessary to check and to reset, if necessary, the pump stroke setting and oil pump cable adjustment, and to bleed any air from the pump. These tasks are described in the two subsequent Sections of this Chapter; when they are completed, refit the oil/water pump cover and tighten securely the four retaining screws.

9   Should it be necessary at any time to inspect the oil pump drive components, the crankcase right-hand cover must be removed as described in Section 8 of Chapter 1. Remove the oil pump and invert the cover. Using a small-bladed screwdriver, prise the drive gear retaining circlip out of its groove in the pump drive shaft and lift the white nylon drive gear away, so that the gear locating pin and the second circlip can be withdrawn. Push the shaft out of its bush in the cover.

10   As the oil pump drive components are lightly loaded and well lubricated, wear is unlikely to occur until a very high mileage has been covered. Damage of any sort should be visible easily and must be rectified by the renewal of the parts concerned. The shaft oil seal may be prised from its housing with a suitably shaped screwdriver. If necessary the drive shaft bush may be driven out using a hammer and a drift of suitable size, after the casing has been heated by immersing it in boiling water. The state of wear of the drive shaft and bush can be assessed by feeling for excessive free play when the shaft is installed in the bush. Renew the shaft or the bush as necessary.

11   When removing or refitting the bush, and also the oil seal, ensure that the casing is well supported on wooden blocks or a clean, flat surface as appropriate. Complete the assembly procedure by following the reverse of the removal procedure.

12   A final note concerns the owners of DT125 LC machines in particular, with especial reference to machines up to engine/frame number 10V-011070 in the case of unrestricted models, or 12W-000122 in the case of the restricted versions. Cases have occurred of a severely restricted supply of oil to the pump due to the oil tank/oil pump feed pipe having been pinched by a raised rib on the inside of the oil/water pump cover. Although it must be stressed that this applies only to the above machines, which should all have been modified by the selling dealer, it is a point worth checking on any DT125 LC machine whenever the oil/water pump cover is removed. Owners of RD125 LC machines are not affected as the cover is different in shape and size, but are advised to bear this point in mind, as it illustrates the degree of care needed when routing the oil pipes.

13   Remove the oil/water pump cover and examine carefully the larger diameter oil feed pipe. If any signs of pinching are seen, or any marks which show that the pipe is chafing on the cover, the following action must be taken. First find the small band of white paint which is applied to the oil tank/oil pump feed pipe. This should be situated next to the sealing grommet as the pipe enters from above. Pull the pipe backwards or forwards through the grommet until the painted band is aligned with the upper edge of the grommet as described. The correct length of pipe will then be in place inside the pump cover; this is essential as while too short a length would risk the pipe being pulled off the oil pump union, too long a length will result in kinks and chafing as it is compressed inside the cover. When the pipe is at the correct length, examine the inside surface of the oil/water pump cover. Yamaha recommend that an area 7 mm (0.27 in) and 1.5 mm (0.06 in) deep be trimmed away from the raised rib at the point where it fouls the oil feed pipe. Finish off with a file or emery paper to round off any sharp edges that may have been left.

14   Make a final check of the pump cover/oil feed pipe clearance using white paint. Check that both feed pipes are correctly routed and of the correct length as described above, then apply a coat of white paint to the edges of the internal rib that have just been trimmed. Before the paint dries, position the pump cover and its gasket on the crankcase right-hand cover. When the cover is withdrawn, there should be no traces of white paint visible on the oil feed pipes. If necessary, trim off some more material and repeat the check until a lack of white paint on the pipes indicates that sufficient clearance now exists.

17.1 Autolube pump is a sealed unit and must be renewed if faulty

17.3a Pump check valve assembly

17.3b Ensure feed pipe union is a tight fit in pump body on reassembly

17.8a Always use a new gasket when refitting oil pump

17.8b Do not omit small thrust washer from end of drive spindle

17.8c Tighten securely the two pump mounting screws

17.9 Pump drive pinion is secured by a circlip

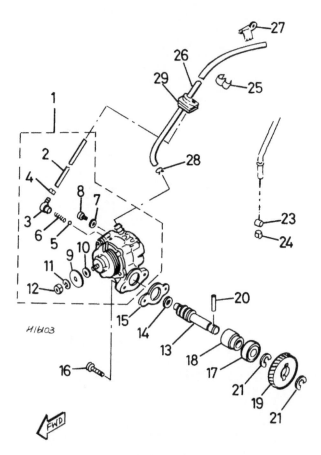

Fig. 3.11 Oil pump

1	Oil pump body	15	Gasket
2	Carburettor feed pipe	16	Screw – 2 off
3	Pipe union	17	Oil seal
4	Spring clip	18	Bush
5	Check valve ball	19	Drive gear
6	Spring	20	Locating pin
7	Sealing washer	21	Circlip – 2 off
8	Oil bleed screw	23	O-ring
9	Adjustment plate	24	Spring clip
10	Shim	25	Pipe securing clip
11	Spring washer	26	Oil pump feed pipe
12	Lock nut	27	Pipe securing clip
13	Drive shaft	28	Pipe clip
14	Thrust washer	29	Casing grommet

18 Oil pump: bleeding

1   It is necessary to bleed the oil pump every time the main feed pipe from the oil tank is removed and refitted. This is because air will be trapped in the oil line, no matter what care is taken when the pipe is removed.

2   Check that the oil pipe is connected correctly, with the retaining clip in position. Then remove the cross-head screw in the outer face of the pump body with the fibre washer beneath the head. This is the oil bleed screw.

3   Check that the oil tank is topped up to the correct level, then place a container below the oil bleed hole to collect the oil that is expelled as the pump is bled. Allow the oil to trickle out of the bleed hole, checking for air bubbles. The bubbles should eventually disappear as the air is displaced by fresh oil. When clear of air, refit the bleed screw. DO NOT replace the oil/water pump cover until the pump setting has been checked, as described in the next Section.

4   Note also that it will be necessary to ensure that the oil pump/carburettor feed pipe is primed if this was disturbed. Unless this is checked the engine will be starved of oil until the pipe fills. The procedure required to avoid this is to start the engine and allow it to idle for a few minutes whilst holding the pump pulley in its fully open position by pulling the pump cable. The excess oil will make the exhaust smoke heavily for a while, indicating that the pump is delivering oil to the engine.

19 Oil pump: checking the minimum stroke setting and cable adjustment

1   The oil pump fitted to the machines described in this Manual is set at the factory to deliver a predetermined amount of oil to the engine at any given engine speed. The actual delivery rate of the pump is governed by the minimum stroke setting; alterations to this setting being made by the use of shims. The manufacturer recommends that the minimum stroke setting be checked at regular intervals, as described in the Routine Maintenance Chapter of this Manual but it should be noted that in practice, actual alteration of the setting is unlikely to be required very often. The work necessary to check and reset the pump stroke setting is the same for both models and is described in the first part of this Section. The next task is to

check that on RD125 LC models the oil pump cable is correctly adjusted so that the oil pump is correctly synchronised with the carburettor. This latter task is neither necessary nor possible on DT125 LC models but it is recommended that the operation of the oil pump cable is checked carefully to ensure that the engine continues to receive the correct amount of oil. Refer to the relevant sub-sections below for the necessary directions. All work on the oil pump or cable must commence with the removal of the oil/water pump cover from the front of the crankcase right-hand cover.

Checking and resetting the pump minimum stroke – both models

2   The pump's minimum stroke adjustment should be checked first. Start the engine and allow it to idle. Observe the front end of the pump unit, where it will be noticed that the pump

adjustment plate moves in and out. When the plate is out to its fullest extent, stop the engine and measure the gap between the plate and the raised boss of the pump pulley using feeler gauges. Do not force the feeler gauge into the gap – it should be a light sliding fit. Make a note of the reading, then repeat the procedure several times. The largest gap is indicative that the pump is at its minimum stroke position. If the pump is set up correctly, the gap found should be 0.20 – 0.25 mm (0.008 – 0.010 in) on both machines described in this Manual.

3   If the pump setting is found to be incorrect, slacken and remove the adjustment plate retaining nut, then withdraw the spring washer and the adjustment plate. The pump stroke is set by adding or removing shims from behind the adjustment plate, these shims being available from Yamaha dealers in thicknesses of 0.3, 0.5 and 1.0 mm (0.0118, 0.0197 and 0.0394 in). When the shim thickness is correct refit the adjustment plate, the spring washer, and the retaining nut, then start the engine and recheck the minimum stroke setting. If necessary, repeat the procedure to ensure that the setting is correct.

Checking operation of oil pump cable – DT125 LC model

4   The DT125 LC model described in this Manual is fitted with an oil pump whose setting is controlled by a cable connected via a junction box to the throttle cable, this junction box being of a new type which automatically compensates for wear in the cables and ensures that the oil pump is synchronised accurately with the carburettor at all times. The junction box is discussed separately in Section 20 of this Chapter. All that can be done at the oil pump is to check that the cable is operating correctly, a task which must be carried out after the adjustment of the throttle cable has been checked, and reset if necessary, as described in Section 8 of this Chapter.

5   When the throttle cable has been checked and is known to be in correct adjustment, and the oil/water pump cover has been removed, start the engine and allow it to idle until the pump adjustment plate moves out to its fullest extent, ie to the minimum stroke position, as described above, so that the pulley is free to rotate throughout its full movement. Stop the engine and allow the twistgrip to rotate forwards to the fully closed position. Slowly rotate the twistgrip to open the throttle and watch closely the pump pulley. The pulley should start to rotate just as the throttle cable takes up its free play and starts to lift the throttle slide. Fully open the throttle. The pulley should rotate smoothly to the fully open position, and should return equally smoothly to its fully closed position when the twistgrip is then released and allowed to snap shut under spring pressure.

If this is not found, refer to the next Section of this Chapter. If all is well, apply a few drops of oil to the exposed length of the oil pump inner cable and to the moving parts of the oil pump, then refit the oil/water pump cover and its gasket, paying careful attention to the routing of the oil feed pipes as described in Section 17 of this Chapter.

Checking adjustment of oil pump cable – RD125 LC model

6   The RD125 LC model described in this Manual is fitted with an oil pump that is connected to the throttle cable in a manner similar to that described above for the DT125 LC model, but differs in that a conventional junction box is employed. This means that the cable must be checked and adjusted at regular intervals to compensate for cable wear so that the oil pump is synchronised correctly with the carburettor at all times. Commence oil pump cable adjustment by checking that the throttle cable is correctly adjusted, using the adjuster on the carburettor top and that on the twistgrip to provide the specified throttle cable free play, this operation being described in full in the relevant paragraph of Section 8 of this Chatpter. Remove the oil/water pump cover.

7   Start the engine and allow it to idle until the pump adjustment plate moves out to its fullest extent, ie. to the minimum stroke position, as described above, so that the pump pulley is free to rotate throughout its full movement. Stop the engine and check that the twistgrip is in the fully closed position, then look closely at the outer face of the pulley. Various marks will be seen, one of which is a round indentation drilled in the pulley at about 90° from the cable end nipple (see accompanying photograph). This is the pulley adjustment mark to be used: all others should be ignored. Slowly rotate the twistgrip until all free play has just been eliminated from the cables, then check that the pump plunger pin aligns exactly with the round alignment mark.

8   If adjustment is necessary, slacken the locknut of the adjuster set in the crankcase right-hand cover, then rotate the adjuster as necessary to achieve the correct setting. Tighten the adjuster locknut and refit the rubber sleeve over the adjuster, then open and close fully the throttle to settle the cables. Check that the adjustment has remained the same, resetting it if necessary. Lubricate the exposed length of the oil pump cable inner and the moving parts of the oil pump itself with a few drops of oil, then refit the oil/water pump cover, taking care that the oil feed pipes are routed correctly as described in Section 17 of this Chapter.

18.1 Screw must be removed to bleed air from pump and feed pipe

19.2 Measuring oil pump minimum stroke setting

19.3a Remove retaining nut, spring washer, and adjustment plate to reach shims

19.3b Shims are available in different thicknesses to adjust setting

19.7a Use only alignment mark arrowed – ignore all others

19.7b Alignment mark must coincide exactly with plunger pin at idle (see text)

20 Throttle/oil pump cable junction box: description and maintenance – DT125 LC only

1   To ensure that the synchronisation of the carburettor and oil pump remains exactly set without the need for constant checking and adjustment, Yamaha have introduced on the DT125 LC model only, a junction box which compensates automatically for wear in either the junction box/carburettor throttle cable or the junction box/oil pump cable.

2   The principle by which this operates is shown in figure 3.13, part 'A' and 'B'. In a normal junction box (part 'A') the throttle cable lifts a one-piece slider to which the carburettor cable and oil pump cable are attached. If all cables are correctly adjusted, and checked regularly to ensure this, the junction box will work very well. If, on the other hand, excessive free play is allowed to creep in through infrequent or incorrect adjustment, the oil pump will no longer be synchronized correctly with the carburettor, and will deliver too much or too little oil, depending on which cable is incorrectly set. Yamaha's new design, shown in basic form in figure 'B', introduces a rotating link instead of the slider. As the twistgrip is rotated, the link also rotates to take up the excess free play in whichever cable is worn.

3   A detailed description of the junction box's method of

operation is as follows. Refer to figure 3.12 for identification of the parts concerned. The twistgrip/junction box inner cable is connected to the slider via the rotor guide, its end nipple being retained in position by the set pin. As the slider moves up the junction box, the rotor turns in the rotor guide taking up the free play in the cables. A small compression spring set in the slider acts on the rotor guide, permitting the rotor to turn until its pressure is overcome by the greater resistance of the throttle valve and oil pump pulley return springs when the cable free play is eliminated, so that a raised locking tab on the rear of the slider comes into contact with serrations on the rotor periphery and locks it solidly. The junction box then opens the throttle valve and oil pump together in the normal way.

4   Referring to figure 3.14, in part A the oil pump cable is shown as having excessive free play. In part B the rotor is turning to take up this free play as the slider moves up the junction box, and in part C the free play has been eliminated and the slider lock has come up to bear on the rotor so that it cannot move.

5   It will be evident from the above description that the whole assembly relies on each cable and component being free from dirt or corrosion if it is to operate correctly. If the checks described in Section 19 of this Chapter and in Routine Maintenance reveal any sign that the assembly is not

functioning correctly, remedial action must be taken at once. While the work can be carried out on the machine, it is very awkward and there is a high risk of dirt getting in, or of one of the smaller components being lost; it is therefore recommended that the complete throttle/oil pump cable assembly be removed before work starts. Disconnect the throttle cable from the carburettor as described in Section 5 of this Chapter, disconnect the oil pump cable from the oil pump as described in Section 17, and free the upper end of the cable from the twistgrip by slackening and removing the two twistgrip clamping screws. Remove the seat, the sidepanels, and the petrol tank as described in Section 2. Disengage the junction box mounting rubber from the bracket on the frame and withdraw the complete assembly from the machine.

6   Slide back the rubber sleeves from the junction box, then slacken and remove the four screws which secure the box cover, and withdraw the cover. Turn the set pin through $\frac{1}{2}$ turn and withdraw it so that the twistgrip cable end nipple can be disengaged from the rotor guide, then very carefully pick out the rotor guide, ensuring that the small spring does not fly out of the slider. Pick out the rotor and disengage the end nipples of the two lower cables from it. The slider can then be picked out.

7   Carefully inspect the cables and junction box components for signs of wear or damage, renewing any individual component which is no longer serviceable; all are available through Yamaha dealers. Carefully clean off all traces of dirt and corrosion. Thoroughly lubricate the cables as described in Routine Maintenance then reassemble the cables and junction box by reversing the dismantling procedure. Do not apply any oil or grease to the junction box components, but use instead a synthetic aerosol lubricant such as WD40, or one of the proprietary silicone based lubricants suitable for use with nylon components. Connect the cables to the carburettor and oil pump, then use the adjuster on the carburettor top to remove all but the slightest trace of free play from the cable between the junction box and the carburettor.

8   Pass the upper end of the throttle cable up to the handlebar ensuring that it is correctly routed with no sharp turns or kinks, then connect it to the twistgrip again and use the adjuster below the twistgrip to set the required amount of free play in the cable. If necessary, unscrew the adjuster on the carburettor top to provide sufficient slack, but remember that the compensating mechanism will work best if it has to take up the free play in the oil pump cable alone.

9   Refit the petrol tank, the sidepanels and the seat, then open and close fully the twistgrip several times to settle the cables. Check the adjustments again and reset them if necessary, but if all is in order, tighten the adjuster locknuts and replace the protecting rubber shoes. Make a final check that the oil pump cable is rotating correctly the pump pulley, as described in Section 19 of this Chapter, before refitting the oil/water pump cover.

20.1a While the RD125 is fitted with a conventional junction box ...

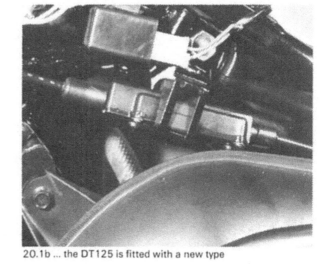

20.1b ... the DT125 is fitted with a new type

20.5 Remove assembly from machine before dismantling (see text)

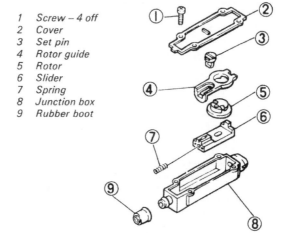

1   Screw – 4 off
2   Cover
3   Set pin
4   Rotor guide
5   Rotor
6   Slider
7   Spring
8   Junction box
9   Rubber boot

Fig. 3.12 Throttle/oil pump cable junction box – DT model

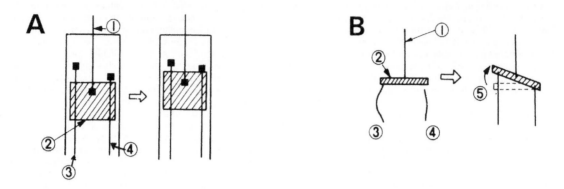

Fig. 3.13 Operation of the throttle/oil pump cable junction box – DT model

1   Throttle cable
2   Slider

3   Carburettor cable
4   Oil pump cable

5   Self adjustment method

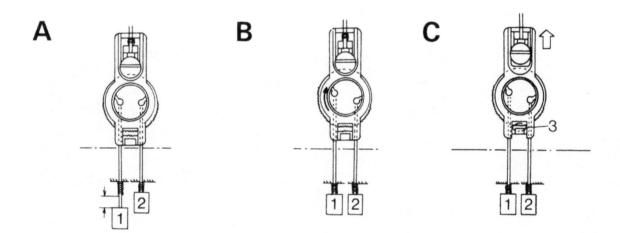

Fig. 3.14 Throttle/oil pump cable junction box self-adjustment method – DT model

1   Oil pump cable

2   Carburettor cable

3   Slider lock

21 Gearbox lubrication: description and maintenance

1   The gearbox oil is contained inside the crankcases and lubricates the clutch and primary drive and the gearbox and ancillary components.

2   The crankcase reservoir is reached through a screwed plug in the crankcase right-hand cover and drained through a single drain plug situated in the underside of the crankcase. The level is checked by referring to a sight glass set in the crankcase right-hand cover, immediately under the kickstart. Maximum and minimum level lines are marked on the periphery of the glass to indicate the correct level.

3   Maintenance is restricted to periodic checks of the oil level and to regular changing of the oil itself. The machine must always be standing upright on level ground whenever work is undertaken on the gearbox oil; it is only in this position that the maximum amount of oil can be removed by draining, and it is also only in this position that the level can be checked accurately. Routine oil changes are made more effective if the machine is fully warmed up to normal operating temperature before the drain plug is removed. The oil is much thinner at high temperatures, enabling more rapid and more complete draining to take place, and any particles of dirt and swarf are held in suspension in the oil and are therefore much more likely to be removed.

4   Once the drain plug has been removed and the oil drained as completely as possible, check the condition of the drain plug sealing washer, renewing it if necessary, and refit the drain plug. Tighten it down to a torque setting of 2.0 kgf m (14.5 lbf ft). On refilling the gearbox, use 600 cc (1.06 pint) after a full rebuild and 550 cc (0.97 pint) at routine oil changes. The difference is the amount of residual oil left in the crankcases after routine draining. Temporarily refit the filler plug, start the engine and warm it up to normal operating temperature. With the machine fully warmed up and placed on its wheels on level ground, stop the engine and leave the machine for a few minutes to allow the oil to settle. The oil level should fall between the maximum and minimum marks on the periphery of the sight glass. Add more oil or drain off the surplus to achieve this.

5   Once the oil level is known to be correct, replace the filler plug and check that both gearbox plugs are securely tightened. Remove any surplus oil and subsequently check that no oil leaks appear.

Chapter 4 Ignition system

For modifications, and information relating to later models, see Chapter 8

Contents

Specifications

	RD125 LC	DT125 LC
Ignition system		
Type	Capacitor discharge (CDI)	Capacitor discharge (CDI)
Voltage	12V	6V
Ignition timing – BTDC:		
At 1300 rpm	17°	Not available
At 3000 rpm	30°	22°
Piston position – BTDC	N/A	2.2 ± 0.15 mm
		(0.087 ± 0.006 in)
Flywheel generator		
Make	Hitachi	Mitsubishi
Model	10W	F3T25271
Source coil resistance	Green-Brown 305 ohms	Black/red-Black 180 ohms
	± 15%	± 15%
	Red-Brown 86 ohms ± 15%	N/A
	White/Red-Black 100 ohms	
Pulser coil resistance	± 15%	N/A
CDI unit		
Make	Hitachi	Mitsubishi
Model	10W	F8T05471
Ignition HT coil		
Make	Hitachi	Mitsubishi
Model	10W	F6T505
Minimum spark gap	6 mm (0.24 in) @ 1350 rpm	6 mm (0.24 in) @ 1350 rpm
Primary winding resistance	0.54 – 0.66 ohm	8.5 – 11.5 ohm
	@ 20°C (68°F)	@ 20°C (68°F)
Secondary winding resistance	4.96 – 7.44 K ohm	5.02 – 6.79 K ohm
	@ 20°C (68°F)	@ 20°C (68°F)
Spark plug		
Manufacturer	NGK	NGK
Type	BR8ES (BR9ES optional)	BR8ES
Gap	0.7 – 0.8 mm	0.7 – 0.8 mm
	(0.028 – 0.031 in)	(0.028 – 0.031 in)

1 General description

1 The machines described in this Manual are equipped with fully electronic capacitor discharge ignition (CDI) systems. This arrangement provides a more powerful and accurate ignition system and can be considered almost maintenance-free.

2 The most important advantage of electronic ignition, is that it removes all mechanical components from the system, the spark being triggered electronically rather than by a contact breaker assembly. Because there are no contact breakers to wear, the owner is freed from the task of periodically adjusting or renewing them. Once the electronic system has been set up, it need not be attended to unless it has been disturbed in the course of dismantling or failure occurs in the electronic components in the system.

3 Note that although the ignition systems fitted to the DT125 LC and RD125 LC models are broadly similar, in that they are both Capacitor Discharge systems, the two systems are very different in construction. Take care to follow the relevant instructions for your machine when working on the ignition system. There is no difference between the ignition systems of the unrestricted and restricted versions of either model.

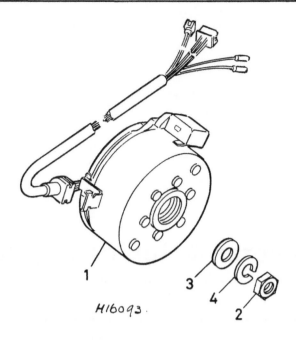

H16093.

Fig. 4.1 Flywheel generator – RD model

1   Flywheel rotor and stator   3   Washer
2   Nut   4   Spring washer

2 Electronic ignition system: principles of operation

RD125 LC models

1 Energy for the ignition system is drawn from the exciter coil. This is mounted on the generator stator, and is integral with the normal alternator windings. It is a two-stage arrangement, having low speed and high speed windings. The low speed windings produce a high output voltage at low engine speeds, this voltage dropping off as the engine builds up speed. The high speed windings, on the other hand, produce little energy at low engine speed, but the output voltage rises along with engine speed. The two outputs are combined, offsetting each other to give a fairly constant output voltage, this being the sum of the output of each set of windings.

2 The exciter coil assembly feeds the CDI unit, a sealed electronic assembly which forms the heart of the system. This unit contains, amongst other things, a capacitor and a thyristor, or silicon controlled rectifier (SCR). The capacitor is charged with the high voltage output from the exciter coil assembly. The thyristor, or SCR, is in effect an electronic switch. When signalled electrically by the pulser, it allows the capacitor to discharge through the primary windings of the ignition coil. This in turn induces a high tension pulse in the secondary windings, which is fed to the spark plug.

3 The pulser, or pickup, comprises a small coil mounted outside the alternator rotor on a projection from the stator. A permanent magnet embedded in the flywheel rotor is arranged to pass beneath the pulser coil. As the magnet passes the pulser coil, a weak current is generated and it is this that is used to trigger the thyristor in the CDI unit.

DT125 LC models

4 The energy for this type of system is drawn from an exciter, or source, coil mounted on the generator stator, the power being fed to the CDI unit behind the steering head. Here it is converted to direct current (dc) and stored in a capacitor.

5 The thyristor is triggered in this case by a weak current from the same source coil, this current rising to a level high enough to provide the necessary signal voltage, whereupon a spark is produced at the plug in the same way as described above for the RD125 LC models. This simple system differs in that the stator plate must be set accurately to determine the ignition timing and the stator mounting holes are slotted to permit this.

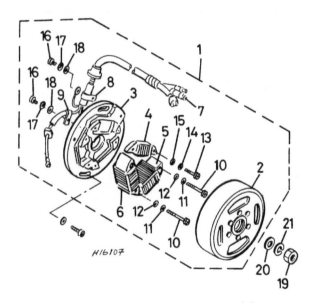

H16107

Fig. 4.2 Flywheel generator – DT model

1   Flywheel generator assembly   11   Spring washer – 4 off
2   Flywheel rotor   12   Washer – 4 off
3   Flywheel stator   13   Screw – 2 off
4   Ignition source coil   14   Spring washer – 2 off
5   Lighting coil   15   Washer – 2 off
6   Lighting coil   16   Screw – 2 off
7   Wiring harness   17   Spring washer – 2 off
8   Wiring clamp   18   Washer – 2 off
9   Wiring clamp   19   Nut
10   Screw – 4 off   20   Washer
    21   Spring washer

a) No spark is produced, or weak spark

```
┌─────────────────────┐
│ Check of connections│ .. Check wiring connections and
└─────────────────────┘    switches. See Section 4
         │
         │ OK                    Faulty ──────► Correct
         │
┌─────────────────────┐
│     Spark test      │ .............. Check by substitution of new
└─────────────────────┘                plug or spark plug cap
         │
      No spark                    Spark ──────► Plug is faulty
         │
┌─────────────────────┐
│   Source coil test  │ ............... Measure coil resistance.
└─────────────────────┘                 See Section 5
         │
         │ OK                    Faulty ──────► Replace
         │
┌─────────────────────┐
│   Pulser coil test  │ ............... Measure coil resistance.
└─────────────────────┘                 See Section 6
         │
         │ OK                    Faulty ──────► Replace
         │
┌─────────────────────┐
│  Ignition coil test │ ............... Ignition test
└─────────────────────┘                 See Section 7
         │
         │ OK                    Faulty ──────► Replace
         │
┌─────────────────────┐
│      CDI unit       │ ............... See Section 8
└─────────────────────┘
         │
                                Faulty ──────► Replace
```

b) The engine starts but will not pick up speed

```
┌─────────────────────┐
│     Spark plug      │ ............... Clean or replace
└─────────────────────┘                 See Section 10
         │ OK
┌─────────────────────┐
│   Source coil test  │ ............... Make continuity test.
└─────────────────────┘                 See Section 5
         │ OK
┌─────────────────────┐
│ Ignition timing check│ ........... See Section 9
└─────────────────────┘
         │ OK
┌─────────────────────┐
│ Ignition system is good│ ........ See Section 9
└─────────────────────┘
         │
┌─────────────────────┐
│      CDI unit       │
└─────────────────────┘
         │
                                Faulty ──────► Replace
```

Testing the electronic ignition

3 Electronic ignition system: testing and maintenance

1 As stated earlier in this Chapter, the electronic ignition needs no regular maintenance once it has been set up and timed accurately. Occasional attention should, however, be directed at the various connections in the system, and these must be kept clean and secure. A failure in the ignition system is comparatively rare, and usually results in a complete loss of ignition. Usually, this will be traced to the CDI unit, and little can be done at the roadside to effect a repair. In the event of the CDI failing, it must be renewed as repair is not practicable.

2 If the CDI unit is thought to be at fault, it is recommended that it be removed and taken to a Yamaha dealer for testing. The dealer will have the use of diagnostic equipment, and will be able to test the unit accurately and quickly. Testing at home is less practical and cannot be guaranteed to be accurate. At best, it will enable the owner to establish which part of the system is at fault, although replacement of the defective part remains the only effective cure.

3 Care must be exercised when dealing with the CDI unit. Wrong connections could cause instant and irreparable damage to the unit, so this must be an important consideration when removing and replacing the unit. If the unit is to be disconnected and removed for testing, use a short length of insulated wire to short-circuit each pair of terminals to avoid any electric shock from residual energy in the capacitor.

4 For those owners possessing a multimeter and who are fully conversant with its use, certain test sequences are given. It is not recommended that the inexperienced attempt to test the system at home, as more damage could be sustained by the system, and an unpleasant shock experienced by the unwary operator. It should be appreciated that the CDI system is capable, in certain circumstances, of producing sufficient current output to be dangerous to the operator. It follows that an attitude to safety should be adopted similar to that applying when testing household mains electricity. Before performing any tests, the following points must be noted. Disconnect all the connectors in the ignition system, thus isolating the various components. When taking readings with any instrument other than the Yamaha test equipment, remember that the resistance measurements taken must be treated only as a guide to the condition of the component being tested as the results gained will vary slightly from one instrument to the next. Readings will also vary with the temperature of the component, all those given in the Specifications Section and throughout the text of this Chapter being correct at 20°C (68°F). In general, therefore, if the readings obtained correspond closely with those given, the tested component may be considered to be in good order, but if any doubt arises because of the results obtained by the test, or if the fault persists, the component should be taken to a local Yamaha dealer for accurate testing.

5 It will be noted that some of the resistance tests require a meter capable of reading in ohms rather than kilo ohms. Many of the cheaper multimeters are only capable of the latter, and thus are of limited use for accurate testing of low-resistance components. When testing the ignition system to isolate faults it is important to follow a logical sequence to avoid wasted effort and time. Use the accompanying flow charts as a guide, referring to the subsequent sections for more detailed information.

4 Wiring, connectors and switches: checking

1 The single most likely cause of ignition failure or partial failure is a broken, shorted or corroded connector, switch contact or wire. Tracing faults of this nature can often prove time consuming but does not require the use of sophisticated test equipment. Although in rare instances a particular fault will not be evident from a physical examination, it is worthwhile checking the obvious points before resorting to expensive professional assistance.

2 Refer to the wiring diagram at the end of the book, trace and check each of the alternator output leads, connector blocks and all connections to the CDI unit, ignition switch and kill switch. Do not ignore the obvious; it can be very frustrating to

spend a long time checking the ignition system, only to discover that the kill switch was set to the 'Off' position. The kill switch and some of the wiring connectors are relatively exposed and may have become contaminated by water. To eliminate this possibility, spray each one with a water dispersing aerosol such as WD40.

3   Check the wiring for breakages or chafing paying particular attention to the wiring in areas like the steering head where steering movement causes flexing. The plastic insulation may appear intact even if the internal conductor has broken. If a wire is suspected of having broken internally or shorted against the frame it should be checked using a multimeter as a continuity tester.

5   Exciter or source coil: testing

RD125 LC models

1   Starting from the sealing grommet set in the crankcase top surface, trace the main generator lead up the frame left-hand front downtube to the various connectors which connect it to the main wiring loom. It may be necessary to remove the radiator cover to gain adequate access to these connectors. Separate the connectors and identify the green, brown, and red wires and their terminals.

2   Set the multimeter on the relevant ohms scale and measure the resistance between the green and the brown wire terminals to check the coil low speed windings. Reset, if necessary, the meter and then check the coil high-speed windings by measuring all resistance between the red and the brown wire terminals. Bearing in mind the notes given in Section 3 of this Chapter, the readings obtained should correspond with those given below.

Exciter (source) coil – resistance
    Green to brown wires 305 ohm $\pm$ 15% (259 – 350 ohm)
    Red to brown wires 86 ohm $\pm$ 15% (73 – 99 ohm)

3   If the coil is found to be suspect by this test, remove the stator plate from the machine as described in Section 7 of Chapter 1 and take the assembly to a Yamaha dealer for accurate testing on Yamaha test equipment. If the coil is then pronounced faulty and a careful inspection does not reveal any obvious damage such as trapped or broken wires, then the entire generator assembly must be renewed.

4   The only alternative to this rather expensive solution is to have the coil removed from the stator plate and rewound by an expert. It must be stressed that such work can be undertaken only by an expert, the Yamaha dealer may be able to assist in locating such a person.

DT125 LC models

5   Remove the sidepanels, the seat, and the petrol tank as described in Section 2 of Chapter 3, then trace the main generator lead from the sealing grommet set in the crankcase top surface across the top of the engine/gearbox unit and up the frame right-hand rear downtube to the various connectors joining the lead to the main wiring loom. Separate the connectors and identify the black/red and black wires and their terminals.

6   Set the multimeter to the relevant ohms scale and measure the resistance between the black/red and the black wire terminals. Bearing in mind the notes made in Section 3 of this Chapter, the reading obtained should correspond with that given below.

Exciter (source) coil and pulser coil – resistance

    Black/red to black wires 180 ohm $\pm$ 10% (162 – 198 ohm)

7   If the coil is found to be suspect by this test, remove the stator plate from the machine as described in Section 7 of

Chapter 1 and take the assembly to a Yamaha dealer for accurate testing on Yamaha test equipment. If the coil is then pronounced faulty it should be renewed unless a careful inspection reveals damage such as trapped or broken wires which may be repaired easily by the owner. While repairs may be undertaken by having an auto-electrical expert rewind the coil, this is unlikely to be an economic proposition as the individual coil is available as a replacement part from a Yamaha dealer.

5.3 Although stator coils can be tested, the complete assembly must be renewed if one is faulty – RD125 only

5.7 DT125 coils can be renewed individually, if necessary

6   Pulser coil: testing

RD125 LC models

1   Find and separate the connectors joining the main generator lead to the wiring loom as described in Section 5 of this Chapter, then identify the white/red and black wires and their terminals.

2   Using a multimeter set to the relevant resistance scale, measure the resistance between the two wires. Bearing in mind

the notes made in Section 3 of this Chapter, the reading obtained should correspond with that given below.

Pulser coil (pick-up) – resistance
     White/red to black wires 100 ohm ± 15% (85 – 115 ohm)

3   If the coil is found to be suspect by this test, remove the stator plate from the machine as described in Section 7 of Chapter 1 and follow the same course of action as described in Section 5 of this Chapter. Once again, if the coil is found to be faulty, the entire generator assembly must be renewed as no individual components are available as replacement parts. The only alternative is to try to find an auto-electrical expert who is prepared to undertake a repair.

DT125 LC models
4   Since the same coil serves the function of both exciter and pulser coils, the test for this component is the same as described in the second half of Section 5 of this Chapter.

7.1 Location of RD125 HT coil

7   Ignition HT coil: testing

1   The ignition HT coils fitted to the machines described in this model are similar in appearance and location. Both are sealed grey or black plastic units with a large external mounting bracket and are readily identified by the HT lead projecting from one end. In the case of both machines the coil is mounted on the frame top tubes to the rear of the steering head and it is therefore necessary to remove the seat, the sidepanels, and the petrol tank as described in Section 2 of Chapter 3 to gain access to the coil before testing can take place.
2   If the ignition coil is suspected of having failed it can be tested by measuring the resistance of its primary and secondary windings. The test can be performed with the coil in place on the frame, having first disconnected the high tension lead at the spark plug and the low tension lead at the connector.
3   Set the meter to the relevant resistance scale and connect the meter negative (–) probe to the coil earth pole, which is formed by the external mounting bracket, or to a suitable earth point on the frame. To test the primary windings, connect the meter positive (+) probe to the terminal of the black wire which forms the low tension lead and measure the resistance between the two.The reading obtained should be within the limits given below.

Primary winding resistance
RD125 LC 0.6 ohm ± 10% (0.54 – 0.66 ohm) @ 20°C (68°F)
DT125 LC 10 ohm ± 15% (8.5 – 11.5 ohm) @ 20°C 68°F

4   To test the HT coil secondary windings, unscrew the spark plug cap from the end of the high tension lead, set the meter to the relevant resistance scale and connect its negative (–) probe to earth as described above. Connect the meter positive (+) probe to the end of the high tension lead and measure the resistance between the two points. Compare the reading obtained with that given below.

Secondary winding resistance
RD125 LC 6.2 K ohm ± 20% (4.96 – 7.44 K ohm) @ 20°C (68°F)
DT125 LC 5.9 K ohm ± 15% (5.02 – 6.79 K ohm) @ 20°C (68°F)

5   If either of the values obtained differs markedly from the specified resistances it is likely that the coil is defective. It is recommended that the suspect coil is taken to a Yamaha dealer who can then verify the coil's condition and supply a replacement unit where necessary. The coil is a sealed unit and therefore cannot be repaired.

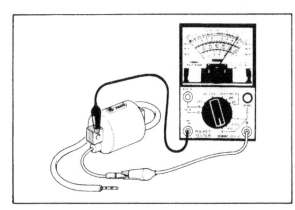
Primary coil test

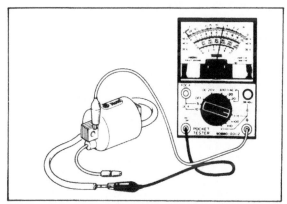
Secondary coil test

Fig. 4.3 Ignition coil resistance tests

8   CDI unit: testing

1   If the tests shown in the preceding Sections have failed to isolate the cause of an ignition fault it is likely that the CDI unit is itself faulty. Whilst it is normally possible to check this by making resistance measurements across the various terminals, Yamaha do not supply the necessary data for the LC models. It

follows that it will be necessary to enlist the help of a Yamaha dealer who will be able to check the operation of the unit by substituting a sound item.

2   In both cases the CDI unit is a sealed black box secured by a rubber mounting to protect it from the effects of vibration. It is connected to the main loom by four pins a block connector and a single snap connector in the case of the RD125 LC model, and by three single and one double snap connector in the case of the ST125 LC model.

3   On the RD125 LC model, the unit is clipped to the bottom forward face of the rear mudguard. Remove the battery as described in Chapter 7, then slacken and remove the two screws which retain the top of the battery tray to the frame. Very carefully lift the battery tray to release the bottom mounting, which is a prong fitted into a rubber grommet set in a bracket on the frame, and lift the battery tray until a hand can be inserted behind it to pick the CDI unit off its mountings. Reverse the above procedure to refit.

4   The CDI unit fitted to the DT125 LC model is much easier to reach. Remove the seat, the sidepanels, and the petrol tank as described in Section 2 of Chapter 3. The CDI unit will be seen clipped to the right-hand side of the frame top tube immediately to the rear of the finned, silver-coloured, voltage regulator and above the ignition HT coil. Disconnect the unit at its snap connectors and pull it carefully off its mounting bracket. Reverse this procedure on refitting.

9   Checking the ignition timing

RD125 LC models

1   The electronic ignition system fitted to this model cannot be adjusted in any way as ignition timing and spark advance are controlled electronically. All that can be done is to check that the system is functioning correctly.

8   The system can only be checked using a stroboscopic timing lamp or strobe, preferably one of good quality. The inexpensive neon lamps should be adequate in theory, but in practice produce a light of such low intensity that the timing marks cannot be easily seen. These lamps can also be affected by the operation of an electronic ignition system in such a way that a spurious reading can result. It is much better to purchase or borrow one of the more expensive but much more precise xenon tube lamps which are powered by an external source, ie from a separate battery or from the household mains supply.

3   Remove the gearchange linkage and the crankcase left-hand cover from the machine, connect the strobe according to its manufacturer's instructions and start the engine. Allow the engine to warm up until it will tick over smoothly at the specified idle speed of 1300 rpm, then aim the strobe at the pulser coil. The raised ignition triggering magnet should be visible immediately adjacent to the coil as shown in the accompanying illustration. Increase the engine speed to 3000 rpm and watch the magnet. It should appear to move smoothly clockwise until it is fully underneath the pulser coil. This is again shown in the accompanying illustration.

4   If any discrepancy is noted when conducting this test, take the machine to an authorised Yamaha dealer for the test to be checked on his accurate equipment. This is because any such discrepancy will indicate a fault in either the generator assembly or possibly the CDI unit, both of which are expensive to renew. As previously stated, no adjustment is possible and the only solution to a fault of any sort is to renew the generator assembly, or the CDI unit, in its entirety. It may be possible, however, to have a faulty coil rewound by an auto-electrical expert who has the necessary skill.

DT125 LC models – checking the ignition timing

5   The DT125 LC models are fitted with a less sophisticated form of CDI ignition in which the spark timing is governed by variation in magnetic flux as the magnets of the rotor pass over the poles of the ignition source coil. It is essential, therefore, that the stator is positioned correctly in relation to the rotor to achieve the necessary standard of accuracy. While the manufacturer recommends that the timing be checked at regular intervals, it is felt that this is the counsel of perfection since the ignition timing will not alter unless the stator/rotor relationship is disturbed by mechanical failure, or in the extremely rare event of a partial breakdown of the ignition source coil or the CDI unit. It should be noted that the most likely form of mechanical failure will be if carelessly tightened stator plate mounting screws work loose and allow the stator plate to move.

6   If the owner wishes to check his machine's ignition timing, whether as part of Routine Maintenance or because symptoms, such as a sudden loss of power, have developed which indicate the possible presence of a fault, the equipment necessary is as follows. In addition to the usual spanners, screwdrivers, and other hand tools, a stroboscopic timing lamp or strobe, will be required which must be one of the better quality types as described above for RD125 LC models, and also an electric tachometer suitable for a single cylinder two-stroke with electronic ignition.

7   The ignition timing is checked as follows. First remove the crankcase left-hand cover and connect the strobe and the electric tachometer according to the instructions of their respective manufacturers. The two fixed timing marks to be aligned are the mark stamped on a raised boss of the crankcase wall at about 10 o'clock from the crankshaft and the mark etched or stamped on the periphery of the rotor. These two marks should be emphasised with paint or a spirit-based felt marker to make them more visible. Start the engine and allow it to warm up until it will tick over smoothly. Check that the tachometer is functioning correctly; it should indicate an engine speed of approximately 1300 rpm. Increase the engine speed to 3000 rpm and aim the strobe at the timing marks. The mark on the rotor should align exactly with the mark on the crankcase wall, and if this is the case, the timing is correct, and the engine may be stopped so that the strobe and tachometer can be disconnected and the crankcase left-hand cover refitted. If such is not the case, the ignition timing is incorrect and the fault must be found and rectified immediately.

8   Possible causes of incorrect ignition timing are mechanical failure or electrical failure. Mechanical faults will take the form of a damaged or distorted crankshaft, a sheared Woodruff key (usually caused by an insufficiently tightened rotor retaining nut), or by the stator plate having been allowed to move due to slackened retaining screws. These are the most probable causes but other faults of a similar nature may occur depending on the use to which the machine has been put or to any work which may have been carried out in the past. Mechanical faults of any sort should be obvious on close inspection and should be rectified by referring to the relevant Sections of Chapter 1. If electrical failure is suspected, the ignition source coil and the CDI unit should be tested as described in the relevant Sections of this Chapter and rectified by the renewal of any faulty component.

DT125 LC models – setting the ignition timing

9   If new components are being fitted such as the crankshaft, flywheel rotor, or the stator plate, or if there is any reason to doubt the accuracy of the existing timing marks, the following procedure must be carried out to set the ignition timing accurately and to provide accurate timing marks for future use. In addition to the equipment listed in paragraph 6 above, the following items will be needed. A flywheel puller, Yamaha Part number 90890-01189, or a cheaper pattern version of the same, to remove the flywheel rotor easily and safely, and a dial gauge set, Yamaha Part number 90890-01252. This latter set comprises the gauge itself, an extension rod with a ball-point tip to ensure accurate measurement, and an adaptor suitable for a 14 mm spark plug thread; its use is essential to establish the correct crankshaft angle by setting the piston in a

predetermined position before top dead centre (BTDC) in the cylinder bore. It should be noted that one or two manufacturers are now advertising good quality dial gauge sets in the national motorcycle press; it is up to the owner to decide whether the expense of purchasing one is justified or not.

10 The first step in setting the ignition timing is the removal of the exhaust system from the machine, a task which is necessary to permit the dial gauge assembly to be installed and easily read. The exhaust system must be removed from the machine as described in Section 14 of Chapter 3, a procedure which will in turn require the removal of the seat, the sidepanels, and the petrol tank as described in Section 2 of the same Chapter. If the machine has just been run, take great care to avoid personal injury through burns when handling the hot exhaust components. Remove the spark plug and install the dial gauge adaptor, ensuring that it is tightened securely. Screw the gauge extension rod firmly into the gauge and insert the assembly into the adaptor, securing it by tightening the grub screws in the adaptor. Rotate the crankshaft anti-clockwise by turning the generator rotor. As the piston approaches top dead centre (TDC) the gauge reading will increase, stopping momentarily as TDC is passed and then decreasing as the piston begins to descend. Rock the crankshaft to and fro until the exact position of TDC is found, then set the gauge to read zero at this point. Move the rotor back and forth a few times to make sure that the needle does not move past zero.

11 Observing the gauge, rotate the rotor clockwise until a reading of 3 mm or 4 mm is shown, then slowly rotate it anti-clockwise again until the reading shows 2.2 mm (0.087 in). This is the piston position BTDC which establishes the correct crankshaft and rotor position to check the timing marks. At this position the mark on the periphery of the rotor must align exactly with the mark on the crankcase wall, although a tolerance of 0.15 mm (0.006 in) is allowed on either side of the set figure and any variation within these limits is permissible. If the marks do not line up, check that the piston is positioned exactly at 2.2 mm (0.087 in) and scribe or punch a new mark on the crankcase wall exactly in line with the mark on the rotor. Once it has been established that accurate timing marks exist, then these marks may be used for all subsequent timing checks.

12 Remove the generator rotor as described in Section 7 of Chapter 1, then slacken the two screws which fasten the stator plate. It will be noted that there is a mark stamped on the stator plate itself which should align exactly with the old mark on the crankcase wall. Rotate the stator plate as necessary until its

mark is exactly aligned with the new mark just punched in the crankcase wall. Tighten the stator plate fastening screws securely. To prevent confusion in the future, erase the old mark on the crankcase wall as much as possible and emphasise the new mark using paint or a felt marker.

13 Refit the generator rotor, tightening its retaining nut to a torque setting of 5.0 kgf m (36 lbf ft) and use the dial gauge to check that the timing marks align exactly. Repeat the above procedure if any discrepancy has crept in. When the timing marks align correctly, remove the dial gauge assembly and refit the spark plug, the exhaust system and the petrol tank, the sidepanels and the seat as described in Sections 2 and 14 of Chapter 3.

14 Connect up the strobe and electric tachometer, start the engine and carry out the ignition timing check described in paragraph 7 above. The timing marks should align exactly at 3000 rpm. Stop the engine, disconnect the strobe and the electric tachometer, and refit the crankcase left-hand cover.

9.3 Note raised ignition trigger on RD125 alternator rotor

9.7 Mark on rotor wall must align with mark on crankcase – DT125 only

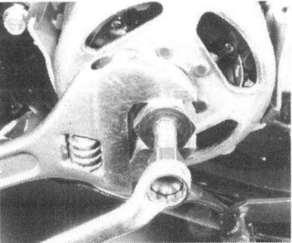

9.12a Flywheel puller being used to remove DT125 rotor ...

9.12b ... the same tool can be used on the RD125

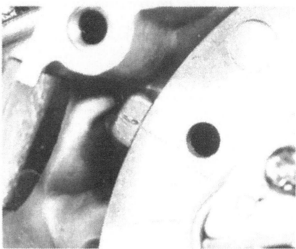

9.12c Slacken stator screws and move stator so that timing marks align

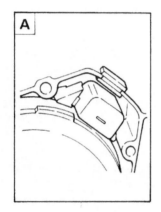

Fig. 4.4 Checking the ignition timing – **RD model**

A   At 1300 rpm, 17° BTDC   B   At 3000 rpm, 30° BTDC

---

10 Spark plug: checking and resetting the gap

1   Both models described in this Manual are fitted as standard with an NGK BR8ES spark plug. In most operating conditions the standard plug should prove satisfactory. However, alternatives are available to allow for varying altitudes, climatic conditions, and the use to which the machine is put. For example constant high-speed riding for extended mileages will require the use of a spark plug of a colder heat range than standard to cope with the excess heat generated, ie the BR9ES. Similarly excessive low-speed riding will require the use of a spark plug of a hotter heat range than standard to burn away the carbon and oil deposits which will result from such use, ie the BR7ES.

2   Note however that the advice of an authorised Yamaha dealer or similar expert should be sought before the plug heat range is altered from standard. The use of too cold, or hard, a grade of plug will result in plug fouling and the use of too hot, or soft, a grade of plug will result in engine damage due to the excess heat being generated. If the correct grade of plug is fitted, however, it will be possible to use the condition of the spark plug electrodes to diagnose a fault in the engine or to decide whether the engine is operating efficiently or not. The accompanying series of   photographs will show this clearly.

3   The final point on the selection of spark plug types and grades concerns the letter 'R' in the prefix mentioned above. This indicates that the plug has a resistor built in to suppress the radio and TV interference produced by the HT pulse, the manufacturers having decided that it is better to suppress the interference as close to its source as possible, ie in the spark plug itself. The ignition system has been designed and constructed with this type of spark plug in mind; its reliability and performance would be adversely affected if a standard, non-resistor spark plug were substituted for the correct type, therefore not only must the spark plug be of the correct heat range, it must also be a resistor type at all times.

4   It is advisable to carry a new spare sparking plug on the machine, having first set the electrodes to the correct gap. Whilst sparking plugs do not fail often, a new replacement is well worth having if a breakdown does occur. Ensure that the spare is of the correct heat range and type.

5   The correct electrode gap is 0.7 – 0.8 mm (0.028 – 0.031 in). The gap can be assessed using feeler gauges. If necessary, alter the gap by removing the outer electrode, preferably using a proper electrode tool. **Never** bend the centre electrode, otherwise the porcelain insulator will crack, and may cause damage to the engine if particles break away whilst the engine is running.

6   Before refitting a spark plug into the cylinder head coat the threads sparingly with a graphited grease to aid future removal.

Electrode gap check - use a wire type gauge for best results

Electrode gap adjustment - bend the side electrode using the correct tool

Normal condition - A brown, tan or grey firing end indicates that the engine is in good condition and that the plug type is correct

Ash deposits - Light brown deposits encrusted on the electrodes and insulator, leading to misfire and hesitation. Caused by excessive amounts of oil in the combustion chamber or poor quality fuel/oil

Carbon fouling - Dry, black sooty deposits leading to misfire and weak spark. Caused by an over-rich fuel/air mixture, faulty choke operation or blocked air filter

Oil fouling - Wet oily deposits leading to misfire and weak spark. Caused by oil leakage past piston rings or valve guides (4-stroke engine), or excess lubricant (2-stroke engine)

Overheating - A blistered white insulator and glazed electrodes. Caused by ignition system fault, incorrect fuel, or cooling system fault

Worn plug - Worn electrodes will cause poor starting in damp or cold weather and will also waste fuel

Use the correct size spanner when tightening the plug otherwise the spanner may slip and damage the ceramic insulator. The plug should be tightened by hand only at first and then secured with a quarter turn of the spanner so that it seats firmly on its sealing ring. If a torque wrench is available, tighten the plug to a torque setting of 2.0 kgf m (14.5 lbf ft).

7   Never overtighten a spark plug otherwise there is risk of stripping the threads from the cylinder head, especially as it is cast in light alloy. A stripped thread can be repaired without having to scrap the cylinder head by using a 'Helicoil' thread insert. This is a low-cost service, operated by a number of dealers.

10.3 Always use recommended make and type of spark plug

11 Spark plug (HT) lead and suppressor cap: examination

1   Erratic running faults and problems with the engine suddenly cutting out in wet weather can often be attributed to leakage from the high tension lead and spark plug cap. If this fault is present, it will often be possible to see tiny sparks around the lead and cap at night. One cause of this problem is the accumulation of mud and road grime around the lead, and the first thing to check is that the lead and cap are clean. It is often possible to cure the problem by cleaning the components and sealing them with an aerosol ignition sealer, which will leave an insulating coating on both components.

2   Water dispersant sprays are also highly recommended where the system has become swamped with water. Both these products are easily obtainable at most garages and accessory shops. Occasionally, the suppressor cap or the lead itself may break down internally. If this is suspected, the components should be renewed.

3   Where the HT lead is permanently attached to the ignition coil, it is recommended that the renewal of the HT lead is entrusted to an auto-electrician who will have the expertise to solder on a new lead without damaging the coil windings.

4   When renewing the suppressor cap, be careful to purchase one that is suitable for use with resistor spark plugs. The traditional method of suppressing HT interference was to incorporate a resistor into the spark plug cap; now that the resistor is sited in the spark plug itself, suppressing any interference at its source, the suppressor caps have much lower resistance values and the ignition system is designed with this in mind. If the original suppressor cap should be replaced by an old-style item with its high resistance, the combination of the heavy resistances in both spark plug and cap would overload the ignition system components, producing hard starting and misfiring initially, and ultimately causing the failure of the ignition system.

Chapter 5 Frame and forks

For modifications, and information relating to later models, see Chapter 8

Contents

Specifications

	RD125 LC	DT125 LC
Frame		
Type	Welded tubular steel	
Front forks		
Type	Oil damped, coil sprung telescopic	
Wheel travel	140 mm (5.5 in)	230 mm (9.1 in)
Fork spring free length:		
Main spring	484.60 mm (19.08 in)	591.49 mm (23.29 in)
Service limit	477.0 mm (18.8 in)	Not available
Secondary spring	Not applicable	51.11 mm (2.01 in)
Service limit	Not applicable	Not applicable
Fork oil capacity – per leg	147 cc (5.17 fl oz)	304 cc (10.7 fl oz)
Fork oil level	180 mm (7.09 in)	(168.4 mm (6.63 in)
Recommended fork oil	SAE 10 fork oil	SAE 10W/30 SE engine oil
Rear suspension		
Type	Cantilever (Yamaha Monocross)	
Wheel travel	120 mm (4.72 in)	200 mm (7.87 in)
Rear suspension unit		
Type	De Carbon, coil sprung, gas/oil damped	
Travel	55 mm (2.17 in)	94 mm (3.70 in)
Spring free length	234.5 mm (9.23 in)	290 mm (11.42 in)
Wear limit	232.5 mm (9.15 in)	N/A
Gas pressure	15 kg/cm² (213 psi)	14 kg/cm² (199 psi)
Rear sub-frame		
Sub-frame/frame maximum clearance (endfloat)	1.0 mm (0.039 in)	
Sub-frame maximum free play	1.0 mm (0.039 in)	

Torque settings

Component	kgf m	lbf ft
Engine mounting bolts	2.5	18.0
Front-wheel spindle nut:		
RD125 LC	7.5	54.0
DT125 LC	4.0	29.0
Top and bottom yoke pinch bolts	2.5	18.0
Damper rod Allen bolt	2.0	14.5
Fork crown bolt:		
RD125 LC	4.0	29.0
DR125 LC	5.4	39.0
Handlebar clamp bolts	1.5	11.0
Brake caliper/fork lower leg mounting bolts	3.5	25.0
Rear suspension unit upper mounting bolt	2.5	18.0
Rear suspension unit preload adjuster locknut –		
DT125 LC only	5.5	40.0
Sub-frame pivot bolt	4.3	31.0
Brake torque arm/sub-frame mounting bolt –		
RD125 LC only	2.3	17.0
Rear wheel spindle nut	8.5	61.5
Brake torque arm/brake backplate nut –		
RD125 LC only	2.0	14.5
Footrest plate mounting bolts – RD125 LC only	6.5	47.0
Pillion footrest mounting bolts – RD125 LC only	3.0	22.0

1 General description

The Yamaha RD/DT125 LC models use conventional welded tubular steel frames. The front wheel is supported by oil-damped coil spring telescopic suspension units. Rear suspension is by a cantilever assembly, pivoting on plain bushes, supported by a single suspension unit. The unit is coil sprung with nitrogen under high pressure as a supplementary springing medium. Damping is hydraulic.

2 Front forks: removal and refitting

1   It is unlikely that the front forks will need to be removed from the frame together with the fork yoke and steering stem. Although feasible, this method is both unwieldy and time consuming. It is far easier to remove the individual fork legs, then the steering head assembly, where necessary, as described in Section 3 of this Chapter.

RD125 LC models
2   Place the machine on its centre stand. Make sure that it is standing securely on a flat surface, then place blocks beneath the front of the engine to raise the front wheel clear of the ground. Unscrew the knurled retaining ring which secures the speedometer drive cable to its gearbox on the front wheel. Pull the cable clear and lodge it clear of the wheel and forks.
3   Straighten and remove the split pin which secures the wheel spindle nut. Slacken and remove the nut, then support the wheel while the spindle is displaced and removed. If necessary, tap the spindle through but take care not to damage its threads. Once the spindle has been removed the wheel can be lowered to the ground and removed from the machine.
4   It is good practice to fit a wooden wedge between the brake pads to prevent the caliper piston from being displaced should the brake lever be squeezed while the wheel is removed. Slacken and remove the two brake caliper/fork lower leg mounting bolts and then the single bolt which secures the brake hose to the upper part of the fork lower leg. Tie the brake caliper and hose to the frame so that both are well clear of the forks. Slacken and remove the four front mudguard mounting bolts and lift the mudguard away.
5   If the forks are to be dismantled it is better to release the fork top plugs at this stage. It can be done with the forks removed, as shown in the photograph accompanying this

Section, but for convenience they should be removed while the fork stanchions are secured in the yokes. Start by prising off the black plastic caps which cover the stanchion tops. This will reveal the fork plugs which will be found to be retained by a wire internal circlip.
6   Try to obtain some assistance at this stage. The plugs are removed by pushing them down against fork spring pressure and the easiest method is to use a cross-point screwdriver in the central depression in the cap. While the cap is held down it will be necessary for the second person to dislodge the retaining clip with a small electrical screwdriver. Once this has been removed the plug can be slowly released and will be pushed out of the stanchion. This operation can be accomplished unaided, but it is best to have two pairs of hands if possible.
7   Slacken the pinch bolts on the top and bottom fork yokes and pull the fork legs downwards out of the yokes. Twist the stanchions by hand only, to free them, and never risk damaging the chromed surface of the stanchions by gripping them with a metal tool. If the stanchions are stuck in the fork yokes by accident damage or corrosion, there are several methods of freeing them. The first is to apply a liberal dose of penetrating fluid at the point where the fork yokes grip each stanchion, and then to rotate each stanchion by hand to free it. If this fails it is permissible gently to tap a screwdriver blade into the slot in the clamps of each yoke, and to open up the clamp very slightly to free the stanchion. If this method is used, the pinch bolts must be removed fully and great care taken not to overstress the clamp by forcing it too far. If all else fails, find a smooth round metal bar of the same diameter as the wheel spindle (but never use the spindle itself), insert this through the wheel spindle lug of the fork leg to be removed, and strike it sharply downwards with a suitable hammer.
8   If drastic measures are needed to remove the fork legs from the yokes, remember to pay particular attention to cleaning the respective mating surfaces of the yokes and the stanchions and to apply a smear of grease to those surfaces on refitting, both as an aid to easy fitting and to prevent the onset of corrosion in the future.
9   The fork legs are refitted by reversing the removal sequence described above. Slide each fork leg into position, ensuring that the top of each stanchion is flush with the top surface of the top yoke. Tighten the top and bottom yoke pinch bolts by just enough to retain each stanchion. When rebuilding the front end, especially if the steering head assembly has been dismantled, it is important to ensure that the forks are correctly aligned when refitted. To this end, assemble the forks loosely, refit the front wheel, then bounce the forks a few times so that the various components assume their correct relative positions. Tighten the

pinch bolts, working from the wheel spindle upwards. Refit the
front wheel as described in Section 4 of Chapter 6.
10 Note the following torque settings when assembling the
forks and calipers.

Fork yoke pinch bolts	2.5 kgf m (18 lbf ft)
Wheel spindle nut	7.5 kgf m (54 lbf ft)
Caliper mounting bolts	3.5 kgf m (25 lbf ft)

DT125 LC models

11 Before any work can be carried out the machine must be
supported securely in the upright position on a suitable stand,
so that the front wheel is clear of the ground. The usual
procedure is to place a stout wooden box underneath the
engine/gearbox unit so that the machine rests on its crankcase
bashplate. Any variation on this theme is permissible, making
use of wooden or concrete blocks, milk crates, or anything else
which comes to hand, but make sure that the machine is quite
secure and will not fall over.
12 Remove the front wheel from the machine as described in
Section 4 of Chapter 6. Release the brake cable from the upper
part of the fork lower leg by slackening and removing the single
clamping bolt, then either remove the cable completely from the
machine or tie it to the frame so that it is clear of the forks.
13 If the forks are to be dismantled it is advisable to release the
fork top plugs at this stage. This task is the same as that
described in paragraphs 5 and 6 for the RD125 LC models and
should be carried out in the same way. Slacken the pinch bolts
in the top and bottom yokes, then slacken the gaiter upper
clamp on each leg so that the gaiter can be pulled down and the
stanchion exposed. Pull the stanchion downwards out of the
yokes, noting that if any difficulty is encountered the same
methods described above for the RD125 LC models can be
employed to release each stanchion.
14 On refitting, smear a light coating of grease over the entire
surface of each stanchion, including that part which will be
covered by a gaiter when installed, and push the stanchions up
through the yokes. Each stanchion should project 10 mm (0.39
in) above the top surface of the top yoke, and the tops of the
two stanchions must be exactly level with each other. Tighten
the top and bottom yoke pinch bolts by just enough to retain
each stanchion and refit the front wheel as described in Section
4 of Chapter 6. Apply the front brake and bounce the forks a few
times so that all the components align correctly and easily;
there can then be no question of a component being stiff to
move or of its wearing unevenly because it is awkwardly
clamped. Tighten the front wheel spindle nut to a torque setting
of 4.0 kgf m (29 lbf ft), taking care to keep the front brake lever
firmly applied whilst so doing, and then working from the
bottom upwards, tighten the fork yoke pinch bolts to a torque
setting of 2.5 kgf m (18 lbf ft), taking care that the brake cable
guides are correctly aligned.
15 Adjust the front brake and secure the cable to the fork lower
leg by the clamp provided then slide each fork gaiter up its
stanchion to butt against the fork bottom yoke and fasten it in
that position by tightening securely the clamp.

All models

16 If work is undertaken at any time on the front forks and
steering head assembly of either of the machines described in
this Manual, the following points must be checked before the
machine is taken out on the road, and any faults found must be
rectified before the machine is used. Make a thorough test of
the brakes and the suspension action, and ensure that all nuts
and bolts are securely fastened and that the split pin securing
the wheel spindle nut is in good condition and correctly fitted.
Check that any control cables that were disturbed have been
correctly connected again and properly adjusted, also that all
electrical circuits have been connected and are now functioning
correctly. Finally, if the steering head has been dismantled, do
not forget to make any final adjustments that may be necessary
to the steering head bearings themselves.

2.4 Slacken and remove mudguard mounting bolts

2.5 Fork spring is secured by a plug and a circlip

2.7a Slacken fork yoke pinch bolts ...

2.7b ... and pull fork leg down with a twisting motion to release

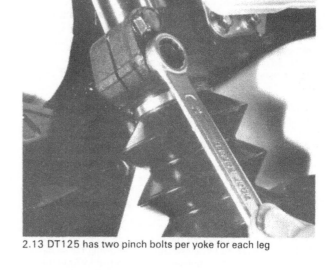

2.13 DT125 has two pinch bolts per yoke for each leg

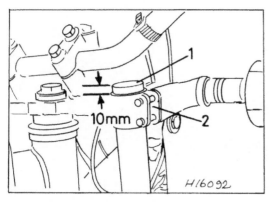

Fig. 5.1 Correct position of front fork leg in top yoke - DT model

1 Fork stanchion	2 Top yoke

3   Steering head assembly: removal and refitting

1   When working on either of the machines described in the Manual, it will be necessary to complete a certain amount of preliminary dismantling before the steering head assembly can be removed. The front wheel must be removed, as described in Section 4 of Chapter 6, followed by the removal of the front forks, as described in the previous Section of this Chapter. It is recommended that the seat, the sidepanels, and the petrol tank be removed also, to protect the painted finish of the petrol tank during the course of the work, this operation being described in Section 2 of Chapter 3. At the very least, the tank should be covered with a thick blanket or similar padding to protect it. Disconnect the battery to prevent short circuits as the wiring is disconnected.

2   Although the full sequence of dismantling will be given below, it should be noted that certain short cuts may be made, depending on the reason for the work being undertaken. For example, if the steering head is being dismantled to grease the bearings during the course of Routine Maintenance, it would not be absolutely necessary to remove the instrument panel from the top yoke, or, in the case of the DT125 LC model, to remove the front mudguard from the bottom yoke. It is, of course, up to the owner to decide how much of the dismantling procedure is to be avoided in this way.

3   When removing components from the machine, take great

care to refit all nuts, bolts, washers and mounting rubbers in their correct respective positions to avoid the loss of any small but important components and to prevent any possibility of confusion on reassembly. Have a pen and paper ready so that notes can be made of the correct position of components or of any other points which must be remembered on rebuilding.

RD125 LC models

4   The fairing fitted to these models is retained by two separate brackets which are bolted to the headlamp bracket. Slacken and remove the two bolts, one on each side, which secure the fairing to its brackets, then gently pull the bottom edge of the fairing forwards off the two rubber grommets which are mounted on lugs protruding upwards from the lower part of the headlamp bracket. Put the fairing to one side.

5   Slacken and remove the two screws which secure the headlamp rim to the headlamp shell, then withdraw the rim and light unit, disconnecting the headlamp wiring so that the unit can be withdrawn completely from the machine. Disconnect all the electrical connectors and connector blocks inside the headlamp shell, withdrawing each wire or lead as it is freed, so that any individual electrical component can be removed easily if desired. Allow the leads from the handlebar switches to hang down clear of the machine. Using a suitable pair of pliers, slacken the knurled rings securing the instrument drive cables to the bases of the instruments and withdraw each cable by pulling it downwards away from its respective instrument. Remove the speedometer cable completely from the machine and allow the tachometer drive cable to hang down clear of the steering head area.

6   Slacken and remove the two bolts which secure the headlamp bracket to the bottom yoke, then pull the bracket out slightly at the bottom and pull it downwards out of the two damping rubbers set in the top yoke. Check the condition of these rubbers and renew them if they are at all worn or perished. Put the headlamp bracket to one side, complete with the fairing brackets, headlamp shell, horn, and front flashing indicator lamp units. These latter components can then be removed from the headlamp bracket.

7   Slacken and remove the two bolts which secure the instrument panel to the top yoke, then withdraw the panel assembly. The ignition switch is also secured to the underside of the top yoke by two bolts, and can be removed at this stage if this is required. Slacken and remove the four bolts which secure the two handlebar clamps to the top yoke, then remove the two clamps and lift the handlebars away from the top yoke. Move the handlebars backwards to rest on top of the frame or on top of the petrol tank, if the latter is suitably covered.

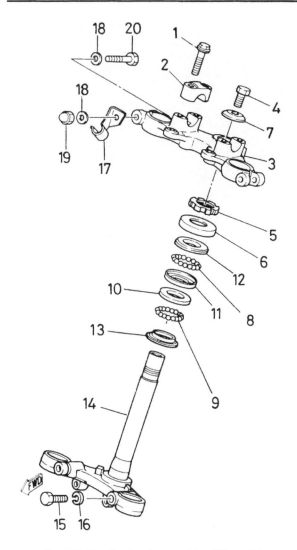

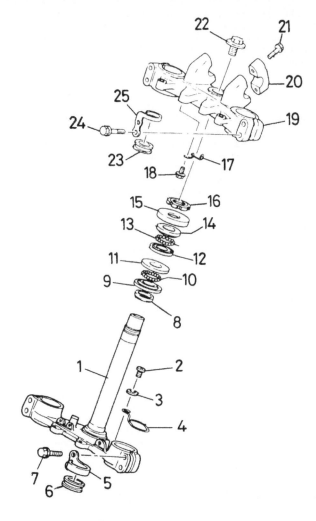

Fig. 5.2 Steering head assembly - RD model

1	Bolt - 4 off	11	Upper bearing cup
2	Handlebar clamp - 2 off	12	Upper bearing cone
3	Top yoke	13	Lower bearing cone
4	Crown bolt	14	Steering stem/lower yoke
5	Adjusting nut	15	Bolt - 2 off
6	Dust cover	16	Spring washer - 2 off
7	Washer	17	Cable guide
8	Upper bearing balls	18	Washer - 4 off
9	Lower bearing balls	19	Domed nut - 2 off
10	Lower bearing cup	20	Bolt - 2 off

Fig. 5.3 Steering head assembly - DT model

1	Steering stem/lower yoke	14	Upper bearing cone
2	Screw	15	Dust cover
3	Spring washer	16	Adjusting nut
4	Cable guide	17	Cable guide
5	Cable guide	18	Bolt
6	Grommet	19	Top yoke
7	Bolt - 2 off	20	Handlebar clamp
8	Dust seal	21	Bolt - 4 off
9	Lower bearing cone	22	Crown bolt
10	Lower bearing balls	23	Grommet
11	Lower bearing cup	24	Bolt
12	Upper bearing cup	25	Cable guide
13	Upper bearing balls		

DT125 LC models

8   Slacken and remove the three bolts which secure the headlamp fairing to its mounting brackets, these latter being bolted in their turn to the top and bottom yokes. Two bolts pass through the upper part of the headlamp fairing, one each side, and the third is situated on the left-hand side just below the light unit. Withdraw the headlamp fairing far enough for the headlamp wiring to be disconnected, so that the fairing can be removed fully, complete with light unit and bulbs. Disconnect all the wiring connectors and connector blocks, withdrawing each lead or wire as it is freed, so that the various electrical components can be removed separately if required. Allow the leads from the handlebar switches to hang down clear of the machine. Using a suitable pair of pliers, unscrew the knurled

ring securing the drive cable to the speedometer and remove the cable from the machine.

9   The instrument panel is secured by two bolts and nuts which pass through the top yoke, and the ignition switch is fastened by two bolts which pass upwards into the top yoke. These two assemblies may be removed, if desired, by unscrewing their respective bolts or nuts. The handlebars are secured to the top yoke by two clamps. Slacken and remove the four bolts which pass through these clamps and into the top yoke. The clamps can then be removed and the handlebars lifted clear, moved backwards clear of the steering head area and secured to the frame so that they are out of the way.

10 This will leave the front mudguard and the horn bolted in place on the bottom yoke, and the front flashing indicator

lamps, the headlamp fairing brackets, and cable guides on the pinch bolts of the top yoke. There is no need to remove any of these components during the course of dismantling, and they should only be unbolted if renewal or attention of any sort is necessary.

All models

11 Once all the necessary preliminary dismantling has been carried out, slacken the fork crown bolts situated centrally in the top yoke, then remove the bolt and the yoke. Before proceeding further, it should be noted that the steering head bearings are of the uncaged cup and cone type, containing a number of steel balls in each race which will drop free as the bottom yoke is removed; some means of catching the balls must be devised therefore. It is a good idea to place a clean plastic bag or a piece of clean rag under the steering head area to catch the balls as they drop.

12 Using a pin spanner or C-spanner, slacken and remove the steering head adjusting nut, supporting the bottom yoke all the time to ensure that it does not drop clear. Remove the thin metal cover which protects the top race from the entry of dust or dirt, and carefully lower the bottom yoke away from the steering head. If sufficient care is taken the top cone will not be disturbed as the steering stem passes through it and attention can be devoted solely to the task of collecting the steel balls of the steering head bottom race. Collect all these in a clean container, noting that both the models described in this Manual have nineteen $\frac{1}{4}$ in (No 6) steel balls fitted to the bottom race.

13 When the bottom yoke has been removed and the balls of the bottom race collected, remove the top cone and collect all the balls of the steering head top race, noting that while the RD125 LC model is fitted with the same as in the bottom race, ie nineteen $\frac{1}{4}$ in (No 6) steel balls, the DT125 LC model differs in having twenty-two $\frac{3}{16}$ in (No 4) steel balls fitted.

14 On reassembly, fit the bottom yoke and adjust the steering head bearings as described in Section 4 of this Chapter. Temporarily slide the two fork legs into position in the bottom yoke, tightening the bottom yoke pinch bolts by just enough to retain the fork legs in position. Install the top yoke, then fit and hand-tighten the fork crown bolt and the top yoke pinch bolts. Check that the fork legs are correctly aligned and tighten the fork crown bolt to a torque setting of 4.0 kgf m (29 lbf ft) on RD125 LC models, or 5.4 kgf m (39 lbf ft) on DT125 LC models.

15 The reassembly procedure is then a straightforward reversal of the removal sequence described in this Section and in Section 2 of this Chapter. Where such settings are given, use a torque wrench to tighten the relevant nuts and bolts. Note that the handlebar upper mounting clamps must be fitted so that they are exactly parallel to the matching surfaces of the top yoke, ie the gaps at the front and the rear between each clamp and the top yoke must be of equal size. Tighten the four handlebar clamp bolts in stages to achieve this, finishing off by tightening each bolt to a torque setting of 1.5 kgf m (11 lbf ft). If no torque wrench is available, these bolts must be tightened just enough to retain the handlebars securely, or else the clamps will be overstressed, with a consequent risk of their cracking in use.

16 Pay careful attention to giving control cables and wiring leads easy smooth runs, this task being made much easier if the handlebar and controls have been correctly aligned. Refer to the colour-coding of the electrical wiring shown in the diagram at the back of this book when connecting the wiring. Do not forget to connect the battery again and to test all the circuits before using the machine on the road.

17 Before tightening the front wheel spindle nut, check that the front forks are correctly aligned, apply the front brake and push up and down on the handlebars to operate the front suspension and to ensure that all the fork components are settled in place. Using a torque wrench where applicable, tighten first the wheel spindle nut, then the mudguard bolts (RD125 LC only), and the fork yoke pinch bolts, working from the bottom upwards as described in Section 2 of this Chapter.

3.4a RD125 fairing is fastened by two bolts ...

3.4b ... and by two grommets

3.9 DT125 instrument panel is secured by two bolts

4 Steering head bearings: examination, refitting and adjustment

1   Before commencing reassembly of the forks examine the steering head races. They are unlikely to wear out under normal circumstances until a high mileage has been covered. If, however, the steering head bearings have been maladjusted, wear will be accelerated. If, before dismantling, the forks had a pronounced tendency to stick in one position when turned, often in the straight ahead position, the cups and cones are probably indented and need to be renewed.

2   Examine the cups and cones carefully; it is not necessary to remove the cups for this. The bearing tracks should be polished and free from indentations, cracks or pitting. If signs of wear are evident, the cups and cones must be renewed. In order for the straight line steering on any motorcycle to be consistently good, the steering head bearings must be absolutely perfect. Even the smallest amount of wear on the cups and cones may cause steering wobble at high speeds and judder during heavy front wheel braking. The cups are an interference fit on their respective seatings and can be tapped from position using a suitable long drift. The top cone, as already mentioned, is lifted away in the course of dismantling and is as easily replaced when rebuilding. The bottom cone, however, is a tight fit on the steering stem. Clamp the bottom yoke in a soft-jawed vice to hold it securely without damaging the painted finish and, using two screwdrivers or tyre levers, carefully lever the bottom cone away from its seating. Take great care not to damage the rubber dust seal which is situated underneath the bottom cone of the DT125 LC model only. Once the bottom cone is removed from its seating, it can be pulled easily off the steering stem. Carefully examine the rubber dust seal and renew it if worn, to prevent the entry of dirt into the steering head bearings.

3   If working on a DT125 LC model, do not forget that the rubber dust seal is fitted on the steering stem underneath the bottom cone. Position them carefully and use a hammer and a long tubular drift to tap the bottom cone on to its seating. Great care must be taken not to damage the bearing track of the bottom cone. Use a drift which will bear only on the inside shoulder of the cone to avoid such damage occurring. Tap the cups firmly on to their respective seatings in the frame using a suitably sized socket as a drift. Again take great care not to damage the bearing track surfaces.

4   The ball bearings themselves should be cleaned using paraffin. If found to be marked, chipped or discoloured in any way they should be renewed as a complete set.

5   On reassembly pack the bottom cone and top cup liberally with grease, and place the bearings in position, using the grease to hold them in place. Both models described in this Manual are fitted with nineteen $\frac{1}{4}$ in (No 6) steel balls in their steering head lower races but whereas the RD125 LC model has the same number and size of balls, ie nineteen $\frac{1}{4}$ in (No 6) fitted to its upper race, the DT125 LC model differs in having twenty-two $\frac{3}{16}$ (No 4) steel balls fitted to its upper race. Note that in all cases, when the set number of balls has been fitted, a gap will be left which is the size of one more ball. This is intended, as some clearance is needed to prevent the balls skidding against one another and accelerating the rate of wear.

6   With the balls held in place as described, very carefully pack the bottom cup with grease and fit the bottom yoke. Support it with one hand and pack the area around the upper bearing with grease. Replace the top cone, ensuring that the balls do not become displaced in the process, and the thin metal collar, then thread the adjusting nut down into position. Using a pin spanner or C-spanner, tighten down the nut until it seats lightly. Do not overtighten it. Turn the nut back through $\frac{1}{8}$ turn from its lightly tightened position. This will serve as the basis for correct adjustment of the steering head bearings, which operation is now described in full. Remember that it can be carried out effectively only with the complete front fork assembly fitted, and should not be forgotten, therefore, during the rebuilding procedure.

7   The steering head bearings, when adjusted correctly, will have all traces of free play eliminated from them without being pre-loaded in any way. To check this, the machine must be placed on its centre stand with the front wheel raised clear of the ground by means of a box or other support underneath the engine. With the forks in the straight ahead position, grasp a fork lower leg in the area of the wheel spindle with one hand and attempt to push and pull the forks backwards and forwards. Any free play will be felt immediately by the fingers of the other hand between the bottom yoke and the frame around the lower steering head bearing. Any such free play must be just eliminated by slackening the fork crown bolt above the top yoke and by tightening the adjusting nut immediately under the top yoke, using a C-spanner. To check for overtightened steering head bearings, position the machine as above, and push lightly on one handlebar end. The front forks should smoothly and easily fall away to the opposite lock. Any signs of stiffness or notchiness will be apparent immediately and should be removed by slackening off the adjusting nut. After adjusting the steering head bearings, tighten the fork crown bolt to a torque setting of 4.0 kgf m (29 lbf ft) on RD125 LC models, and to 5.4 kgf m (39 lbf ft) on DT125 LC models, then recheck the bearing adjustment as described above to ensure that it has remained the same.

8   Note that if it is impossible to adjust the bearings correctly, it must be assumed that the steel balls or the bearing tracks are damaged in some way. The bearings should then be stripped for examination and renewal of the affected components.

4.7 Steering head bearings are adjusted using a C-spanner on adjusting nut

5 Fork yokes: examination and renovation

1   To check the top yoke for accident damage, push the fork stanchions through the bottom yoke and fit the top yoke. If it lines up, it can be assumed the yokes are not bent. Both must also be checked for cracks. If they are damaged or cracked, fit new replacements.

6 Fork legs: dismantling and reassembly

RD125 LC models

1   Having removed the forks from the yokes as described in Section 2 withdraw the fork springs and invert each leg over a drain tray until the damping oil has emptied. It is assumed that the fork top plugs were removed as described prior to removal

of the stanchions from the fork yokes. If the plugs are still in position they can be removed by clamping the stanchion in a vice fitted with soft jaws. Take care not to crush or score the stanchion. Depress the plug with a cross-point screwdriver and prise out the wire circlip which retains it. Release the pressure on the plug, allowing it to be displaced by the fork spring. The fork leg can now be removed and drained as described above.

2   Once the oil has been drained, slacken the Allen bolt which passed up through the bottom of the lower leg and into the damper rod. It is quite likely that the damper rod will tend to rotate in the lower leg and thus impede the removal of the bolt. If this problem arises, clamp the assembly in a vice using soft jaws to hold the lower leg by the caliper mounting lugs. Obtain a length of wooden dowel about $\frac{1}{2}$ in in diameter and form a taper on one end.

3   Pass the dowel down the stanchions, having first withdrawn the fork spring. Push the dowel hard against the head of the damper rod to lock it in position while an assistant slackens the retaining bolt. If the dowel proves difficult to hold, a self-grip wrench or similar can be used to obtain sufficient leverage and pressure. Once the retaining bolt has been removed the fork stanchion can be pulled out of the lower leg. Invert the lower leg to tip out the damper rod seat.

4   Invert the stanchion to release the damper rod and rebound spring, placing these components to one side to await examination. Unless new oil seals have been fitted recently they should be renewed as a matter of course to prevent failure at a later date. Lever out the dust seal, using a suitably-shaped screwdriver and taking great care not to damage the seal or the fork leg, then prise out the wire retaining clip using a small screwdriver.

5   The oil seal can be levered out of its recess in the top of the lower leg as described above for the dust seal. The oil seal lip should be lubricated prior to installation, and then tapped into place using a large tubular drift to ensure that it enters the leg squarely. A large socket is ideal for this purpose. Remember to fit the retaining clip, ensuring that it locates correctly. Note that if the seal is a particularly tight fit, whether on removal or refitting, the task can be made easier by carefully pouring boiling water over the fork leg exterior in the area of the seal. This will cause the alloy of the lower leg to expand, loosening the seal in its housing. Take great care not to damage the fork leg during removal or refitting of the seals. When the retaining circlip has been refitted to locate the fork oil seal, tap the dust seal into place using the same tubular drift or socket.

6   After the fork components have been examined for wear as described in Section 7, reassembly can commence. Make sure that all components are fitted to the leg from which they were removed. Slide the damper rod and rebound spring into the stanchion, then lubricate the stanchion surface with oil. Slide it into the lower leg making sure that the seal lip does not become distorted or damaged. Lock the damper rod, then fit the retaining bolt and its sealing washer, tightening it securely to a torque setting of 2.0 kgf m (14.5 lbf ft). It is recommended that the threads of the Allen bolt be degreased and coated with Loctite to provide an additional measure of security against the bolt's working loose.

7   Fill each leg with 147 cc (5.17 fl oz) of fork oil. The grade of oil recommended by Yamaha is SAE 10 fork oil with SAE 10W engine oil being an acceptable alternative. Variations in the damping effect of the oil can be achieved by using a lighter or a heavier grade of oil as desired. Most of the proprietary brands of fork oil are sold in different grades to enable the owner to tune the suspension of his machine in this way.

8   When the fork leg has been filled with the correct amount of oil, pump the stanchion gently up and down to distribute the oil around the leg, finishing by compressing the stanchion as far as possible into the lower leg. Measure the level of the oil below the top edge of the stanchion using a length of welding rod and a ruler, as shown in the accompanying photographs. The correct level is 180 mm (7.09 in). Add or remove oil to raise or lower the level if it is not correct. Measurement of the oil level makes it possible for the oil level in the fork legs to be checked easily

and topped up, if leakage has occurred, without the necessity for dismantling and draining the forks.

9   When the oil level is known to be correct, pull the stanchion out to its fullest extent and refit the fork spring, noting that the end which must be fitted first, ie at the bottom of the leg, is the end with the wider-spaced coils which are tapered inwards slightly. Secure the spring with the fork top plug and its retaining circlip, then refit the black plastic cap.

DT125 LC models

10   If the fork top plug was not released during fork leg removal, it must be removed now. Lever out the black plastic cap, push down the plug with a cross-point screwdriver and prise out the wire circlip, noting that it will be necessary to hold the stanchion while this is done. The easiest method is to clamp the leg in the bottom fork yoke, or failing this, to use a strap wrench. It is not advisable to clamp the stanchion in a vice because of the risk of damage due to scoring or overtightening. Use of this method is permissible, however, if soft jaw padding is used and if the vice is not overtightened. Be very careful when removing the plug as the fork spring is under tension and will push the plug clear with some force. It follows that firm hand pressure must be applied to the screwdriver to counter this.

11   Lift out the fork secondary, or upper spring, noting precisely which way round it is fitted. It is useful to degrease and to mark the top surface of the spring, using paint or a spirit-based felt marker, so that the spring is refitted the correct way round on reassembly. Remove the thick washer which acts as a spacer between the two fork springs, marking its top surface as described above, and then pull out the fork main, or lower, spring markings its top surface in the same fashion. Put the springs, spacer and top plug to one side and release the gaiter top securing clamp. The gaiter can then be pulled off the fork lower leg and slid off the stanchion. Invert the fork leg over a drain tray and allow the fork oil to run out. Pump the stanchion in and out to assist this until all the oil has fully drained. Repeat the above procedure on the remaining leg, but note that each leg should be dismantled and reassembled separately to avoid interchanging components which would cause increased wear.

12   Wrap some rag around the lower leg and clamp the assembly in a vice, taking care not to overtighten and thus distort the lower leg. Using an Allen key, slacken and remove the damper bolt from the recessed hole in the bottom of the lower leg. This will often cause some difficulty because the bolt threads are coated in a locking compound and once slightly loose there is a tendency for the damper rod to turn in its seat. To overcome this problem a length of wooden dowel can be employed to hold the damper rod. Grind a coarse taper on one end of the dowel and insert it down the bore of the stanchion so that it engages in the recessed head of the damper rod.

13   It will now be necessary to push on the dowel to obtain grip on the damper. The dowel must not be allowed to turn, and to this end it is recommended that a self-locking wrench is clamped across its end to provide a handle. With an assistant restraining the damper rod, the bolt should now unscrew.

14   Lever out and slide off the dust seal which is fitted around the top of the lower leg. Using a suitable screwdriver, prise out the large circlip which retains the oil seal in the top of the lower leg. Remove the lower leg from the vice and clamp it at the wheel spindle lug with the leg and stanchion horizontal. Grasp the stanchion firmly, push it in to the lower leg as far as possible, and pull it sharply out. Repeat the process until the oil seal is driven out of its housing by the slide-hammer action of the stanchion and bushes. If the oil seal is particularly tight fit, pour some boiling water over the upper end of the fork lower leg, causing the alloy of the leg to expand, thus loosening the seal in its housing.

15   Once the stanchion and the lower leg are separated, the oil seal, the stamped washer underneath it, and the top bush can be pulled off over the top of the stanchion. Ensure that the upper length of the stanchion is quite clean before doing this to

avoid unnecessarily damaging the seal or the bush. The removal and refitting of the bottom bush is described in the next Section of this Chapter.

16 Invert the stanchion to tip out the damper rod and rebound spring, and invert the fork lower leg to tip out the damper rod seat. The rebound spring is pulled easily off the damper rod and the piston ring fitted around the head of the damper rod can be prised off with a suitable tool such as an electrical screwdriver. The latter should only be disturbed if renewal is necessary.

17 On reassembly, check that all components are clean and dry and that they are being refitted to the leg from which they were removed. Refit the piston ring (if removed) to the groove in the damper rod head, and slide the rebound spring over the other end of the damper rod. Insert the damper rod assembly into the stanchion and allow it to drop so that the damper rod protrudes beyond the lower end of the stanchion. Insert the damper rod seat into the fork lower leg so that its tapered surface points upwards and allow it to drop. Check that it is seated correctly in the recess machined in the bottom of the fork lower leg.

18 Insert a long rod into the stanchion so that it rests on the head of the damper rod, and hold the long rod so that the damper rod is locked in place at the bottom of the stanchion. Coat the bottom bush and the stanchion with fork oil and insert the two as a single unit into the fork lower leg until the damper rod can be felt to contact the seat. Check that the damper rod has fitted fully into the bore of the seat. Apply a few drops of thread locking compound to the threads of the damper rod Allen bolt and insert the bolt, complete with its copper sealing washer, into the base of the fork lower leg. Screw the bolt upwards into the damper rod. Lock the damper rod by the method used on dismantling and tighten the damper rod Allen bolt securely to a torque setting of 2.0 kgf m (14.5 lbf ft).

19 Coat the upper surface of the stanchion with fork oil and push it into the lower leg as far as possible. Fix the fork leg assembly in an upright position by clamping the wheel spindle lug lightly in a soft-jawed vice. Coat both surfaces of the fork top bush with fork oil and slide this down the length of the stanchion into its recess in the fork lower leg. Take great care that the bush is fitted squarely, it should be possible to fit it by hand alone, but a few gentle taps with a hammer and a suitable wooden drift will be permissible. In the case of the machine featured in the photographs, a short length of tubing was found which fitted over the stanchion and which had one edge square-cut and undamaged. This edge was used to position the bush and the tube used to press the bush into place, with its top surface flush with the bottom of the larger diameter recess that forms the oil seal housing.

20 The spacer between the top bush and the oil seal takes the form of a thick metal washer that has been stamped out so that it has a raised lip around its outer edge. Fit the washer over the top of the stanchion with this raised lip facing upwards and drop it into place over the bush. Lightly grease the inner and outer surfaces of the new oil seal and push it down over the stanchion into its recess in the lower leg. Tap it squarely and evenly into place using a hammer and a wooden drift on the hard outer ring only. If it can be found, a long metal tube that extends beyond the top end of the stanchion can be used to tap the seal into place, in the same way as a socket is used on the engine/gearbox oil seals. If such a tube is used, it must be sufficiently large in diameter to bear only on the hard outer ring of the seal, and the edge that comes into contact with the seal must be square-cut, absolutely clean, and undamaged.

21 When the seal has been tapped into its housing far enough to reveal the groove machined to accept the large circlip, cease tapping to avoid distorting the seal. Fit the circlip, pack the recess with grease to provide additional protection against dirt and corrosion, then refit the dust seal. This is easily pressed home by hand only, provided care is taken to keep it square to the fork lower leg. Pull the stanchion out to its fullest extent.

22 Fill each fork leg with 304 cc (10.7 fl oz) of fork oil. The recommended grade for this model is SAE 10W/30SE engine oil, but, as stated above for the RD125 LC models, proprietary

brands of fork oil are now available which will give better damping qualities and last longer as they are designed specifically for use in telescopic forks. They have the added advantage of being available in a variety of grades, enabling the owner to tune the suspension as he wishes. When the fork leg has been filled, pump the stanchion gently up and down to distribute the oil, finishing by compressing the stanchion into the lower leg as far as possible.

23 Using a length of welding rod or some other suitable item, and a ruler, measure the distance between the top of the stanchion and the top of the oil. The correct distance should be 168.4 mm (6.63 in). Add or remove oil to raise or lower the level if it is not correct. When the correct level has been established, pull the stanchion out to its fullest extent and fit the main fork spring, the thick spacer, and the secondary fork spring, using the marks made on removal to identify the top surface of each component so that it is fitted the correct way up. Insert the fork top plug and hold it against spring pressure while the retaining circlip is fitted. With the top plug secure, refit the black plastic cap.

6.1 Withdraw fork springs noting the way they were fitted

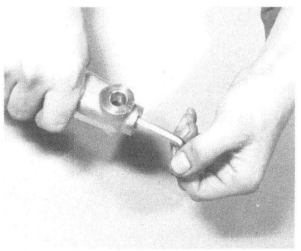

6.2 Allen bolt secures damper rod to fork lower leg

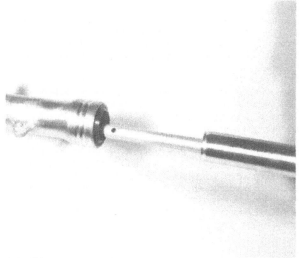

6.3a Withdraw stanchion from fork lower leg

6.3b Invert lower leg to tip out damper rod seat

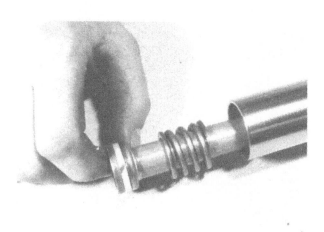

6.4a Tip damper rod and rebound spring out of stanchion

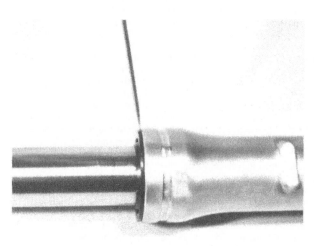

6.4b Prise out dust seal with great care – DT125 shown ...

6.4c ... but RD125 is similar

6.4d Prise out oil seal retaining circlip

6.5a Carefully lever oil seal out of fork lower leg

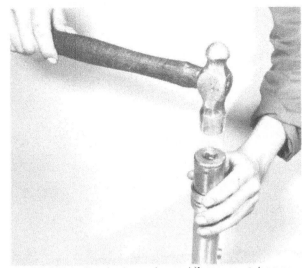

6.5b Refitting oil seal using socket as drift to prevent damage

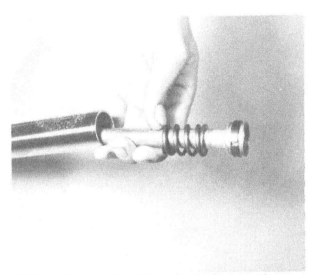

6.6a Insert damper rod and rebound spring into stanchion top

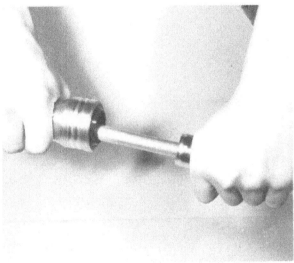

6.6b Be careful not to damage oil seal when inserting stanchion

6.6c Apply thread locking compound to damper rod bolt

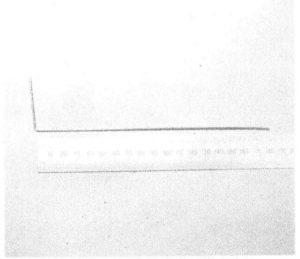

6.8a Dipstick fabricated from length of welding rod ...

6.8b ... and used to measure fork oil level

6.9 Ensure circlip is correctly seated in its groove

6.10 DT125 fork springs are retained in same way as RD125

6.11 Springs and spacer must be marked to show top surface

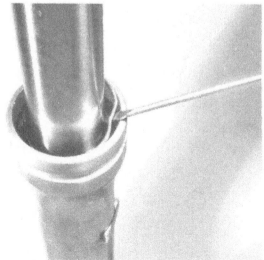

6.14 Same method is used to remove dust seal and circlip

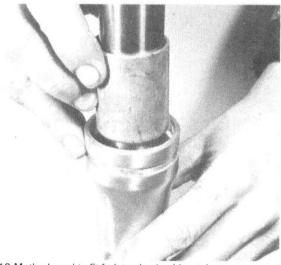

6.19 Method used to fit fork top bush without damage

6.20a Fit spacer as shown, noting raised lip

6.20b Coat stanchion with fork oil before sliding seal into place

6.20c Press seal squarely into fork lower leg

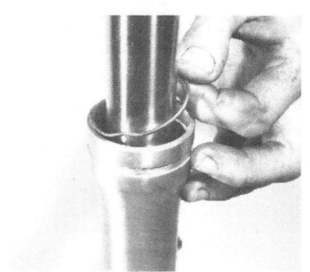

6.21a Do not omit oil seal retaining circlip

6.21b Dust seal can be pressed home by hand

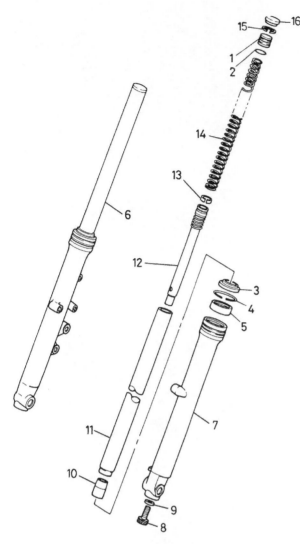

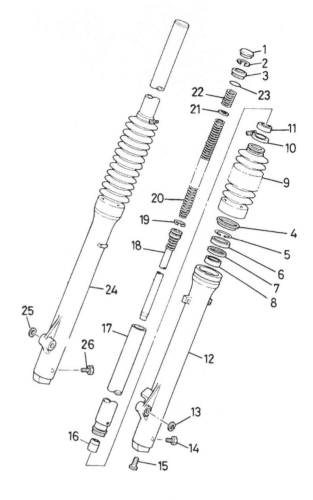

Fig. 5.4 Front forks - RD model

1	Top plug	9	Sealing washer
2	O-ring	10	Damper rod seat
3	Dust seal	11	Stanchion
4	Retaining clip	12	Damper rod and rebound
5	Oil seal		spring
6	Complete right-hand	13	Damper rod piston ring
	fork leg	14	Spring
7	Lower left-hand leg	15	Wire circlip
8	Allen bolt	16	Plastic cap

Fig. 5.5 Front forks - DT model

1	Plastic cap	15	Allen bolt
2	Wire circlip	16	Damper rod seat
3	Top plug	17	Stanchion
4	Dust seal	18	Damper rod and rebound
5	Retaining clip		spring
6	Oil seal	19	Damper rod piston ring
7	Washer	20	Main spring
8	Top bush	21	Washer
9	Gaiter	22	Secondary spring
10	Gaiter clamp	23	O-ring
11	Lower bush	24	Complete right-hand
12	Lower right-hand leg		fork leg
13	Washer	25	Washer
14	Oil drain plug	26	Drain plug

7 Fork legs: examination and renovation

1   Carefully clean all the fork components, removing all traces of rust, dirt, and old fork oil. Check that the oil control orifices in the damping rod are clear. Examine all the components, looking for excessive wear, hairline cracks or any other damage. The parts most likely to wear are the sealing lips of the dust seal and oil seal and the sliding surfaces of the stanchion or bushes, and the lower leg. Unfortunately no specifications are given by the manufacturer for any of these components and any assessment of wear will have to be by time-honoured trial and error methods. Any part found to be damaged, worn, or otherwise defective, must be renewed at once.

2   With all the components clean and dry, fit the top bush temporarily back on the stanchion. Slide the stanchion assembly into the lower leg and clamp the wheel spindle lug firmly in a vice. Move the stanchion backwards and forwards in an attempt to feel for any play which may be present. As the lubricating film normally present between the sliding surfaces has been removed, any such free play should be immediately apparent. Repeat the test at several points, pulling the stanchion gradually out, until it is in the fully extended position. If any free play is felt, and the visual check revealed no serious signs of wear, take the fork leg assembly to an authorised Yamaha dealer for the various components to be compared with new parts. While it is inevitable that some free play will be present, especially in the fully extended position, a degree of

experience will be necessary to assess whether any components will have to be renewed. Note that if this amount of wear has taken place, it will normally be evident due to areas of scoring on the component concerned. While the above test applies mainly to the owners of DT125 LC machines, it should be used also by owners of RD125 LC machines. Although the much greater working surface of the forks fitted to the latter will result in a slower rate of wear than on the trail bike version, when such wear does reach the stage where repairs are necessary it will be more expensive to put right. For this reason, care must be taken by RD125 LC owners to check the fork legs for wear whenever they are dismantled so that any problems can be rectified as soon as possible, thus prolonging as much as possible the life of the forks.

3   On DT125 LC models only, if it is decided that the bottom bush is so seriously worn that renewal is necessary, the course of action recommended by Yamaha is to renew the stanchion and the bush as a single component. It should be noted that the stanchion is sold complete with the bush in accordance with this policy. The reason for this is that the method chosen to secure the bush to the stanchion renders it impossible to remove the bush without damaging it, and makes fitting a new item equally difficult. In view of the above, it is recommended that the average owner complies with the suggested policy and takes the various fork leg components to an authorised Yamaha dealer for an expert second opinion. The dealer may pronounce the bush still serviceable, but if not, he can only confirm your worst fears and supply a new stanchion. For the more knowledgeable owner, it should be noted that a separate bush is listed as a replacement part by Yamaha, but it will require a considerable amount of skill and equipment to fit. The old bush may be cut open with a hacksaw, taking great care not to damage the stanchion chromed surface, and then prised off using a suitable flat-bladed screwdriver. The new bush must then be heated in an oven so that it will expand sufficiently to be pressed into its correct position on the stanchion. It is recommended that a press of some sort is employed, rather than a hammer and a tubular drift, to ensure that no damage is done to the bush.

4   If any stiffness is encountered when moving the stanchion in the above test, it must be assumed that the stanchion is bent or distorted. To check this, remove the oil seal and the top bush (where fitted) from the stanchion and roll the stanchion on a dead flat surface. Any misalignment will be immediately obvious. If the forks have been removed to rectify damage incurred in an accident, this check will not, of course, be necessary. Bent or distorted stanchions may be straightened once if the tubing has not been cracked or creased by the impact. If the stanchion has been straightened previously, or if the tubing is damaged, the stanchion must be renewed. As a general rule, it is always best to err on the side of safety and fit new stanchions, especially since there is no easy means to detect whether the metal has been overstressed or fatigued. Remember that any attempt at straightening the stanchion must be made only by a qualified expert and should on no account be made by anyone without the equipment necessary and the skill and experience to use it.

5   If the stanchions have been found to be straight and unworn by the test described, carefully examine the parts of them which are exposed when in position on the machine. Any signs of rust pitting must be removed. Deep pits can be filled with Araldite and then smoothed to match the original surface. It should be remembered that machines with fork stanchions pitted badly enough to damage the oil seals, and cause oil leakage, will be failed as unroadworthy when submitted for the annual MOT certificate. This is due to the reduction in damping efficiency which would result from such oil leakage. On RD125 LC models it is worth fitting a pair of gaiters, available from any good motorcycle dealer, to cover the exposed stanchions and to prevent such damage occurring in the future.

6   Inspect the lower legs carefully, looking for cracks around the mudguard mountings on RD125 LC models, and around the wheel spindle lug. Any such damage may be reclaimed by welding, but such a task must be left to an expert as alloy welding is a skilled operation. If damage is severe, it would be better to renew the lower leg on the grounds of safety.

7   It is advisable to renew the oil seals whenever the forks are dismantled, even if the seals appear to be in good condition. The seals are relatively cheap and their renewal at this stage will save the inconvenience of having to strip the forks at a later date, should leakage occur past the disturbed seals. They should be discarded and new ones fitted on reassembly. Similarly examine, and renew if necessary, the O-rings which seal the fork top plugs as these play an equally important part in preventing oil leakage. The external dust seals should be checked for splits, wear, or other damage and renewal if necessary. Note that if gaiters are to be fitted to an RD125 LC model, the dust seals should be left in place to provide a second line of defence against dirt and corrosion.

8   After a very high mileage has been covered the fork springs will take a permanent set, resulting in a spongy fork action. On RD125 LC models, the overall length of the spring must not be less than 477 mm (18.8 in). On DT125 LC models, however, no wear limits are set; assessment is therefore a matter of personal experience. Measure the length of each spring and compare the measurement obtained with that given in the Specifications Section of this Chapter. If the springs are found to be appreciably shorter they must be considered faulty and renewed. Note that in all cases, the only answer to fork springs that are ground to be shorter than the set figure, and therefore weaker, is the renewal of the springs concerned, although it is usually recommended that suspension components be renewed as a set to ensure consistent suspension action. While it is possible to add spacers to preload the springs, this is not a satisfactory solution to damaged or weakened springs.

9   Having examined those components of each fork leg which are most likely to suffer damage or wear, spend some time checking the remaining items. The damper rod should be clean, straight and undamaged and should fit securely into its seat, with little or no free play apparent between the two, and the taper formed on the outside of the seat should be undamaged. The Allen bolt and its copper washer should be in good condition, as should the fork top plug and its retaining circlip. The gaiters fitted to DT125 LC models must not be split or cracked, but note that the breather holes pierced in the lower part of each gaiter must be clean and free from blockages. If there is any doubt about any of these components, take them to a local Yamaha dealer for comparison with new replacement parts. Renew any item that is found to be worn or damaged.

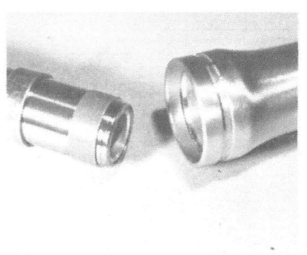

7.2 Bushes can be checked for wear, but renewal may not be easy

7.9a Check damper components for wear

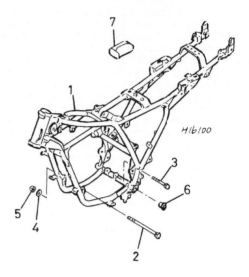

Fig. 5.6 Frame - RD model

1	Frame	5	Nut
2	Front engine mounting bolt	6	Plug - 2 off
3	Rear engine mounting bolt	7	Tool roll
4	Washer		

7.9b Sealing washer and threads of damper rod bolt must be unworn

8 Frame: examination and renovation

1   The frame is unlikely to require attention unless accident damage has occurred. In some cases, renewal of the frame is the only satisfactory remedy if the frame is badly out of alignment. Only a few frame specialists have the jigs and mandrels necessary for resetting the frame to the required standards of accuracy, and even then there is no easy means of assessing to what extent the frame may have been overstressed.

2   After the machine has covered a considerable mileage, it is advisable to examine the frame closely for signs of cracking or splitting at the welded joints. Rust corrosion can also cause weakness at these joints. Minor damage can be repaired by welding or brazing, depending on the extent and nature of the damage.

3   Remember that a frame which is out of alignment will cause handling problems and may even promote 'speed wobbles'. If misalignment is suspected, as a result of an accident, it will be necessary to strip the machine completely so that the frame can be checked, and if necessary, renewed.

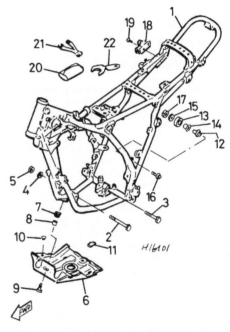

Fig. 5.7 Frame - DT model

1	Frame	12	Mounting bolt
2	Front engine mounting bolt	13	Collar
3	Rear engine mounting bolt	14	Bush
4	Washer	15	Washer
5	Nut	16	Bolt
6	Engine protection plate	17	Nut
7	Grommet - 3 off	18	Helmet lock
8	Spacer - 3 off	19	Screw
9	Bolt - 3 off	20	Tool roll
10	Damping rubber	21	Securing band
11	Damping rubber	22	Special spanner

9   Rear suspension assembly: general information

1   On the LC models the conventional pivoted rear fork, or 'swinging arm' is replaced by a welded tubular sub-frame which allows a single large suspension unit to control the rear wheel movements. The unit is mounted between the uppermost point of the sub-frame and the frame top tube, running almost horizontally between the dual seat. This cantilever suspension arrangement is known by Yamaha as 'Monocross'.

2   The suspension unit is of conventional construction. An oil-filled damper unit is fitted inside the main spring and is designed to provide control of rear wheel movement by the usual expedient of allowing oil to be forced through a small orifice. A quantity of nitrogen is contained in a high pressure chamber at one end of the unit and this allows for oil movement without the risk of aeration of the damping oil and the consequent loss of damping effect. A free-floating piston and O-ring keep the oil and nitrogen separate.

3   The damper incorporates a floating valve arrangement which is designed to reduce damping effect when the damper is moved suddenly. This allows the suspension to move rapidly under the forceful input of a large rut for example, and allows a higher degree of suspension compliance than is normally possible. The floating valve is effectively a spring steel disc, much like a thin washer, and is trapped between two seating faces of the damper piston. Its operation is best illustrated by referring to the accompanying figure in which various damping rates are shown.

4   In A, the unit is under fairly gently compression and the damping oil flows through the four notches in the valve. If a large bump is encountered the damping rate is too high to allow the unit to react quickly so the valve is arranged to deform, as shown in B, to allow a much greater oil flow through the piston. The same system applies under rebound as shown in C, during gentle movement and D, where the valve has deformed to allow rapid movement at the rear wheel.

5   In the case of both models, the spring preload is adjustable. Adjustment is made via a collar which incorporates a cam arrangement on RD125 LC models, and by a locknut and threaded adjuster on DT125 LC models. While the spring preload on RD125 LC models has six positions, that on DT125 LC models is adjustable to any position desired, within the limits of spring length set by the manufacturer. No provisions for the adjustment of the damping effect are made.

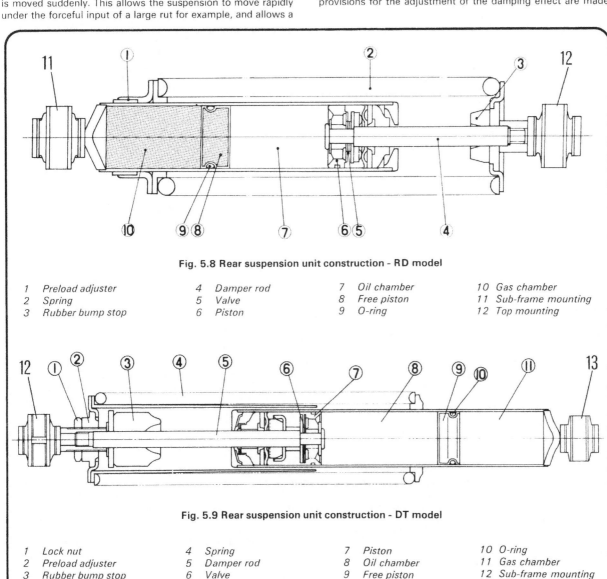

Fig. 5.8 Rear suspension unit construction - RD model

1 Preload adjuster	4 Damper rod	7 Oil chamber	10 Gas chamber
2 Spring	5 Valve	8 Free piston	11 Sub-frame mounting
3 Rubber bump stop	6 Piston	9 O-ring	12 Top mounting

Fig. 5.9 Rear suspension unit construction - DT model

1 Lock nut	4 Spring	7 Piston	10 O-ring
2 Preload adjuster	5 Damper rod	8 Oil chamber	11 Gas chamber
3 Rubber bump stop	6 Valve	9 Free piston	12 Sub-frame mounting
			13 Top mounting

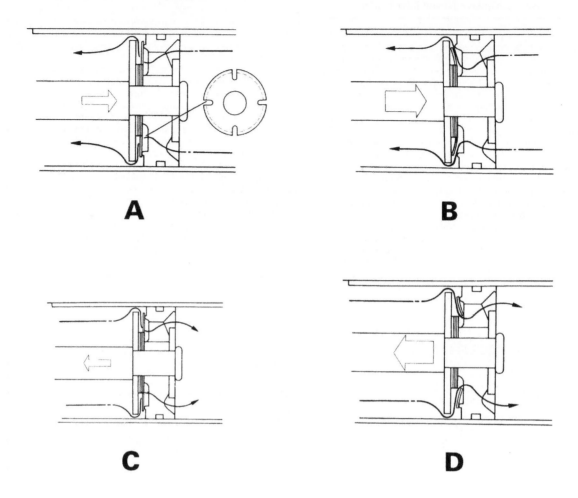

Fig. 5.10 Rear suspension unit operation

10 Rear sub-frame: removal, examination and refitting

1   The rear sub-frame pivot is formed by two large lugs which fit into the gaps between a large lug on the rear of the crankcase castings and the two sides of the main frame. A headed bush of synthetic material is pressed into each lug, these bushes bearing on two hardened steel sleeves. The sleeves are themselves retained by the sub-frame pivot bolt which passes through both sides of the frame, through both sub-frame pivots, and through the crankcase lug. The sub-frame pivots are sealed at both ends to prevent the entry of dirt, and are shimmed to take up any endfloat that may be present.

2   To assess the condition of the sub-frame pivot bearings, it will be necessary to remove any restrictions upon its movement, which will involve the removal of the rear wheel and of the suspension unit. Rear wheel removal is described in Section 15 of Chapter 6. Remove the seat, the sidepanels, and the petrol tank as described in Section 2 of Chapter 3, and the air filter cover (RD125 LC only) to gain access to the suspension unit top mounting. Straighten and withdraw the split pin which secures the suspension unit/sub-frame mounting pin, then remove the plain washer and tap the mounting pin out until the sub-frame is released from the suspension unit and can drop clear. Support the sub-frame as the pin is removed so that it is not damaged by dropping sharply on to another component or the floor. Slacken and remove the nut which secures the suspension unit top mounting bolt, then withdraw the spring or plain washer and tap out the bolt. Pull gently the suspension unit upwards and forwards to withdraw it from the machine.

3   When the sub-frame is free of any restrictions, check first for endfloat by grasping it firmly at the pivot and by pulling and pushing from left to right along the line of the pivot bolt. Ideally, no free play should be found, but a maximum of 1.0 mm (0.039 in) is permissible, this measurement being obtained by using feeler gauges in the gap between the main frame and either of the sub-frame pivot sealing caps. If excessive, this clearance must be reduced by inserting shims between the sealing cap and the steel sleeve of each pivot, a task which will require the removal of the sub-frame. The state of the bushes can be assessed by grasping the fork ends of the sub-frame and attempting to move them from side to side; again there should be no free play apparent, but a maximum of 1.0 mm (0.039 in) is permissible. If movement in excess of this figure is found, the sub-frame must be removed for the bushes to be renewed. This operation is described below.

4   When working on a DT125 LC model, the rear brake pedal assembly must be removed to allow the pivot bolt to be withdrawn; disconnect the stop lamp rear switch extension road and use a small flat-bladed screwdriver to prise off the circlip which retains the pedal on its pivot. Unhook the pedal return spring and remove the pedal assembly. On both models, the sub-frame is now ready to be removed. Slacken and remove the

pivot bolt retaining nut and the plain or spring washer fitted underneath it. Tap out the pivot bolt using a hammer and a long metal drift. If corrosion has taken hold this task will be very difficult. Spray or pour a liberal quantity of penetrating fluid over those areas concerned, allow some time for the fluid to work, and then resume tapping. Exercise extreme caution and some patience to prevent damage to the machine. When the pivot bolt is fully removed, disconnect the chain at its connecting link, if this was not done during wheel removal, and withdraw the sub-frame from the machine.

5   Place the sub-frame on a convenient working surface and remove the sealing caps from the outside of each pivot lug, noting the location and number of any shims found beneath them. Tap out the hardened steel sleeves, then lever out the seals fitted on the inside edge of each pivot lug. The seals should be renewed as a matter of course, as they are relatively cheap and play a vital part in keeping dirt out of the pivot bearings. If the check described above revealed excessive wear in the synthetic bushes, these must be removed and new ones fitted, but it should be noted that there is no way to remove the bushes without damaging them in the process. If renewal is necessary, pass a long drift through the opposite pivot lug to rest against the inside edge of the bush to be removed, then tap out the bush. Repeat the operation for the remaining bush.

6   Thoroughly clean the entire sub-frame assembly and all the components of the pivot bearings, removing all traces of dirt, grease, and corrosion. Use emery paper to remove rust deposits from the pivot bolt and from the hardened steel sleeves, taking care to produce a clean polished surface on each. Any component which is found to be visibly worn or damaged after cleaning should be renewed, unless the owner has the expertise to attempt repairs as described below.

7   The sub-frame assembly should be checked for cracks or for other signs of wear or damage, or for distortion due to any accident damage. Although some repairs may be possible, depending on the nature of the damage, these must be entrusted to an expert and not attempted by an unskilled person. On the grounds of safety alone, it would be preferable to renew the sub-frame assembly, however expensive this might prove.

8   The sub-frame pivot bolt must be absolutely straight and unworn, and its threads must be in good condition. Roll it on a flat surface to check for straightness; any distortion should be evident immediately. Check that the threads are undamaged and that the self-locking retaining nut is working properly. It should grip the threads securely as the nut is screwed down. If the threads are damaged, most good motorcycle dealers will have the necessary taps and dies with which they can be reclaimed, and a bent pivot bolt is as easily straightened, provided it is taken to an expert. Check the clearance between the pivot bolt and the steel sleeves by sliding the sleeves into their correct respective locations over the bolt and feeling for any signs of free play; none should be discernible. If any free play is felt, compare the bolt and the sleeves with new components and renew any item that is found to be worn.

9   Check the clearance between the steel sleeves and the synthetic bushes by inserting the sleeves into the bushes and feeling for free play. Again, none should be discernible in an ideal situation, but it is inevitable that manufacturing tolerances will produce a limited amount. Compare the used items with new components to gain some idea of permissible tolerances. Note that if heavy scoring is visible on any of the sub-frame pivot components, there will be no need for such comparisons as such scoring will show which component has worn. It is not an economic proposition to attempt to reclaim worn components; renewal is the only practical solution.

10   If the synthetic bushes were removed, reassembly must start by fitting the new components. Yamaha insist that the new bushes must not be driven into place, but must be pushed home using a suitable press. Few owners will have the use of such equipment, but bush fitting is possible by employing a drawbolt arrangement. Obtain a bolt or a length of threaded rod that is about 1 inch longer than the combined width of either of the sub-frame pivot lugs and one bush. Also required are a suitable nut (two if a threaded rod is used) and two thick and robust plain washers which are large enough in diameter to fit over the head of the bush and over the inside end of the pivot lug. In the case of the threaded rod being used, fit one nut to one end of the rod and, if required, stake it in place for convenience.

11   Lightly grease the outside of one of the new bushes to assist fitting, and push it into a pivot lug from the outside (frame side) by hand, ensuring that it is fitted squarely into the lug. Slide one of the washers over the bolt or rod so that it rests against the head of the bolt or the nut, then pass the assembly through the bush and pivot lug. As the thread appears on the inside edge of the pivot lug, fit the remaining washer and the nut. Screw the nut down to tighten the drawbolt assembly on to the bush and the lug. Make a final check that the bush is kept square, then tighten slowly the nut so that the bush is drawn into the lug. Keep checking that the bush is entering squarely into the lug, and stop tightening as soon as the head of the bush contacts the outside edge of the pivot lug. Once the bush is fully home, remove the drawbolt assembly and repeat the sequence to fit the remaining bush to the other pivot lug.

12   Grease liberally the new seals and push them squarely into the inside edge of each pivot lug. While this operation can be done easily by hand, if any difficulty is encountered the drawbolt assembly just described can be brought into service again, but care must be taken not to damage the seal. Pack the interior of each pivot lug with grease, apply a generous amount of grease to the inside and outside of each of the hardened steel sleeves, and insert a sleeve into each pivot lug from the outside, taking care not to damage the lips of the seal as the sleeve passes through them.

13   Refit the shims inside each sealing cap, ensuring that each shim is fitted in its original location. If, when conducting the checks described at the beginning of this Section, excessive endfloat was found, shims must be added as necessary to take up the clearance. If the shims were disturbed and cannot be refitted correctly, or if a new sub-frame has been fitted, the endfloat must be measured and shimmed from scratch. Fit the sealing caps without the shims and replace the sub-frame in the machine as described subsequently in this Section, tightening the pivot bolt to the recommended torque setting. Using feeler gauges, measure the endfloat between one of the sealing caps and the frame, then purchase or fabricate shims of sufficient thickness to remove the clearance completely. Remove the sub-frame again and fit the new shims inside the sealing caps. Note that it is better to use one or two thick shims than several thin ones, and that the shims must be divided as equally as possible between the two sealing caps so that the sub-frame is positioned centrally in the frame. Refit the sub-frame to the machine and check the clearance again, remembering that ideally there should be no clearance at all, but that a tolerance of up to 1.0 mm (0.039 in) is permissible.

14   If the above procedure was not necessary on the machine being worked on, grease the inside of each sealing cap and refit them over the outside edge of each pivot lug. Pack grease into the pivot lugs, the crankcase lug and the holes in the frame to prevent corrosion forming on the pivot bolt, and smear grease over the bolt itself. Offer up the sub-frame, aligning the pivot lugs, and push the pivot bolt through from right to left on RD125 LC models, and from left to right on DT125 LC models. Refit the spring or plain washer to the pivot bolt end, removing any surplus grease, and refit the self-locking nut. Tighten the nut to a torque setting of 4.3 kgf m (31 lbf ft) and wipe away any surplus grease. Pass the final drive chain over the pivot and back through the sub-frame itself, then on DT125 LC models, grease the brake pedal pivot and refit the brake pedal assembly.

15   Refit the suspension unit as described in the next Section of this Chapter and complete the remainder of the reassembly by reversing the instructions followed on dismantling. Before taking the machine out on the road, check the rear brake and chain adjustment and the rear suspension operation. Check that all nuts and bolts are securely fastened, and that all split pins, where fitted, are renewed and correctly secured.

10.1 Sub-frame pivots fit each side of large crankcase lug

10.5a Note location and number of any shims found behind end caps

10.5b Seals should be levered out and renewed

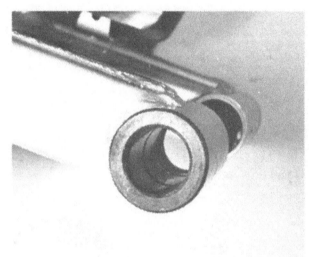

10.5c Worn bushes can be drifted out of sub frame

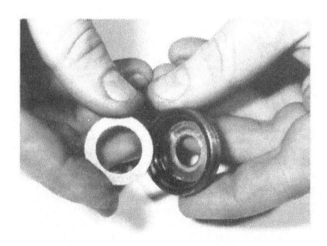

10.13 Add shims, if necessary, to take up endfloat

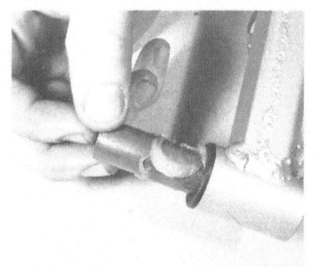

10.14a Grease bushes liberally before fitting

10.14b Do not omit rubber chain protector before refitting end caps

10.14c Offer up sub frame and align pivots

10.14d Tighten sub frame pivot bolt to recommended torque setting

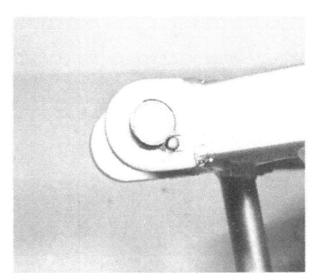

10.15 Note flat on mounting pin head located against raised stop

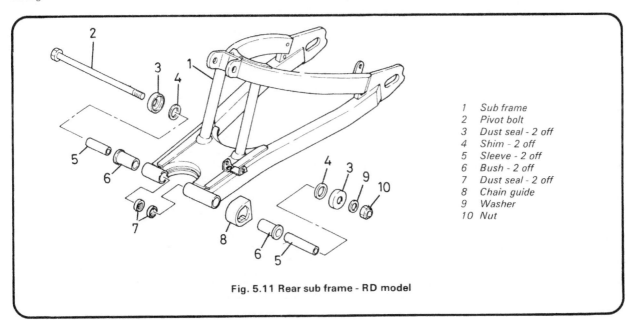

1   Sub frame
2   Pivot bolt
3   Dust seal - 2 off
4   Shim - 2 off
5   Sleeve - 2 off
6   Bush - 2 off
7   Dust seal - 2 off
8   Chain guide
9   Washer
10 Nut

Fig. 5.11 Rear sub frame - RD model

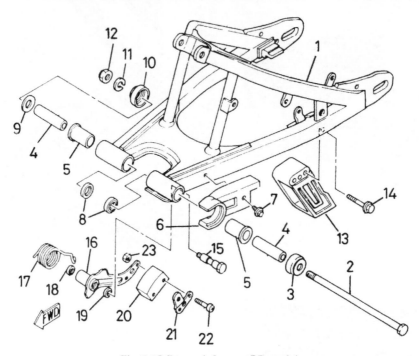

Fig. 5.12 Rear sub frame - DT model

1	Sub frame	7	Bolt and washer	13	Chain guide	19	Oil seal
2	Pivot bolt	8	Dust seal - 2 off	14	Bolt - 2 off	20	Rubbing block
3	Dust seal	9	Shim	15	Pivot bolt	21	Plate
4	Sleeve - 2 off	10	Dust seal	16	Jockey arm	22	Screw - 2 off
5	Bush - 2 off	11	Spring washer	17	Return spring	23	Nut
6	Guard	12	Nut	18	Oil seal		

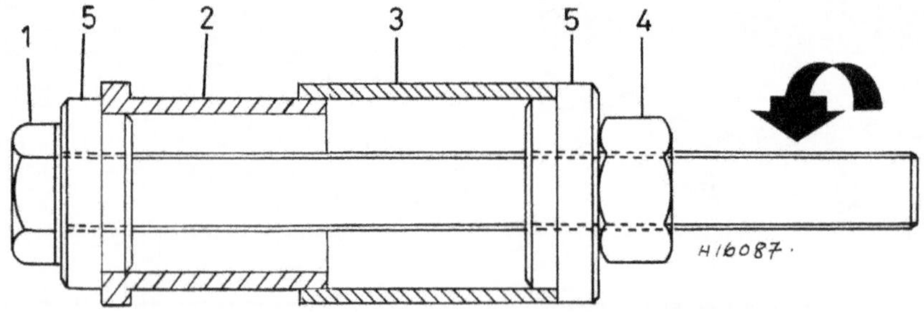

Fig. 5.13 Fitting the sub frame pivot arm bushes

1	Drawbolt	4	Nut
2	Bush	5	Thick washer
3	Pivot lug		

11 Rear suspension unit: removal, examination and refitting

1   Place the machine on a suitable stand so that it is supported securely in the upright position with the rear wheel well clear of the ground. Remove the seat, the sidepanels, and the petrol tank as described in Section 2 of Chapter 3, and the air filter cover (RD125 LC only) to gain access to the suspension unit top mounting. With the rear wheel in place, it is easiest to release the top mounting first so that the sub-frame can drop to the point where access is more easily gained to the lower mounting; slacken and remove the nut which secures the top mounting bolt, then remove the washer and tap out the bolt, supporting the back wheel on a wooden block or similar. Remove the support and by holding on to the top of the suspension unit, restrain the sub-frame and rear wheel as they

drop, to prevent accidental damage. Straighten and withdraw the split pin which secures the suspension unit bottom mounting, then remove the plain washer and tap out the pin. Lift the unit away from the sub-frame mounting, noting on DT125 LC models only the positions of the two additional spacers and the thrust covers, and pass it up through the aperture in the rear mudguard.

2   Having removed the suspension unit, there is little that can be done to it. The damper itself is a sealed unit and should be renewed if defective, but while the spring is removable on either of the machines described in this Manual, new components are not available as replacement parts as Yamaha only offer the spring and damper as a single assembly. This policy renders pointless any attempts to dismantle the unit further. In the event that springs or damper units of proprietary manufacture become available, however, the procedure of dismantling is described as far as possible.

3   When removing a spring, note that it is under quite a lot of compression and no attempt should be made to disturb it unless a safe method of compressing it is available. The method employed in practice was as follows. Slacken the adjuster locknut on DT125 LC models, then unscrew fully the adjuster to reduce the pressure. On RD125 LC models, rotate the cam adjuster so that the preload is in its softest (No 1) position. Clamp the suspension unit in a vice by the mounting eye attached to the body of the damper itself, and not by the eye attached to the piston rod. The safest spring compressors that are readily available are those used by car garages or sold at car accessory shops for use on the MacPherson struts that are fitted to the front suspension of many modern cars, particularly Fords. Following the manufacturer's instructions, clip the compressors on to the coils of the spring, one on each side as shown in the accompanying photograph, and tighten each one by equal amounts until the spring pressure is released safely and the two retaining collets can be picked out. The spring can then be pulled off the suspension units and the pressure gradually released by unscrewing the compressors in stages. The spring is refitted by reversing the above process.

4   Once removed, the spring can be measured to assess its condition. Those fitted to RD125 LC models must not be less than 232.5 mm (9.15 in) long, when measured in the uncompressed state, and those fitted to DT125 LC models must not be appreciably shorter than the 290 mm (11.42 in) set by the manufacturer.

5   The only other components of the suspension unit to be inspected are the bonded rubber bushes fitted in each mounting eye. If cracked, perished, or worn in any other way, these must be renewed. The removal of bonded rubber bushes is always a difficult operation, the most effective method being to make up an extractor based on a drawbolt arrangement, as shown in the accompanying illustration. Find a length of tubing that is at least as long as the total width of the bush to be removed, and whose diameter is the same as that of the suspension unit mounting eye. Also required will be two thick plain washers, one of which is large enough to fit over one end of the tubing and the other being of the same outside diameter as the bush itself, a bolt or length of threaded rod that is about 1 inch longer than the combined width of the bush and the tubing, and a nut to fit the bolt (two nuts if threaded rod is used). If threaded rod is used fit one nut to one end of the rod and stake it in place.

6   Fit the smaller washer over the bolt or rod so that it rests against the head of the bolt or the nut on the end of the rod. Pass the bolt or rod through the centre of the bush and fit the tubing over the protruding end so that it butts against the suspension unit mounting eye, then fit the large washer and secure the whole assembly by tightening down the remaining nut. Proceed to tighten the nut, thus compressing the tool and drawing the smaller washer through the mounting eye, taking the bush with it. The bush will distort quite markedly in the early stages, until the smaller washer comes firmly into contact with the metal outer and starts to exert pressure on it. Note that if, as is likely, corrosion has taken hold it will be useful to soak the complete mounting eye in penetrating fluid to render a difficult operation as easy as possible. Never apply any form of heat to any part of the suspension unit while removing or refitting these bushes as this would damage the seals and rubber components of the unit, rendering it useless, and would be also a very dangerous exercise due to the fact that the unit contains pressurised gas, and would therefore incur a risk of explosion.

7   Fitting new bushes is slightly easier, but will require the use of a vice, or of the drawbolt assembly used on removal. Two lengths of tubing will be needed, one of which is the same diameter as the suspension unit mounting eye and is at least as long as the width of the protruding metal centre, and another which is of the same length, but is of the same diameter as the metal outer of the bush. Thoroughly clean the inside of the mounting eye and the outside of the new bush, removing all traces of corrosion and any small burrs or raised edges. Smear a small quantity of grease over both surfaces and insert the bush as far as possible into the mounting eye by hand, ensuring

that it is fitted squarely. Place the larger diameter tube against one jaw of the vice and press the suspension unit mounting eye against it, ensuring that the two are aligned exactly. Place the smaller diameter tube against the metal outer of the bush and clamp the assembly together by closing the jaws of the vice. While this task can be carried out by one person exercising a degree of dexterity, it is much easier if the aid of an assistant is enlisted. Check that all the components are square and that the metal centre of the bush is under no pressure at all. If any pressure is applied to the bush metal centre or to the rubber itself, the rubber will distort and split, rendering useless the bush, and making the whole operation pointless. When all is correctly assembled, tighten down the vice jaws, thus pressing the bush into the mounting eye. Stop when the metal outer of the bush is flush with each edge of the mounting eye. If no vice is available, the operation can be carried out by substituting the drawbolt and plain washers used on removal, the bolt being passed through the centre of the tubes and bush to simulate the clamping action of the vice.

8   If the suspension unit is so worn as to be causing handling problems due to the reduction in damping efficiency, or if signs of damage are visible, such as a bent piston rod, oil leakage, or a dented unit body, the complete assembly must be renewed, and the old unit discarded. Do not just throw it away. It is first necessary to release the gas pressure and the manufacturers recommend that the following procedure is followed.

9   Refer to the accompanying figure and mark a point 10 – 15 mm above the bottom of the cylinder. Place the unit securely in a vice. Wearing proper eye protection against escaping gas and/or metal particles, drill a 2 – 3 mm hole through the previously marked point on the cylinder.

10 On refitting the suspension unit, follow the reverse of the dismantling instructions, noting the following points. In the case of both models, the suspension units are fitted so that the spring preload adjusters are to the rear, or towards the sub-frame mounting bracket, and the damper bodies must be aligned so that the large caution label is facing upwards. Smear a liberal amount of grease over the top mounting bolt and the bottom mounting pin, and into the centre of each bush, to prevent the onset of corrosion. Push the unit down through the aperture in the rear mudguard and then downwards into its front mounting. Push the mounting bolt through from left to right to retain the unit. Raise the sub-frame and rear wheel so that the unit bottom mounting eye fits between the sub-frame mounting brackets, noting that on DT125 LC models only, the two additional spacers and the two thrust covers must be greased and assembled on each side of the bush. Push the mounting pin through from left to right and rotate it so that one of the flats on the pin head aligns with the small raised lug on the left-hand bracket.

11 Place the plain washer over the protruding right-hand end of the pin and secure the pin by pushing a new split-pin through the hole in its right-hand end. Spread the ends of the split-pin as shown in the accompanying photograph. Place the plain or spring washer over the right-hand end of the mounting bolt, refit the nut, and tighten it securely to a torque setting of 2.5 kgf m (18 lbf ft). If the spring preload setting was altered for any reason, do not forget to reset it before the machine is used on the road. Refit the air filter cover (RD125 LC only) and the petrol tank, the sidepanels, and the seat.

12 A final note on the rear suspension unit concerns the spring preload adjustment. The RD125 LC model employs a conventional cam arrangement by which a collar fitted around the body of the unit can be rotated to any one of six positions using a special C-spanner provided in the machine's toolkit. It is necessary to remove the seat to gain access to the toolkit and to the collar. The standard position is the second from softest (No 2) and the collar must be rotated to the left (viewed from above when installed) to raise the spring preload for stiffer suspension, or to the right for the reverse effect. The DT125 LC model is different in that a threaded adjuster and locknut are fitted to the head of the piston rod, these being accessible from underneath the rear mudguard when the unit is installed.

Slacken the locknut and rotate the adjuster clockwise (viewed from the rear) to raise the spring preload, or anti-clockwise to lower it. The operation is made easier if the special spanner is used. This has thin jaws and a suitably cranked handle and should be provided in the machine's toolkit but if not, a new one can be ordered as a separate item from a Yamaha dealer. The spring preload is altered by changing the installed length of the spring as described. The manufacturer has specified maximum and minimum lengths for the spring to act as guidelines when setting up the suspension and to ensure maximum life for the spring and damper unit. To measure the spring length, the seat must be removed and while this is not strictly necessary, life is made easier if the petrol tank is removed as well. Using a ruler, measure the overall length of the spring from one collar to the other. The spring should be no more than 285 mm (11.2 in) long and no shorter than 270 mm (10.6 in). The standard length on delivery is 280 mm (11.0 in). Note that one full turn of the adjuster in either direction will alter spring length by 1 mm (0.039 in), and that adjustments should be made in increments of 2 mm (0.08 in) to produce any noticeable effect. When the adjustment is made, hold the adjuster and tighten securely the locknut to prevent any alteration in the setting. A torque setting of 5.5 kgf m (40 lbf ft) is given for the adjuster locknut, and this should be adhered to if the necessary equipment is available.

11.1a Slacken and remove top mounting bolt and allow sub frame to drop ...

11.1b ... so that lower mounting can be reached easily

11.3 Spring must be compressed safely before it can be removed

11.5 Check mounting bushes for damage or wear – renewal is not easy

11.10 Grease mounting bolt liberally before fitting

11.11 Spread ends of split pin as shown to secure mounting pin

11.12a While RD125 is fitted with conventional preload adjustment ...

11.12b ... DT125 employs threaded adjuster and locknut

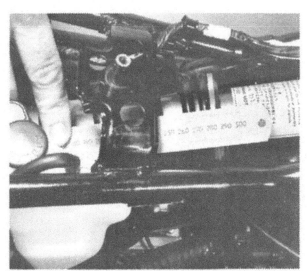

11.12c Measure carefully spring fitted length

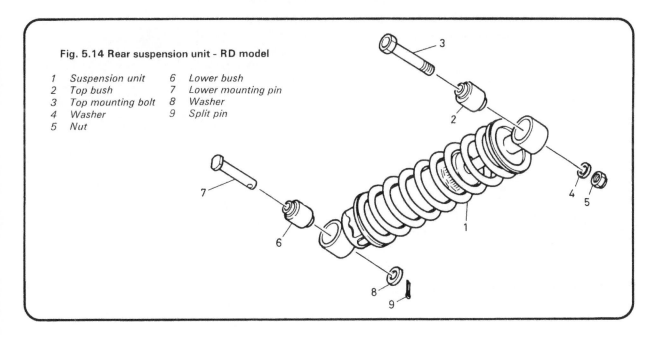

Fig. 5.14 Rear suspension unit - RD model

1 Suspension unit
2 Top bush
3 Top mounting bolt
4 Washer
5 Nut
6 Lower bush
7 Lower mounting pin
8 Washer
9 Split pin

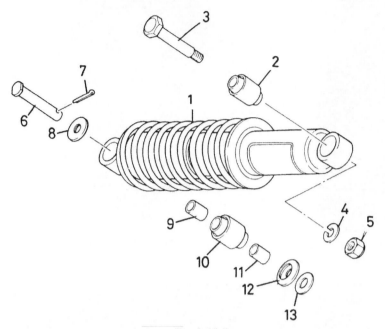

Fig. 5.15 Rear suspension unit - DT model

1   Suspension unit
2   Top bush
3   Top mounting bolt
4   Spring washer
5   Nut
6   Lower mounting pin
7   Split pin
8   Washer
9   Metal insert
10 Lower bush
11 Metal insert
12 Side cover - 2 off
13 Washer

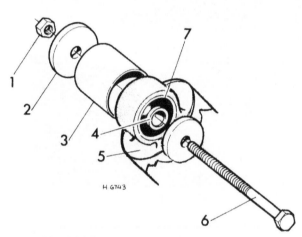

Fig. 5.16 Rear suspension unit bush removal

1   Nut   5   Rear suspension unit eye
2   Thick washer   6   Drawbolt
3   Tubing   7   Bush
4   Spacer

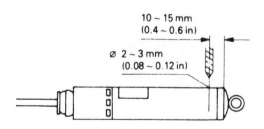

Fig. 5.17 Position of drilling on rear suspension unit

12 Centre stand: examination and maintenance – RD125 LC

1   The centre stand is an important but largely neglected feature of most motorcycles. It is important to check the stand for wear or damage from time to time, as failure of the stand can result in costly repair bills. Check that the stand pivot pin is secure and in good condition, and that it is kept adequately lubricated.
2   Check that the return spring is in good condition. A broken or weak spring may cause the stand to fall whilst the machine is being ridden, and catch in some obstacle, unseating the rider.
3   To remove the stand, place the machine on its prop stand, prise out the circlip which secures the right-hand end of the stand pivot pin, tap out the pin, and release the stand from its return spring. Reverse the process to refit the stand, making sure that the pivot pin is well greased.

12.1 Inspect and lubricate stand pivot at regular intervals

13 Prop stand: examination

1   The prop stand is attached to a lug welded to the left-hand lower frame tube. An extension spring anchored to the frame ensures that the stand is retracted when the weight of the machine is taken off the stand.

2   Check that the pivot bolt is secured and that the extension spring is in good condition and not overstretched. An accident is almost certain if the stand extends whilst the machine is on the move.

14 Footrests: examination and renovation

RD125 LC models

1   The front and rear footrests are mounted on alloy brackets on each side of the machine, these doubling as mounting points for the silencer, and for the brake pedal. The alloy plates may be removed after the latter parts have been detached, though this should rarely prove necessary.

2   The footrests are of the folding type and are unlikely to require frequent attention. Little can be done to repair them, and in the event of extreme wear or damage, the affected parts should be renewed. If the footrests are damaged in an accident, they may be removed and straightened after they have been heated to a dull cherry red using a welding torch or blowlamp. The rubber covers are available separately and can be pulled off the footrest bar if damaged. Use petrol or soapy water to aid refitting.

3   When refitting the alloy footrest plates, tighten the mounting bolts to a torque setting of 6.5 kgf m (47 lbf ft) and tighten the pillion footrest mounting bolts to a torque setting of 3.0 kgf m (22 lbf ft).

DT125 LC models

4   This model employs individual cleated-pattern metal foot-rests which are pivoted by clevis pins on separate brackets to each side of the frame. A small spring is fitted to each footrest to ensure that it returns to its correct position if pivoted away from the horizontal. The pillion footrest assemblies are bolted on to the rear of the sub-frame, and are also pivoted. The elastic qualities of the footrest rubber retain each footrest in the up or down position. It is worth noting that for this model, all the individual components of the front and rear footrests are available as replacement parts, as are those of the brake pedal with its folding tip.

5   If the footrests become damaged in an accident, it may be possible to straighten them after removal from the machine. The area around the deformed portion should be heated to a dull red before any attempt is made to bend the footrest back into shape. If required, the hinged portion of the footrest may be separated from the frame or bracket after removing the split-pin and clevis pin. The rubber, where fitted, should be removed prior to application or heat, for obvious reasons.

6   If there is evidence of failure of the metal either before or after straightening, it is advised that the damaged component is renewed. If a footrest breaks in service, loss of machine control is almost inevitable.

7   Refitting of the footrest assemblies is a reversal of the removal procedure. When refitting the clevis pin, ensure that the return spring is correctly located and a new split-pin fitted.

15 Rear brake pedal: examination and renovation

1   The rear brake pedal is constructed of a malleable metal which is easily straightened in the event of an accidental impact bending it, although the application of heat beforehand is good practice so that the minimum of strain is imposed on the metal. Adjustment is provided by means of a bolt and locknut by which the height of the pedal can be altered to suit the rider's preference.

2   To remove the pedal fitted to the RD125 LC model, disconnect the brake rod from the brake operating arm by unscrewing the adjusting nut. Slacken and remove the single bolt which secures the exhaust system to the right-hand footrest mounting plate, then slacken and remove the two bolts which secure the plate itself to the frame. Withdraw the complete assembly, unhooking the stop lamp rear switch extension spring from the pedal. Prise out the circlip which secures the brake pedal pivot to the footrest plate, pick out the plain washer behind the circlip, and disengage the pedal return spring. Push out the pedal from the footrest plate and remove all traces of dirt and corrosion from the pivot and from that part of the plate through which it passes. The pedal can be separated from the brake rod by removing the split-pin and clevis pin which join the two.

3   To remove the pedal fitted to the DT125 LC model, disconnect the brake rod from the brake operating arm by unscrewing the adjusting nut, then disconnect the stop lamp rear switch extension rod from the pedal itself. Prise out the large circlip which secures the pedal to its pivot, unhook the pedal return spring and pull the pedal away, noting the large plain washer which fits over the pivot between the pedal and the frame.

4   In the case of both models, refitting is a straightforward reversal of the removal instructions given above. Renew any part of the brake linkage that is worn, to preserve the efficient and safe operation of the rear brake. Apply a liberal quantity of grease to the pivots, to prevent the onset of corrosion or wear. If the pedal pivot is worn badly, this can only be reclaimed by the renewal of the part concerned as no separate bushes are fitted. Unfortunately this is likely to prove expensive, and it may be worthwhile attempting to find an engineering firm who are prepared to make up a bush and to press it into place to take up the free play and restore the brake operation. This will involve reaming out the footrest plate on RD125 LC models, and the pedal itself on DT125 LC models, so that the bush can be inserted. The expense of the only alternative solution should make this course a practical proposition.

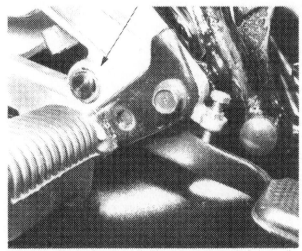

15.2a RD125 brake pedal is retained by large circlip (arrowed)

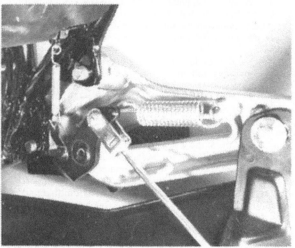

15.2b Check brake linkage for wear and corrosion

16.2 Instrument heads are very delicate

16 Speedometer and tachometer heads: removal, examination, renovation and reassembly

1   Before operations are described in detail, there are a few general observations to be made about these instruments. They must be carefully handled at all times and must never be dropped or held upside down. Dirt, oil, grease and water all have an equally adverse effect on them, and so a clean working area must be provided if they are to be removed.

2   The instrument heads are very delicate and should not be dismantled at home. In the event of a fault developing, the instrument should be entrusted to a specialist repairer or a new unit fitted. If a replacement unit is required it is well worth trying to obtain a good secondhand item from a motorcyle breaker in view of the high cost of a new instrument.

3   Remember that a speedometer in correct working order is a statutory requirement in the UK. Apart from this legal necessity, reference to the odometer readings is the most satisfactory means of keeping pace with the maintenance schedules.

RD125 LC models

4   The instrument heads are removed as a complete assembly as described in Section 3 of this Chapter. Remove the fairing to provide sufficient clearance, and the headlamp unit so that the relevant electrical leads can be disconnected. Disconnect the drive cables at their upper ends by unscrewing the knurled retaining rings. Slacken and remove the two bolts securing the instrument panel to the top yoke, and withdraw the panel assembly. Separate the two halves of the panel assembly by unscrewing the four black-painted screws which pass upwards into the panel top half and by lifting the panel top half away. Each instrument is secured to the panel bottom half by two plated screws. Slacken and remove these and lift the instrument away. Refitting is the reverse of the above.

DT125 LC model

5   The speedometer head can be removed as a complete assembly with the warning light cluster, as described in Section 3 of this Chapter or on its own. Remove the headlamp fairing, then disconnect the drive cable by unscrewing the knurled retaining ring. Using a suitable pair of pliers pull out the two clips (and catch the washer underneath each) that secure the speedometer to the mounting bracket. Pull the speedometer carefully upwards, withdraw the illuminating bulb from its socket, and remove the speedometer from the machine. Again, refitting is a straightforward reversal of the above procedure.

16.4a Instrument panel is secured by two bolts to top yoke

16.4b Panel mounting bracket is retained by two nuts

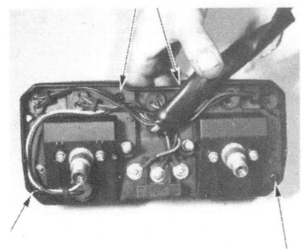

16.4c Unscrew four black screws (arrowed) ...

16.4d ... to separate panel halves

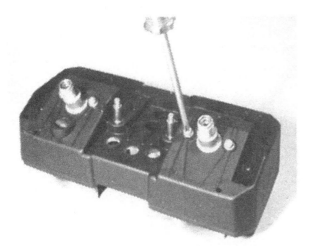

16.4e Each instrument is retained by two screws

17 Speedometer and tachometer drive cables: examination and maintenance

1   If the operation of the speedometer or tachometer becomes sluggish or jerky it can often be attributed to a damaged or kinked drive cable. Remove the cable from the machine by releasing the circlip or the knurled ring at the lower end and the knurled ring at the instrument head. Turn the inner cable to check for any tight spots. If the cable tends to snatch as it is rotated it is likely that the inner has become kinked and will require renewal.

2   Check that the outer cable is in sound condition with no obvious damage. The inner and outer cables are not available separately. When fitting a new cable or refitting an old one, remove the inner and grease all but the upper six inches or so. This will ensure that the cable is adequately lubricated without incurring the risk of grease working up into the instrument head. Ensure that the cable is routed in a smooth path between the drive and instrument panel, avoiding any tight bends which would result in the cable becoming kinked.

17.1a Drive cables are secured by a circlip – DT125 ...

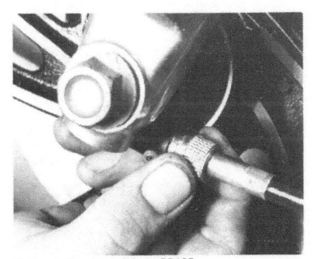

17.1b ... or by a knurled ring – RD125 ...

17.1c ... and by a knurled ring at the upper end

6   Remove all old grease from the drive gear, thrust washers, brake backplate housing and worm gear by wiping the components with a clean rag. Inspect the gears for broken teeth and signs of excessive wear due to lack of lubrication. Renew any components that are found to be worn or damaged as repairs are not possible.

7   Refit the speedometer drive components by reversing the order of dismantling, taking care to pack the recesses with a suitable high melting-point grease. Fit the new dust seal, if required, into its recess in the brake backplate by pressing it into position evenly and squarely. The brake shoes and springs may now be refitted and the brake backplate assembly re-inserted into the wheel hub. Ensure that the speedometer drive tabs are aligned with the corresponding slots in the wheel hub boss.

19 Dualseat: examination and maintenance

As the dualseat is a plastic moulding to which the seat cushion and cover are attached, it will require little maintenance in the life of the machine other than periodic lubrication of the seat catches and lock to minimise wear and corrosion. It is worth noting, however, that new seat covers can be ordered through any Yamaha dealer should the original split or be damaged in an accident.

18 Speedometer and tachometer drives: location and examination

RD125 LC models

1   The speedometer drive gearbox is situated on the left-hand side of the front wheel hub. It is a single component which cannot be stripped easily. Note that there are no parts available to recondition it and any damage or malfunction can only be corrected by renewing the gearbox. The only maintenance required is that a small quantity of grease be packed into it whenever the front wheel is removed for wheel bearing examination or replacement.

2   The tachometer drive consists of a worm mechanism driven by the crankshaft through the clutch outer drum and a driven gear mounted vertically in the crankcase. This mechanism is lubricated by the gearbox oil and is fully enclosed, therefore requiring no maintenance at all. Again any damage or malfunction can only be corrected by renewing the parts concerned. Inspect the mechanism whenever the right-hand outer cover is removed.

DT125 LC models

3   The speedometer drive assembly is contained within the front wheel brake backplate and should be examined and repacked with grease whenever work is carried out on the wheel bearings or brake assembly.

4   To remove the main drive gear from its housing, place the brake backplate assembly, inner side uppermost, on a work surface. Remove the brake shoes and springs, as described in Section 13 of Chapter 6, and inspect the large dust seal for signs of damage or deterioration. To renew this seal, carefully lever it out of position using the flat of a screwdriver; great care must be taken not to damage the surrounding alloy casting. The new seal may be fitted after removal, examination and fitting of the drive gear assembly.

5   Using a suitable pair of circlip pliers, remove the circlip which retains the main drive gear, then lift out the first thrust washer, the drive piece, the drive gear itself, and the second thrust washer. Invert the backplate. The worm gear is retained by a threaded bush which is screwed into the recess in the backplate itself. Using a thin-nosed punch in the slots cut in the outer rim of the bush, unscrew the bush by tapping carefully. Alternatively a peg spanner may be cut from a length of tubing of suitable diameter, so that protrusions are left on one end of the tube which will engage with the slots in the bush. With the bush unscrewed, pull out the worm gear, noting the thrust washer fitted to locate it correctly.

18.1 Speedometer drive gearbox – RD125

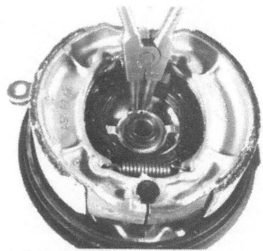

18.3 DT125 speedometer drive is in front brake backplate

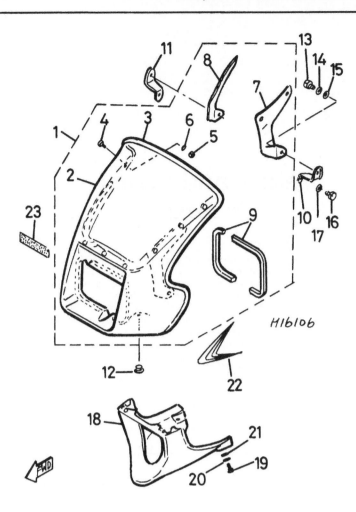

HIbIO6

Fig. 5.18 Fairing

1   Fairing assembly
2   Fairing
3   Screen trim
4   Screw - 9 off
5   Nut - 9 off
6   Washer - 9 off
7   Left-hand mounting bracket
8   Right-hand mounting bracket
9   Headlamp trim
10 Left-hand bracket
11 Right-hand bracket
12 Grommet
13 Bolt - 2 off
14 Spring washer - 2 off
15 Washer - 2 off
16 Bolt - 2 off
17 Spring washer - 2 off
18 Belly fairing
19 Bolt - 2 off
20 Spring washer - 2 off
21 Washer - 2 off
22 Emblem
23 Front emblem

20 Fairing: RD125 LC model

1   The fairing fitted by Yamaha to these machines is a simple plastic moulding retained by two metal brackets which are bolted to the headlamp bracket, the whole assembly being steadied and further supported by two grommets which are fitted over lugs protruding upwards from the lower part of the headlamp bracket. It has a transparent plastic windscreen which is secured by nine screws and nuts. All the component parts, including the various pieces of trim, are available separately if required.

2   Maintenance is restricted to a periodic check that all fixings are in good condition and to cleaning and polishing regularly so that the finish is preserved for as long as possible. Be especially careful when cleaning plastic mouldings as the surface will soon become abraded and dull, attracting road dirt and grime very quickly, if not cared for.

Chapter 6 Wheels, brakes and tyres

For modifications, and information relating to later models, see Chapter 8

Contents

Specifications

Wheels

	RD125 LC	DT125 LC
Type	Cast alloy	Steel wire spoked
Size:		
Front	1.60 x 18	1.60 x 21
Rear	1.85 x 18	1.85 x18
Rim runout limit:		
Radial and axial	2.0 mm (0.08 in)	2.0 mm (0.08 in)

Brakes

	RD125 LC	DT125 LC
Type:		
Front	Hydraulic disc	Drum
Rear	Drum	Drum
Brake disc:		
Thickness	4 mm (0.16 in)	
Wear limit	3.5 mm (0.14 in)	
Maximum runout	0.15 mm (0.006 in)	
Brake pad:		
Friction material thickness	10.6 mm (0.42 in)	
Wear limit	0.8 mm (0.03 in)	
Master cylinder bore ID	12.7 mm (0.49 in)	
Brake caliper bore ID	38.1 mm (1.49 in)	
Brake drum ID	130 mm (5.12 in)	
Wear limit	131 mm (5.16 in)	
Brake shoe:		
Lining thickness	4 mm (0.16 in)	
Wear limit	2 mm (0.08 in)	
Return spring free length	33 mm (1.30 in)	

Tyre size

	RD125 LC	DT125 LC
Front	2.75 x 18-4PR	2.75 x 21-4PR
Rear	3.00 x 18-6PR	4.10 x 18-4PR

Tyre pressure

	RD125 LC	DT125 LC
Front - solo	26 psi (1.82 kg/cm^2)	18 psi (1.26 kg/cm^2)
Rear - solo	28 psi (1.96 kg/cm^2)	22 psi (1.54 kg/cm^2)
Front - pillion	26 psi (1.82 kg/cm^2)	22 psi (1.54 kg/cm^2)
Rear - pillion	32 psi (2.24 kg/cm^2)	26 psi (1.82 kg/cm^2)

For continuous high-speed riding, the rear tyre pressure should be increased to:

Solo	32 psi (2.24 kg/cm^2)	Not applicable
Pillion	40 psi (2.80 kg/cm^2)	Not applicable

Torque settings

Component	kgf m	lbf ft
Front wheel spindle nut:		
RD125 LC	7.5	54.0
DT125 LC	4.0	29.0
Rear wheel spindle nut	8.5	61.5
Brake torque arm/backplate nut - RD125 LC only	2.0	14.5
Sprocket mounting studs - DT125 LC only	3.9	28.0
Sprocket retaining nuts:		
RD125 LC	2.3	17.0
DT125 LC	4.5	32.5
Brake disc mounting bolts	2.0	14.5
Brake caliper/fork lower leg mounting bolts	3.5	25.0
Brake hose union bolts	2.5	18.0
Brake caliper bleed nipple	0.6	4.0

1 General description

The RD125 LC models are fitted with cast aluminium alloy wheels carrying conventional tubed tyres. The front brake is a hydraulically-operated disc and the rear is a single leading shoe (sls) drum, operated by a rod linkage from the brake pedal.

The DT125 LC models are fitted with wheels of conventional wire-spoked construction, using chromed steel rims of 21 inch diameter at the front, and 18 inch diameter at the rear. The tyres are again of the conventional tubed type. Drum brakes of the single leading shoe (sls) type are fitted front and rear, the front being operated by a cable and the rear by a rod linkage.

2 Front wheel: examination and renovation – RD125 LC models

1 Carefully check the complete wheel for cracks and chipping, particularly at the spoke roots and the edge of the rim. As a general rule a damaged wheel must be renewed as cracks will cause stress points which may lead to sudden failure under heavy load. Small nicks may be radiused carefully with a fine file and emery paper (No 600 - No 1000) to relieve the stress. If there is any doubt as to the condition of a wheel, advice should be sought from a reputable dealer or specialist repairer.

2 Each wheel is covered with a coating of lacquer, to prevent corrosion. If damage occurs to the wheel and the lacquer finish is penetrated, the bared aluminium alloy will soon start to corrode. A whitish grey oxide will form over the damaged area, which in itself is a protective coating. This deposit however, should be removed carefully as soon as possible and a new protective coating of lacquer applied.

3 Check the lateral runout at the rim by spinning the wheel and placing a fixed pointer close to the rim edge. If the maximum runout is greater than 2.0 mm (0.080 in) the manufacturer recommends that the wheel be renewed. This is, however, a counsel of perfection; a runout somewhat greater than this can probably be accommodated without noticeable effect on steering. No means is available for straightening a warped wheel without resorting to the expense of having the wheel skimmed on all faces. If warpage was caused by impact during an accident, the safest measure is to renew the wheel complete. Worn wheel bearings may cause rim runout. These should be renewed.

3 Front wheel: examination and renovation – DT125 LC models

1 Wire spoked wheels are often viewed as being prone to problems when compared to the increasingly popular cast alloy and composite types. Whilst this is true to some extent, it is also true that wire spoked wheels are relatively easy and inexpensive to adjust or repair. Spoked wheels can go out of true over periods of prolonged use and like any wheel, as the result of an impact. The condition of the hub, spokes and rim should therefore be checked at regular intervals.

2 For ease of use an improvised wheel stand is invaluable, but failing this the wheel can be checked whilst in place on the machine after it has been raised clear of the ground. Make the machine as stable as possible, if necessary using blocks beneath the crankcase as extra support. Spin the wheel and ensure that there is no brake drag. If necessary, slacken the brake adjuster until the wheel turns freely. In the case of rear wheels it is advisable though not essential, to remove the final drive chain.

3 Slowly rotate the wheel and examine the rim for signs of serious corrosion or impact damage. Slight deformities, as might be caused by running the wheel along a curb, can often be corrected by adjusting spoke tension. More serious damage may require a new rim to be fitted, and this is best left to an expert. Whilst this is not an impossible undertaking at home, there is an art to wheel building, and a professional wheel builder will have the facilities and parts required to carry out the work quickly and economically, Badly rusted steel rims should be renewed in the interests of safety as well as appearance. Where light alloy rims are fitted corrosion is less likely to be a serious problem, though neglect can lead to quite substantial pitting of the alloy.

4 If it has been decided that a new rim is required some thought should be given to the size and type of the replacement rim. In some instances the problem of obtaining replacement tyres for an oddly sized original rim can be resolved by having a more common rim size fitted. Do check that this will not lead to other problems, fitting a new rim whose tyre fouls some other part of the machine could prove a costly error. Remember that changing the size of the rear wheel rim will alter the overall gearing. In most cases it should be possible to have a light alloy rim fitted in place of an original plated steel item. This will have a marginal effect in terms of weight reduction, but will prove far more corrosion resistant.

5 Assuming the wheel to be undamaged it will be necessary to check it for runout. This is best done by arranging a temporary wire pointer so that it runs close to the rim. The

wheel can now be turned and any distortion noted. Check for lateral distortion and for radial distortion, noting that the latter is less likely to be encountered if the wheel was set up correctly from new and has not been subject to impact damage.

6   The rim should be no more than 2.0 mm (0.1 in) out of true in either plane. If a significant amount of distortion is encountered check that the spokes are of approximately equal tension. Adjustment is effected by turning the square-headed spoke nipples with the appropriate spoke key. This tool is obtainable from most good motorcycle shops or tool retailers.

7   With the spokes evenly tensioned, any remaining distortion can be pulled out by tightening the spokes on one side of the hub and slackening the corresponding spokes from the opposite hub flange. This will allow the rim to be pulled across whilst maintaining spoke tension.

8   If more than slight adjustment is required it should be noted that the tyre and inner tube should be removed first to give access to the spoke ends. Those which protrude through the nipple after adjustment should be fitted flat to avoid the risk of puncturing the tube. It is essential that the rim band is in good condition as an added precaution against chafing. In an emergency, use a strip of duct tape as an alternative; unprotected tubes will soon chafe on the nipples.

9   Should a spoke break a replacement item can be fitted and retensioned in the normal way. Wheel removal is usually necessary for this operation, although complete removal of the tyre can be avoided if care is taken. A broken spoke should be attended to promptly because the load normally taken by that spoke is transferred to adjacent spokes which may fail in turn.

10   Remember to check wheel condition regularly. Normal maintenance is confined to keeping the spokes correctly tensioned and will avoid the costly and complicated wheel rebuilds that will inevitably result from neglect. When cleaning the machine do not neglect the wheels. If the rims are kept clean and well polished many of the corrosion related maladies will be prevented.

4   Front wheel: removal and refitting

RD125 LC models

1   Place the machine on its centre stand and support the front wheel clear of the ground with a wooden block under the engine. Unscrew the knurled ring retaining the speedometer cable and withdraw the cable. Remove the split pin from the wheel spindle nut and unscrew the nut. Remove the wheel spindle. It may be necessary to use a hammer and soft metal drift gently to tap out the spindle. Withdraw the front wheel. Wedge a piece of wood between the brake pads to prevent ejection of the caliper piston should the front brake lever be inadvertently applied while the wheel is removed.

2   Refitting is a straightforward reversal of the removal procedure. Guide the brake disc carefully between the brake pads to avoid damage. Make sure that the wheel spindle is clean and completely free of corrosion, then grease it lightly before fitting to ease dismantling in the future. Ensure that the speedometer drive gearbox is correctly located on the driving piece in the hub, and ensure that the ribs on the gearbox are located correctly against the lug on the fork lower leg. Tighten the wheel spindle nut to a torque setting of 7.5 kgf m (54 lbf ft). Replace the split pin with a new one and spread its ends correctly. Lastly apply the front brake lever repeatedly until the pads are moved back against the disc and full lever pressure is restored, check for free front wheel rotation, for correct speedometer operation, and that the front brake works properly.

DT125 LC models

3   Place the machine on a strong wooden box or similar support to hold the machine securely so that its front wheel is clear of the ground. Although it is possible to speed up the process of wheel removal by not disconnecting the speedometer and front brake cables, the complete process is described below for reference. Note that if the shorter method

is adopted, the brake backplate must be removed from the wheel hub as soon as the wheel is clear of the forks, and that great care must be taken not to twist, kink, or otherwise damage the cables.

4   Using a suitable pair of pliers, remove the circlip which retains the lower end of the speedometer drive cable, then remove the cable by pulling it gently out of the recess in the brake backplate. Before the brake cable can be disconnected, enough free play must be gained for this to be possible. Starting at the handlebar end of the cable, pull back the rubber cover, slacken fully the adjuster locknut, and screw the adjuster fully in. Working at the cable lower end, slacken the adjuster lower nut (the locknut) and screw it down off the threaded length of the cable outer, displacing the grommet to permit this, then slacken the cable clamp at the top of the fork lower leg. Pull the cable outer straight up until the cable inner can be slipped out through the slot in the adjuster boss cast in the brake backplate. If insufficient free play exists for this to tbe possible, disconnect the cable at the handlebar end. Disengage the cable lower end nipple from the brake operating arm, then straighten and remove the split pin securing the wheel spindle nut. Slacken and remove the spindle nut, then gently tap out the spindle and remove the front wheel from the machine.

5   On refitting, ensure that the tabs on the speedometer drive piece in the brake backplate are aligned correctly with the corresponding slots in the wheel hub boss, and check that the spacer and dust-excluding collar are correctly inserted in the hub right-hand side. Note that a plain washer is fitted into a recess in the wheel spindle lug of each of the fork lower legs; while these are glued in place at the factory, they may work loose. Check that they are secure whenever the wheel is removed and never omit them. Insert the wheel between the fork lower legs, ensuring that the large lug on the fork left-hand lower leg fits into the slot cast in the brake backplate. Check that the spindle is clean and completely free from corrosion, smear grease over it to ease dismantling in the future, and push it through from right to left. Refit the spindle nut and tighten it by hand only at first. Connect the brake cable by reversing the above disconnecting sequence and return the adjusters to their original positions. Spin the wheel and apply the brake lever hard to centralise the brake shoes and backplate on the drum. While keeping firm pressure on the brake lever, tighten the spindle nut to a torque setting of 4.0 kgf m (29 lbf ft), then release the brake lever and fit a new split-pin to secure the spindle nut, spreading correctly the ends of the split-pin. Refit the speedometer drive cable, applying a smear of grease to the metal abutment on the cable lower end to prevent corrosion. Secure the cable with its circlip and adjust the front brake as described in Section 13 of this Chapter. Finally check for free wheel rotation and correct speedometer operation, and that the front brake works properly.

4.2a Do not omit wheel right-hand spacer

4.2b Ensure that tangs on speedometer drive ring engage in gearbox cut-outs

4.2c Protruding lug on fork lower leg must fit between ribs on gearbox ...

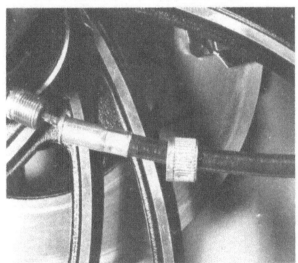

4.2d ... before spindle is refitted

4.2e Connect speedometer cable to gearbox

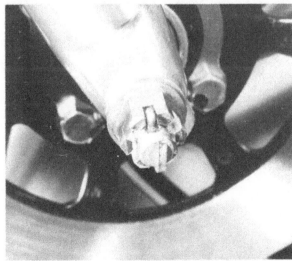

4.2f Tighten spindle nut and secure as shown with split-pin

4.4a Release circlip and withdraw speedometer cable

4.4b Release brake cable from clamp

4.5a Ensure that speedometer drive tangs engage with slots in hub boss

4.5b Do not omit wheel right-hand spacer

4.5c Push spindle through from right-to left, ensuring that ...

4.5d ... large lug on fork lower leg fits into groove in brake-backplate. Note loose plain washer

4.5e Tighten spindle nut only lightly until brake is centralised ...

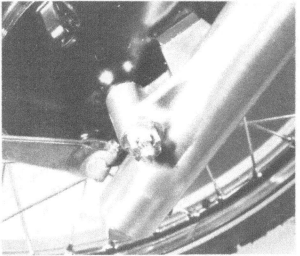

4.5f ... then tighten securely and fit split pin as shown

4.5g Connect speedometer cable and secure circlip

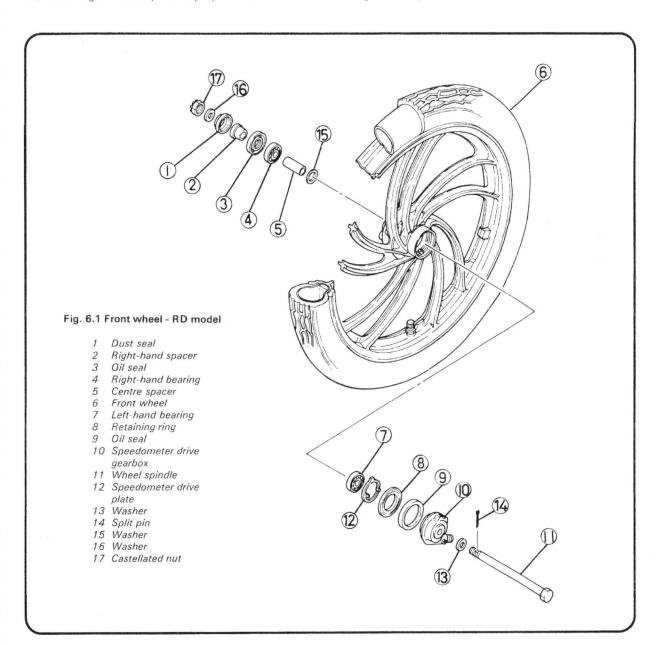

Fig. 6.1 Front wheel - RD model

1 Dust seal
2 Right-hand spacer
3 Oil seal
4 Right-hand bearing
5 Centre spacer
6 Front wheel
7 Left-hand bearing
8 Retaining ring
9 Oil seal
10 Speedometer drive
   gearbox
11 Wheel spindle
12 Speedometer drive
   plate
13 Washer
14 Split pin
15 Washer
16 Washer
17 Castellated nut

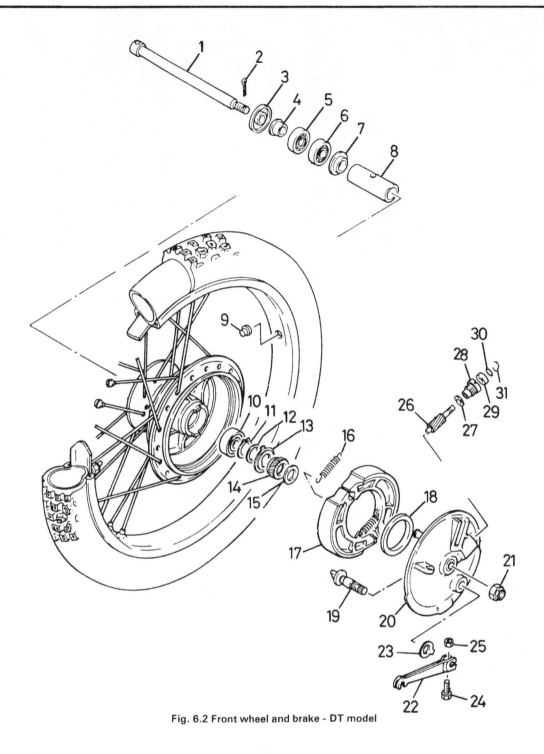

Fig. 6.2 Front wheel and brake - DT model

1	Wheel spindle	12	Thrust washer	22	Brake operating lever
2	Split pin	13	Speedometer drive plate	23	Wear indicator plate
3	Dust seal	14	Speedometer drive gear	24	Bolt
4	Right-hand spacer	15	Thrust washer	25	Nut
5	Oil seal	16	Return spring - 2 off	26	Speedometer driven gear
6	Right-hand bearing	17	Brake shoe - 2 off	27	Washer
7	Spacer flange	18	Oil seal	28	Bush
8	Centre spacer	19	Brake operating cam	29	Oil seal
9	Plug	20	Brake backplate	30	O-ring
10	Left-hand bearing	21	Castellated nut	31	Circlip
11	Circlip				

5 Front wheel bearings: removal, examination and refitting

1 Before the front wheel bearings can be examined, the front wheel must be removed as described in Section 4 of this Chapter. On the RD125 LC models, the speedometer drive gearbox must be withdrawn, followed by the oil seal, the speedometer drive ring retainer, and the speedometer drive ring. Heat the end of an old flat-bladed screwdriver and bend the tip into a slightly curved shape with no sharp edges. This will give a useful tool for levering out the oil seal without damaging the sealing lip or the spring beneath it. The spacer and dust seal on the disc side of the hub can then be withdrawn, if desired, in a similar manner, or they can be driven out with the bearings. For DT125 LC models the brake backplate must be withdrawn. The dust seal and spacer on the opposite side of the hub can be removed as described above or driven out with the bearings as desired.

2 Although the two types of hub are vastly dissimilar in external appearance, they are essentially the same in design. Bearing removal procedure is the same in both cases. Support the wheel firmly on two wooden blocks as close to the hub centre as possible to prevent distortion, ensuring that enough space is allowed to permit bearing removal. Place the end of a small flat-ended drift against the upper face of the lower bearing and tap the bearing downwards out of the wheel hub. The spacer located between the two bearings may be moved sideways slightly in order to allow the drift to be positioned against the face of the bearing. Move the drift around the face of the bearing whilst drifting it out of position, so that the bearing leaves the hub squarely. The end spacer and dust seal will be driven out along with the bearing.

3 With the one bearing removed, the wheel may be lifted and the spacer withdrawn from the hub. Invert the wheel and remove the second bearing, using a similar procedure to that used for the first.

4 Wash the bearings thoroughly in clean petrol to remove all traces of the old grease. Check the bearing tracks and balls for wear or pitting or damage to the hardened surfaces. A small amount of side movement in the bearing is normal but no radial movement should be detectable. Check the bearings for play and roughness when they are spun by hand. All used bearings will emit a small amount of noise when spun but they should not chatter or sound rough. If there is any doubt about the condition of the bearings they should be renewed.

5 Carefully clean the bearing recesses in the hub and the centre space of the hub. All traces of the old grease, which may be contaminated with dirt, must be removed. Examine the oil seals which are removed and renew them if any damage or wear is found.

6 Before replacing the bearings pack them with high melting point grease. This applies equally to the original bearings, if refitted, and to new ones, if the originals are to be renewed. With the wheel firmly supported on the two wooden blocks, tap a bearing into place in the hub noting that the sealed surface must face outwards. Use a hammer and a tubular metal drift or socket spanner which bears only on the outer race of the bearing to drive the bearing into position. If the inner race or sealed surface of the bearing is used to drive it into place, severe damage will be done to the bearing due to the high side loadings thus imposed.

7 Once one bearing has been installed, invert the wheel, fit the central spacer and pack the remaining space no more than $\frac{2}{3}$ full of grease. This is important as although some grease must be present, it will expand when hot and if too much grease is in the centre space, it will force its way past the seals and out onto the brake components. Once the grease is packed in, fit the second bearing in the same manner as the first. The dust seal in the right-hand side of the hub can be pressed into position by hand. The wheel spacer can then be greased and pushed into the dust seal.

8 On RD125 LC models the speedometer drive ring should

now be fitted, ensuring that the drive tangs are located in the slots provided for them in the hub. Insert the drive ring retainer, then use a hammer and a tubular drift or socket spanner which bears on the outside diameter of the oil seal to drive the oil seal gently into position. Install the speedometer drive gearbox and fit the front wheel assembly back into the front forks as described in the relevant paragraphs of Section 4. On DT125 LC models replace the brake backplate in the hub and replace the front wheel in the forks as described in Section 4.

5.1 Use an old screwdriver to lever out oil seals as shown

6 Front disc brake: general

1 As already mentioned, the front brake is of the hydraulically-operated disc type. It should be noted that a number of precautions should be taken when dealing with this system, because brake failure can have disastrous consequences.

2 The hydraulic system must be kept free from air bubbles. Any air in the system will be compressed when the brake lever is operated instead of transmitting braking effort to the disc. It follows that efficiency will be impaired, and given sufficient air, this can render the brake inoperative. If any part of the hydraulic system is disturbed, the system must always be bled to remove any air. See Section 12 for details. It is vital that all hoses, pipes and unions are examined regularly and renewed if damage, deterioration or leakage is suspected.

3 Hydraulic fluid is specially formulated for given applications, and must always be of the correct type. Any fluid conforming to SAE J1703 or DOT 3 may be used; other types may not be suitable. On no account should any other type of oil or fluid be used. Old or contaminated fluid must be discarded. It is dangerous to use old fluid which may have degraded to the point where it will boil in the caliper, creating air bubbles in the system. Note that brake fluid will attack and discolour paintwork and plastics. Care must be taken to avoid contact and any accidental splashes must be washed off immediately.

4 Cleanliness is more important in hydraulic brake system than in any other part of a motorcycle. Dirt will rapidly destroy the seals, allowing fluid to leak out or air to be drawn in. Water, even in the form of moist air, will be absorbed by the fluid which is hygroscopic, the water degrading the fluid, lowering its boiling point until it can boil in use. The master cylinder and any cans of fluid must be kept closed to prevent this.

5 It should be emphasised that repairs to the master cylinder and to the caliper are best entrusted to a Yamaha Service

Agent, or alternatively, that the defective parts should be replaced by a new unit. Dismantling and reassembly requires a certain amount of skill and it is imperative that the entire operation is carried out under surgically clean conditions.

7 Front disc brake: pad renewal

1   Although the front brake pads must be removed from the machine at regular intervals for cleaning and close inspection, it is not actually necessary to remove them for the degree of wear to be assessed. With the machine supported on its stand, view the caliper from the rear, looking upwards to see the pads, then look downwards to see the pads from in front of the fork right-hand lower leg. It will be seen that the two outermost corners of the metal backing of each pad are bent inwards towards the disc, forming wear indicator tabs. When the friction material is worn down to the point where these tabs contact the disc, the pads must be considered worn out and be renewed as a set. Always renew the pads as a set to ensure maximum braking efficiency is maintained; they are usually available only as sets to make sure of this. When the wear indicator tabs touch the disc, a pronounced squeal should be heard whenever the brake is applied; if the brake is used for a sustained period in this state, not only will the available braking effort be reduced severely, but the disc will also be scored heavily. To prevent this possibility from arising the following course is recommended; it will be seen that a deep groove is cut down the centre of each pad. When the friction material is worn down to the point where the groove disappears, the pad should be considered worn out and renewed. This will provide a small safety margin of wear before the wear indicator tabs come into contact with the disc.
2   While it is just possible to remove the pads without disturbing the caliper, the task is very awkward and tricky to accomplish successfully. It is therefore recommended that the caliper is removed, as described below, to carry out this operation. Using an Allen key of suitable size, slacken the pad retaining bolt, then slacken and remove the two brake caliper/fork lower leg mounting bolts and pull the complete caliper assembly away from the fork leg and disc, noting that while it is not necessary to disconnect the brake hose during this operation, great care must be taken to ensure that the hose is not twisted, kinked, or otherwise strained in any way. Unscrew and remove fully the pad retaining bolt, then lift out the pads. Note carefully the exact position of the anti-squeal shim fitted to the back of the pad nearest the piston. This shim must not be omitted and must be refitted in the same position on reassembly; the same applies to the anti-rattle spring fitted in the body of the caliper.
3   Carefully clean both pads, the anti-squeal shim, the anti-rattle spring, and the caliper body, removing all traces of road dirt and any deposits of friction material. If there are any traces of grease or oil on the friction material of either pad, the two must be renewed, regardless of the state of wear. Oil will reduce severely the braking efficiency and cannot be removed satisfactorily. Clean out the centre groove of each pad, and pick out any particles of foreign matter that may be embedded in the friction material, using any sharp-pointed instrument that may be available. Areas of glazing may be removed by rubbing carefully with emery paper to roughen the pad surface and break the glaze.
4   Obtain some copper-based very high melting-point grease that is designed specifically for use in brake calipers. Check that the caliper body slides easily on the two mounting bracket pins, taking remedial action as described in Section 11 of this Chapter if any stiffness is encountered. It will be seen that the metal backing of each pad comes into contact with the caliper body or the caliper mounting bracket when the caliper is assembled; apply a thin smear of grease to the sides of each pad's metal backing at these points to prevent the onset of corrosion. Similarly, apply a thin smear of grease to the holes in the metal backings through which the pad retaining bolt passes,

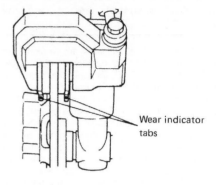

Fig. 6.3 Front brake pad wear check

Wear indicator tabs

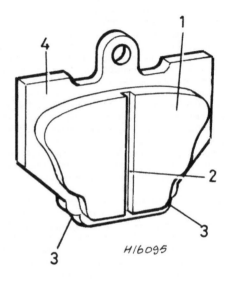

H16095

Fig. 6.4 Front brake pad

| 1 | Friction material | 3 | Wear indicator tabs |
| 2 | Central groove | 4 | Metal backplate |

and to the whole of the back of the pad which is to be fitted against the caliper piston. Take great care to apply only a very thin smear of grease to each of the areas mentioned, and do not allow any grease on to the friction material. Lastly, thoroughly clean the shank of the pad retaining bolt, removing all traces of corrosion, and apply a thin smear of grease to ensure the pads can move easily along it.
5   On reassembly, refit the anti-rattle spring in the caliper body, and ensure that the caliper mounting bracket is pushed hard against the caliper itself. If new pads are to be fitted, the caliper piston must be pushed back to provide sufficient clearance; if this cannot be done by hand, the caliper must be stripped to find the reason, as described in Section 11 of this Chapter. While pushing back the piston, watch the fluid level in the master cylinder reservoir. If this has been overfilled, the level may rise to the point where leakage or spillage may occur, and the surplus fluid should be drained off, therefore, before the machine is used. Refit the anti-squeal shim to the back of one

pad, then insert the pad into the caliper, pushing through the pad retaining bolt to secure it and remembering that if the pads have not been renewed, they must be fitted in their original positions. Insert the remaining pad, push through the pad retaining bolt, and tighten the bolt as hard as is possible with the Allen key alone.

6   Refit the caliper assembly to the machine, ensuring that the disc passes correctly between the pads, and tighten the two caliper/fork lower leg mounting bolts to a torque setting of 3.5 kgf m (25 lbf ft). Check that the pad retaining bolt is securely fastened, remembering that the leverage provided by the longer length of the key will be enough to ensure this. Repeatedly apply the front brake lever to bring the pads into firm contact with the disc, then check the fluid level in the master cylinder reservoir.

7   If the original pads have been refitted, it will be sufficient to check that the fluid level has not fallen below the 'Lower' mark on the sight glass set in the rear face of the fluid reservoir. If necessary, fluid should be added to raise the level to this mark. If new pads have been fitted, remove the reservoir cover by unscrewing the two retaining screws, lift out the rubber

diaphragm, and check that the level has not risen to above the cast level mark which protrudes from the inside face of the reservoir. Remove any surplus fluid or, if necessary, top up to the level mark. Carefully dry the diaphragm on a clean cloth, removing the moisture which tends to gather in its folds, then fold it carefully into its compressed state. Refit the diaphragm and the cover, tightening securely the two screws. If it is necessary to add fluid at any time, be careful to use the same type and make of fluid at all times, and use only fluid which conforms to SAE J1703 or DOT 3. Remember to fasten securely the cap of the fluid container after use, and never use fluid that has come from an unsealed container. Check that all bolts and screws are securely fastened, that there are no leaks from the system, and that the front brake works efficiently before taking the machine out on the road. Wash off immediately any surplus fluid.

8   If new pads have been fitted, they must be bedded-in before they reach full braking efficiency. Use the brake carefully for the first 100 miles, trying to avoid fierce applications and using firm but gentle pressure so that the pads bed in properly without the glazing that would result from over-light use.

7.2a Slacken and remove two mounting bolts to release caliper

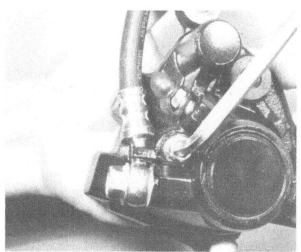

7.2b Unscrew pad retaining bolt ...

7.2c ... to release brake pads

7.2d Note the position of anti-squeal shim ...

7.2e ... and of anti-rattle spring

7.5a Refit pad with anti-squeal shim as shown

7.5b Push back piston to make clearance for new pads

7.7a Fluid must not be above cast level mark (arrowed). Note diaphragm in correctly folded state

8   Hydraulic hose: examination

1   The handlebar-mounted master cylinder assembly and the caliper assembly are connected by a flexible hose, fastened at each end by a conventional banjo union. It is secured at the top of the fork lower leg and at the headlamp brackets by metal clamps which engage on rubber sleeves wrapped around the hose.

2   While the hose is very strong in construction it must withstand considerable pressures in use, and must also endure a considerable degree of flexing due to suspension movement. In addition to the above, the hose is prone to decay due to the natural process of ageing, this manifesting itself on the outside in the form of splits and cracking due to the perishing of the hose material. For the above reasons the manufacturer recommends that the hose be renewed every four years irrespective of its apparent condition. This recommendation is made on the grounds of safety alone and must be backed up by a close inspection of the hose at regular intervals, with the hose being renewed as soon as any damage is found, whether it is four years old or not.

7.7b Tighten securely master cylinder cover

3   Starting at the top and working downwards along the length of the hose, check for any signs of damage such as cracks or splits, and for any signs that the hose is chafing against another component. Actual fluid leakage will be revealed by dark stains on the hose surface. If any signs of damage are found, the hose must be renewed immediately, as repairs are not possible.

4   On refitting the hose, renew the sealing washers at each end as a safety precaution, and ensure the hose is routed correctly, with no sharp bends or kinks. Refit the supporting clamps so that the hose cannot chafe against another component, and tighten the union banjo bolts to a torque setting of 2.5 kgf m (18 lbf ft).

8.4 Secure hose with clamps provided to prevent chafing

9   Front brake disc: removal, examination and refitting

1   The brake disc is machined from a single piece of stainless steel and is bolted to the hub by four bolts; these bolts being secured by two tab washers. Examination of the disc can be carried out with the front wheel installed.

2   Look first for signs of excessive scoring. While some degree of scoring is inevitable, deep grooves or marks will mean that the disc must be renewed to restore full braking efficiency. The only alternative is to locate an engineering company who are willing to skim the disc to remove the marks. If this is the case, remember that the disc must not be skimmed down below the minimum thickness specified by the manufacturer, ie 3.5 mm (0.14 in). Check the disc for warpage, which can often result from overheating or accident damage and may cause brake judder. This is best checked using a dial gauge mounted on the fork leg and should not exceed 0.15 mm (0.006 in).

3   The disc thickness should be measured using a vernier caliper or a micrometer in several places around the disc surface. The nominal thickness is 4.0 mm (0.16 in) and the disc must be renewed if it is worn to less than 3.5 mm (0.14 in) at any point.

4   To remove the disc, the front wheel must be removed as described in Section 4 of this Chapter. Knock back the locking tabs of the two tab washers, then slacken and remove the four bolts. Lift the disc away, but note that if it is to be refitted, it is considered good practice to mark both disc and hub so that the disc can be refitted in its original position. On reassembly, tighten the four bolts evenly and in stages to a torque setting of 2.0 kgf m (14.5 lbf ft). Do not omit to secure each bolt by bending the locking tabs of the tab washers against its head.

9.3 Measuring thickness of brake disc

10   Front disc brake: overhauling the master cylinder

1   The master cylinder forms a unit with the hydraulic fluid reservoir and front brake lever, and is mounted by a clamp to the right-hand side of the handlebars.

2   The unit must be drained before any dismantling can be undertaken. Place a suitable container below the caliper unit and run a length of plastic tubing from the caliper bleed screw to the container. Unscrew the bleed screw one full turn and proceed to empty the system by squeezing the front brake lever. When all the fluid has been expelled, tighten the bleed screw and remove the tube.

3   Select a suitable clean area in which the various components may be safely laid out, a large piece of white, lint-free cloth or white paper being ideal.

4   Remove the locknut and the brake lever pivot bolt to free the lever. As it is lifted away note the small spring which is fitted into the end of the lever blade. Release the front brake switch by pushing a small screwdriver blade into the hole beneath the master cylinder extension which houses it. The switch can now be withdrawn.

5   Remove the two bolts which hold the master cylinder clamp half to the body and then lift the master cylinder away.

Remove the cover and empty the reservoir. If it is still connected, remove the banjo bolt and free the hydraulic hose from the master cylinder body.

6   Pull off the dust seal from the end of the piston bore to expose the piston end and the circlip which retains it. Remove the circlip to free the piston. If the piston tends to stick in the bore it can be pulled clear using pointed-nose pliers. As the piston is removed the main seal and spring will be released.

7   Examine the piston and seals for scoring or wear and renew if imperfect. Excessive scoring may be due to contaminated fluid, and if this is suspected, it is probably worth checking the condition of the caliper seals and piston. It is recommended that the piston seal is renewed as a matter of course because leakage often occurs once it has been disturbed. Note that the various components are available only as a complete master cylinder reconditioning kit and cannot be purchased separately. Use only clean hydraulic fluid to clean the various components during the overhaul. If the master cylinder bore is worn appreciably it must be renewed as repairs are not possible.

8   Soak the new seals in hydraulic fluid for about 15 minutes prior to refitting, then reassemble the master cylinder using the reversal of the dismantling procedure. Take great care that the seals are not damaged as they are inserted.

9   Clamp the master cylinder to the handlebars so that it will be horizontal when the machine is being ridden, and ensuring that the 'Up' mark on the handlebar clamp is facing upwards. Refit the hydraulic hose, tightening the banjo bolt to a torque setting of 2.5 kgf m (18 lbf ft), and ensuring that the rubber boot is correctly positioned.

10   Refill the reservoir and system with clean, unused, hydraulic fluid, and bleed any air from the system as described in Section 12 of this Chapter. Check that the brake is operating correctly before using the machine on the road and use the brake gently for the first 50 miles or so to allow the new components to bed in correctly. Remember also to wash away any spilt fluid before it has a chance to damage the finish of any component.

10.1 Master cylinder and fluid reservoir form one unit clamped to handlebars

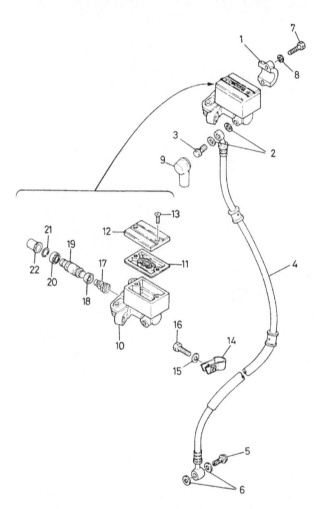

Fig. 6.5 Front brake master cylinder - RD model

1	Handlebar clamp	12	Reservoir cap
2	Sealing washer - 2 off	13	Screw - 2 off
3	Banjo union bolt	14	Hose guide
4	Hydraulic hose	15	Spring washer
5	Banjo union bolt	16	Bolt
6	Sealing washer - 2 off	17	Spring
7	Bolt - 2 off	18	Piston cup
8	Spring washer - 2 off	19	Piston
9	Rubber boot	20	Seal
10	Master cylinder reservoir	21	Circlip
11	Diaphragm	22	Boot

11 Front disc brake: overhauling the caliper unit

1   When working on the caliper unit, remember that cleanliness is essential. Wash the caliper and surrounding components to remove as much road dirt as possible before dismantling commences. Have ready a sufficient quantity of new brake fluid and obtain some very high melting-point brake caliper grease.

2   Slacken the caliper hose union bolt by just enough to permit it to be unscrewed when the caliper is removed from the machine, but do not remove it or slacken it too much as hydraulic pressure will be required to displace the piston. Remove the caliper from the machine and withdraw the pads as described in Section 7 of this Chapter. Withdraw the caliper mounting bracket by sliding it out of the caliper body, away from the piston. Obtain a large, clean, polythene bag, place the caliper inside it and hold the neck of the bag closed around the hose. Apply repeatedly the brake lever, using normal hydraulic pressure to push the piston out of the caliper. The bag will prevent the piston from dropping clear to the ground, where it might be damaged, and will restrict the inevitable shower of brake fluid. Remove the caliper from the bag, unscrew the brake hose union bolt, and allow the remaining fluid to drain into a suitable container. If the caliper body and the piston are so badly corroded or damaged that the above method is not effective in displacing the piston, do not waste time proceeding any further with dismantling. A new caliper assembly should be purchased and fitted immediately and the old one discarded as it will be of no further use.

3   Lay all the components of the caliper assembly out on a clean working surface. Carefully remove, if necessary, the two piston seals and the two sealing grommets from the caliper body, noting that while the two grommets are available separately, the piston seals can be obtained only as part of a kit with the piston and must be treated very carefully if unnecessary expense is to be avoided. The components removed should be cleaned thoroughly, using only brake fluid as the cleaning medium. Petrol, oil, or paraffin will cause the various seals to swell and to degrade, and should not be used under any circumstances. When the various parts have been cleaned, they should be stored in polythene bags until re-assembly to keep them dust free.

4   Examine all the components for signs of wear. Excluding accident damage, which should be obvious, the only areas likely to give trouble are the mounting bracket pins on which the

caliper slides, and the piston and caliper bore. Insert the mounting bracket into the caliper body and feel for free play; if the fit is excessively sloppy, or if the caliper cannot move freely and easily, the complete assembly must be renewed as the individual components are not available as separate items. Similarly examine the caliper bore and the outside diameter of the piston. Any damage, however small, will mean that the affected component must be renewed. Look for score marks, small scratches or nicks, and for pitting due to corrosion. Any one of these could lead to a sudden loss of braking pressure due to fluid leakage. Corrosion on that part of the piston which projects beyond the caliper body must be polished away using fine emery paper.

5   Reassemble the caliper under clinically clean conditions. Insert the seals into their grooves in the caliper bore, then smear hydraulic fluid over the inner (piston) seal and over the caliper bore inside that seal. Apply a very thin smear of caliper grease to the outer (dust) seal, and apply some hydraulic fluid to the piston outside diameter. Insert the piston, flat side first, into the caliper bore, taking the greatest care not to damage the seals. Push the piston as far in as it will go. Refit the two rubber grommets on to the caliper body, pack a small quantity of grease into the two mounting pin holes in the caliper body, and apply a thin smear of grease to each of the pins on the caliper mounting bracket. Insert the caliper mounting bracket into the caliper itself, taking care not to damage the two grommets.

6   Refit the brake pads and other components as described in Section 7 of this Chapter, and refit the caliper assembly to the machine. Tighten the brake caliper/fork lower leg mounting bolts to a torque setting of 3.5 kgf m (25 lbf ft), then connect the brake hose to the caliper, tightening the banjo bolt to a torque setting of 2.5 kgf m (18 lbf ft).

7   Refill the fluid reservoir and system with new hydraulic fluid and bleed any air from the system as described in the next Section of this Chapter. When all air has been removed, apply repeatedly the brake lever to bring the brake pads into firm contact with the disc, and check that the brake is operating correctly before using the machine on the road. Remember to wash away any spilt fluid before it has a chance to damage the finish of any component, and remember also to use the brake as gently as possible for the first 50 miles or so to allow the new components to bed in correctly.

11.2a Slide out caliper mounting bracket

11.2b Use hydraulic pressure to displace caliper piston

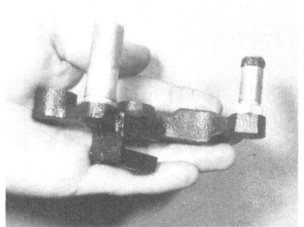

11.4a Check mounting bracket pins for wear or corrosion ...

11.4b ... also matching bores in caliper body. Renew sealing rubbers

Fig. 6.6 Front brake caliper - RD model

1 Bolt - 2 off
2 Bleed nipple cap
3 Piston
4 Anti-squeal shim
5 Brake pads
6 Tab washer - 2 off
7 Brake disc
8 Bolt - 4 off
9 Anti-rattle spring
10 Piston seal
11 Piston seal
12 Caliper
13 Bleed nipple
14 Sealing grommet - 2 off
15 Spring washer - 2 off
16 Washer - 2 off

11.5 Components must be absolutely clean on reassembly

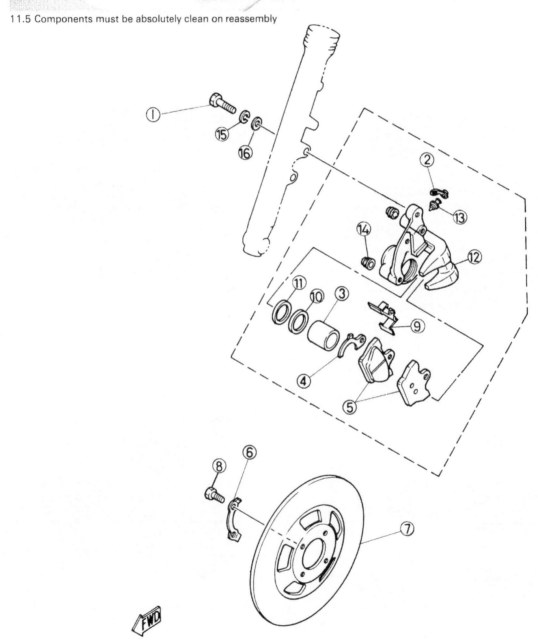

12 Bleeding the hydraulic brake system

1   The method of bleeding a brake system of air and the procedure described below apply equally to either a front brake or rear brake of the hydraulically actuated type.

2   If the brake action becomes spongy, or if any part of the hydraulic system is dismantled (such as when the hose is replaced) it is necessary to bleed the system in order to remove all traces of air. The procedure for bleeding the hydraulic system is best carried out by two people.

3   Check the fluid level in the reservoir and top up with new fluid of the specified type if required. Keep the reservoir at least half full during the bleeding procedure; if the level is allowed to fall too far air will enter the system requiring that the procedure be started again from scratch. Refit the cover on the reservoir to prevent the ingress of dust or the ejection of a spout of fluid.

4   Remove the dust cap from the caliper bleed nipple and clean the area with a rag. Place a clean glass jar below the caliper and connect a pipe from the bleed nipple to the jar. A clear plastic tube should be used so that air bubbles can be more easily seen. Place some clean hydraulic fluid in the glass jar so that the pipe is immersed below the fluid surface throughout the operation.

5   If parts of the system have been renewed, and thus the system must be filled, open the bleed nipple about one turn and pump the brake lever until fluid starts to issue from the clear tube. Tighten the bleed nipple and then continue the normal bleeding operation as described in the following paragraphs. Keep a close check on the reservoir level whilst the system is being filled.

6   Operate the brake lever as far as it will go and hold it in this position against the fluid pressure. If spongy brake operation has occurred it may be necessary to pump rapidly the brake lever a number of times until pressure is achieved. With pressure applied, loosen the bleed nipple about half a turn. Tighten the nipple as soon as the lever has reached its full travel and then release the lever. Repeat this operation until no more air bubbles are expelled with the fluid into the glass jar. When this condition is reached the air bleeding operation should be complete, resulting in a firm feel to the brake operation. If sponginess is still evident continue the bleeding operation; it may be that an air bubble trapped at the top of the system has yet to work down through the caliper.

7   When all traces of air have been removed from the system, top up the reservoir and refit the diaphragm and cover. Check the entire system for leaks, and check also that the brake system in general is functioning efficiently before using the machine on the road.

8   Brake fluid drained from the system will almost certainly be contaminated, either by foreign matter or more commonly by the absorption of water from the air. All hydraulic fluids are to some degree hygroscopic, that is, they are capable of drawing water from the atmosphere, and thereby degrading their specifications. In view of this, and the relative cheapness of the fluid, old fluid should always be discarded.

9   Great care should be taken not to spill hydraulic fluid on any painted cycle parts; it is a very effective paint stripper. Also the plastic glasses in the instrument heads, and most other plastic parts, will be damaged by contact with this fluid.

13 Front drum brake: examination, renovation and adjustment

1   The front brake assembly complete with the brake backplate can be withdrawn from the front wheel hub after removing the front wheel from the forks. With the wheel laid on a work surface, brake backplate uppermost, the brake backplate may be lifted away from the hub. It will come away quite easily, with the brake shoe assembly attached to its back.

12.4 Connect bleed tube arrangement as shown

2   Examine the condition of the brake linings. If they are thin or unevenly worn, the brake shoes should be renewed. The linings are bonded on and cannot be supplied separately. The linings are 4 mm (0.16 in) thick when new and should receive attention when worn to the wear limit thickness of 2 mm (0.08 in).

3   If fork oil or grease from the wheel bearings has badly contaminated the linings, the brake shoes should be renewed. There is no satisfactory way of degreasing the lining material.

4   Examine the drum surface for signs of scoring or oil contamination. Both of these conditions will impair braking efficiency. Remove all traces of dust, preferably using a brass wire brush, taking care not to inhale any of it, as it is of an asbestos nature, and consequently harmful. Remove oil or grease deposits, using a petrol soaked rag.

5   If deep scoring is evident, due to the linings having worn through to the shoe at some time, the drum must be skimmed on a lathe, or renewed. Whilst there are firms who will undertake to skim a drum whilst fitted to the wheel, it should be borne in mind that excessive skimming will change the radius of the drum in relation to the brake shoes, therefore reducing the friction area until extensive bedding in has taken place. Also full adjustment of the shoes may not be possible. If in doubt about this point, the advice of one of the specialist engineering firms who undertake this work should be sought.

6   Note that it is a false economy to try to cut corners with brake components; the whole safety of both machine and rider being dependent on their good condition.

7   Removal of the brake shoes is accomplished by folding the shoes together so that they form a 'V'. With the spring tension relaxed, both shoes and springs may be removed from the brake backplate as an assembly.

8   Before fitting the brake shoes, check that the brake operating cam is working smoothly and is not binding in its pivot. The cam can be removed by withdrawing the retaining bolt on the operating arm and pulling the arm off the shaft. Before removing the arm, it is advisable to mark its position in relation to the shaft, so that it can be relocated correctly. Lightly grease both the shaft and the faces of the operating cam and pivot prior to reassembly.

9   Before refitting existing shoes, roughen the lining surface sufficiently to break the glaze which will have formed in use. Glasspaper or emery cloth is ideal for this purpose but take care not to inhale any of the asbestos dust that may come from the lining surface.

10 Fitting the brake shoes and springs to the brake backplate is a reversal of the removal procedure. Some patience will be needed to align the assembly with the pivot and operating cam whilst still retaining the springs in position; once they are correctly aligned though, the shoes can be pushed back into position by pressing downwards in order to snap them into position. Do not use excessive force, or there is risk of distorting the brake shoes permanently.

11 Adjusting the front brake is best accomplished with the front wheel free to rotate, and is made at the cable lower adjuster. Spin the wheel and carefully screw the adjusting nut down until you hear a rubbing sound which shows that the brake shoes are lightly in contact with the drum surface. Turn the adjuster nut back by $\frac{1}{2}$ - 1 turn until the noise stops. Spin the wheel and apply the brake hard once or twice to settle the brake. Check that the wheel is still free to rotate and that the adjustment has remained the same. This setting should give you 5 - 8 mm (0.20 - 0.30 in) free play measured between the butt end of the brake lever and its handlebar clamp when the brake is firmly applied. Note that the handlebar lever adjuster should

be screwed in as far as possible while adjustment is being made. It should be used only for fine adjustments or for very quick ones.

12 If brake adjustment cannot be made correctly due to insufficient movement at the adjuster, and yet the brake wear indicator or measurement has revealed the linings to be in good order, the following course of action may be taken. Slacken fully both adjuster locknuts and screw in both adjusters so that the maximum free play is gained in the cable. Slacken and remove the pinch bolt retaining the operating arm on the camshaft, lever the arm off the camshaft end, and rotate the operating arm anti-clockwise about the camshaft end as far as the cable will allow. Refit the operating arm on to the camshaft splines as close to that position as possible, and secure the arm with its pinch bolt. This will take most of the surplus free play out of the cable, thus permitting the adjustment procedure described above to be carried out more successfully. This method should be used only to extend the useful life of the cable, and should be employed only when the brake shoes are known to be in good condition.

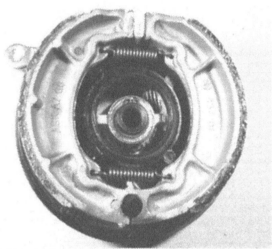

13.1 Brake backplate assembly can be withdrawn as one unit

13.2 Measuring thickness of brake linings

13.4 Brake drum must be clean and unworn

13.7 Remove brake shoes from backplate as shown

13.8a Smear camshaft lightly with grease

13.8b Note notch in camshaft splines to locate pointer correctly

13.8c Refit operating arm in position noted on dismantling

13.11 Brake is adjusted at brake backplate adjuster

14 Rear wheel: examination and renovation – all models

In the case of both machines described in this Manual the rear wheel is exactly the same in design and construction as the front, and differs only in size and hub design. Examination and renovation techniques are exactly the same as those given in Sections 2 and 3 of this Chapter. Refer to the Section appropriate to the machine being worked on, noting only that it is advisable to disconnect the rear chain so that the wheel can be spun freely.

15 Rear wheel: removal and refitting

RD125 LC models

1   Place the machine securely on its centre stand so that the rear wheel is raised clear of the ground. Unscrew the brake adjuster nut from the brake rod end and displace the rod from the brake arm. Push out the trunnion from the end of the brake arm and fit it, the return spring, and the nut to the end of the

brake rod for safe keeping. Detach the brake torque arm from the brake backplate lug.
2   Straighten and remove the split pin from the end of the wheel spindle and slacken the wheel spindle nut. Pull the wheel rearwards slightly to allow the adjusters to be pushed downwards through 90°. The wheel can now be pushed forward and the chain disengaged from its sprocket. Alternatively, the chain can be disconnected at its connecting link. Hang the chain over the rear fork end of the subframe to prevent its trailing in the dust or dirt on the ground. Remove the wheel spindle, using a hammer and a long metal drift gently to tap it out, withdraw the wheel right-hand spacer, and manoeuvre the wheel clear of the machine.
3   The wheel is installed by reversing the removal sequence. The chain adjusters must be fitted so that their stamped alignment marks can be seen aligning with those marks on the subframe. Do not forget to fit the wheel left-hand spacer, having first greased the oil seal lips. Check that the wheel spindle is straight, clean, and free from corrosion, then smear a liberal quantity of grease over it to assist dismantling in the future. Push the spindle through, ensuring that the two chain adjusters and the two spacers are correctly fitted, then fit the spindle nut finger-tight, install the chain and check that the chain tension

and wheel alignment are correct before final tightening. The chain tension should be checked with the machine *off* its centre stand, the correct amount of free play being 35 - 40 mm (1.4 - 1.6 in) measured at the centre of the lower run. Repeat the check several times with the wheel moved to reposition the chain. There will usually be one point at which the chain is at its tightest, and the chain tension should be set at this point.

4   Adjustment is effected by moving each of the adjuster drawbolts by an equal amount to preserve wheel alignment. As a guide, a row of alignment marks is provided on each side of the wheel spindle slot. Further information on chain adjustment and wheel alignment will be found in Section 19.

5   When chain adjustment is correct, refit the brake torque arm, tightening by hand only its retaining nut. Connect the brake rod again, apply firmly the back brake to centralise the brake shoes and backplate on the drum, and tighten the rear wheel spindle nut to a torque setting of 8.5 kgf m (61.5 lbf ft). Maintain pressure on the back brake pedal and tighten the torque arm retaining nut to a torque setting of 2.0 kgf m (14.5 lbf ft). Secure both nuts by fitting new split-pins through each, spreading the ends of the split-pins securely.

6   Finally, adjust the back brake by rotating the adjusting nut as necessary to give 20 - 30 mm (0.8 - 1.2 in) of free play, measured at the brake pedal tip, before the brake shoes come into firm contact with the drum. Adjust the stop lamp rear switch height so that the lamp lights just as the brake shoes are coming into firm contact with the drum. Check that all nuts and bolts are securely fastened, that new split-pins have been correctly fitted where applicable, that the rear wheel is free to rotate easily, and that the rear brake and stop lamp are functioning efficiently.

DT125 LC models

7   Support the machine securely in an upright position so that the rear wheel is clear of the ground, using a stand or a strong wooden box placed under the engine/gearbox unit. Unscrew the brake adjusting nut from the brake rod end and displace the rod from the brake operating arm. Push out the trunnion from the operating arm end and fit it, the return spring, and the nut to the end of the brake rod for safekeeping.

8   Straighten and remove the split-pin which secures the spindle nut then slacken and remove the nut itself. Rotate the snail cam chain adjusters so that the wheel can be pushed forward and the chain disengaged from its sprocket and looped over the fork end of the subframe. Alternatively the chain can be disconnected at its connecting link and the free ends hung over the subframe so that they do not trail in any dirt and debris on

the ground. Remove the wheel spindle, using a hammer and a long metal drift gently to tap it out. Withdraw the wheel right-hand spacer and manoeuvre the wheel clear of the machine.

9   On reassembly, grease the lips of the oil seal and insert the spacer/dust cover into the hub left-hand side. Check that the wheel spindle is straight, clean, and free from corrosion, then smear a liberal quantity of grease over it to assist dismantling in the future and slide the right-hand snail cam along it to fit against the spindle head. Insert the wheel between the fork ends of the rear subframe ensuring that the large tongue on the inside of the right-hand fork end fits into the slot in the brake backplate. Fit the wheel right-hand spacer and push the spindle through from right to left. Refit the left-hand snail cam, the plain washer, and the spindle nut, then loop the chain back over the rear sprocket rotating both snail cams so that the spindle is as far forward as possible to permit this. Tighten the spindle nut as much as possible by hand only so that the wheel is held securely but the snail cams can be rotated. Connect the rear brake rod to the operating arm again, then push the machine off its support so that both wheels are on the ground.

10   The chain tension must be set so that with the machine resting on its wheels only and with the sprung-loaded chain tensioner not touching the chain, there is 45 - 55 mm (1.77 - 2.17 in) free play measured midway between the sprockets on the lower run of the chain, with the chain at its tightest point. Rotate the two snail cams by an equal amount to achieve this. Note that the cut-outs in the perimeter of each cam are numbered, and that there are two cut-outs between each numbered one. The larger the cut-out number, the further back the wheel spindle is positioned. Ensure that the same cut-out in each cam is engaged on their respective raised lugs on each fork end to ensure accurate wheel alignment. When the chain adjustment is correctly set, apply firmly the back brake to centralise the brake shoes and backplate on the drum, and maintain the pressure while the wheel spindle nut is tightened to a torque setting of 8.5 kgf m (61.5 lbf ft).

11   Release the pressure on the brake and fit a new split-pin to secure the spindle nut, spreading correctly the ends of the pin. Rotate the brake adjusting wing nut as necessary to give 20 - 30 mm (0.8 - 1.2 in) of free play, measured at the brake pedal tip, before the brake shoes come into firm contact with the drum. Adjust the stop lamp rear switch height so that the lamp lights just as the brake shoes are coming into firm contact with the drum. Check that all nuts and bolts are securely fastened, that a new split-pin has been correctly fitted, that the rear wheel is free to rotate easily, and that the rear brake and stop lamps are functioning efficiently.

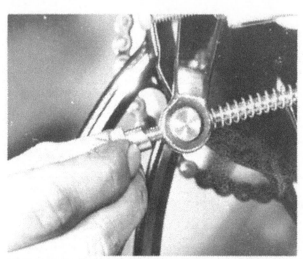

15.1a Unscrew brake adjusting nut and disconnect brake rod

15.1b Slacken and remove rear wheel spindle nut

15.1c Manoeuvre wheel clear of machine

15.1d Do not forget wheel left-hand spacer

15.1e Fit wheel and insert wheel right-hand spacer

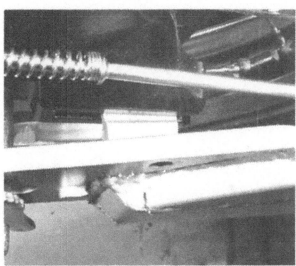

15.7a Lug on sub frame must engage in backplate groove – DT125

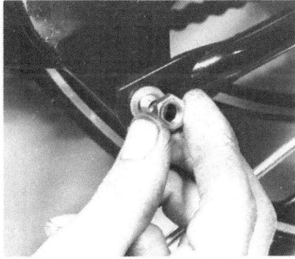

15.7b Refit torque arm, secure with spring washer and nut – RD125

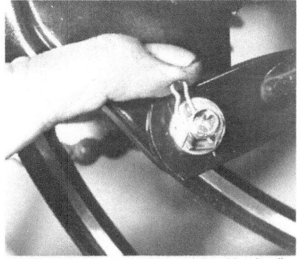

15.7c Tighten nut to torque setting and secure with spring clip

Fig. 6.7 Rear wheel and brake - RD model

1	Wheel spindle	19	Circlip
2	Right-hand spacer	20	Left-hand chain adjuster
3	Brake backplate	21	Castellated nut
4	Brake operating cam	22	Bolt - 2 off
5	Brake shoe - 2 off	23	Lock nut - 2 off
6	Return spring - 2 off	24	Spacer - 4 off
7	Spacer flange	25	Left-hand spacer
8	Right-hand bearing	26	Oil seal
9	Centre spacer	27	Torque arm
10	Tyre	28	Bolt - 2 off
11	Left-hand bearing	29	Spring washer - 2 off
12	Cush drive rubber - 4 off	30	Nut - 2 off
13	Rear wheel	31	R-pin - 2 off
14	Sprocket	32	Brake operating lever
15	Final drive chain	33	Bolt
16	Tab washer - 2 off	34	Wear indicator plate
17	Nut - 4 off	35	Right-hand chain adjuster
18	Washer	36	Split pin

15.7d Connect brake rod to operating arm

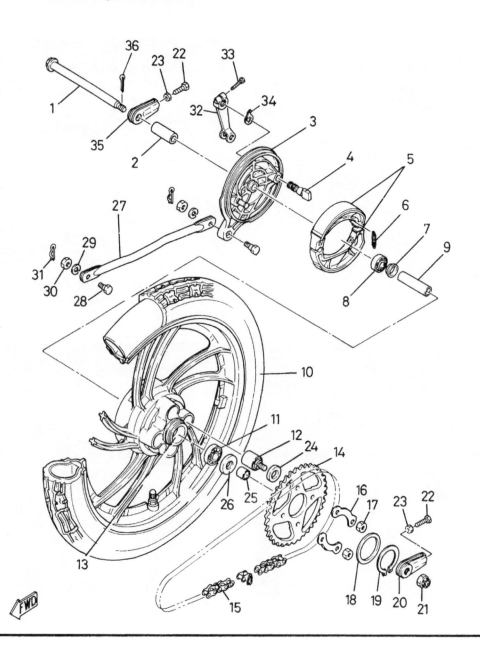

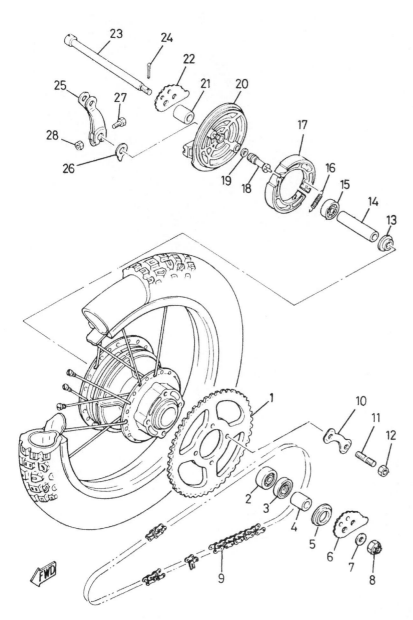

Fig. 6.8 Rear wheel and brake - DT model

1 Sprocket
2 Left-hand bearing
3 Oil seal
4 Left-hand spacer
5 Dust seal
6 Left-hand chain adjuster
7 Washer
8 Castellated nut
9 Final drive chain
10 Tab washer - 2 off
11 Stud - 4 off
12 Nut - 4 off
13 Spacer flange
14 Centre spacer
15 Right-hand bearing
16 Return spring - 2 off
17 Brake shoe - 2 off
18 Brake operating cam
19 Sealing ring
20 Brake backplate
21 Right-hand spacer
22 Right-hand chain adjuster
23 Wheel spindle
24 Split pin
25 Brake operating lever
26 Wear indicator plate
27 Bolt
28 Nut

16 Rear wheel bearings: removal, examination and refitting

Remove the rear wheel as described in the previous Section of this Chapter, then withdraw the brake backplate and the wheel left-hand spacer to gain access to the rear wheel bearings. Due to the similarity in design and construction between the front and rear hubs of each of the machines described in this Manual, the procedures for removal, examination, and refitting of the rear wheel bearings are exactly the same as those given in Section 5 of this Chapter. Refer to that Section, therefore, when working on the rear wheel.

17 Rear brake: examination, renovation, and adjustment

1 The rear brake fitted to both machines described in this Manual is identical in design and construction to that fitted to the front of DT 125 LC models. The rear wheel must be removed, as described in the relevant part of Section 15 of this Chapter, to gain access to the brake components.

2 An external wear indicator is provided on both models so that the state of wear of the brake shoes can be assessed without dismantling the brake assembly. A pointer fitted to the brake camshaft moves within an arc cast in the brake backplate as the brake shoes wear and adjustment is made to compensate. If the pointer reaches the wear limit line at the end of the arc when the brake is firmly applied, the shoes must be considered worn out and renewed immediately.

3 Before adjustment is made on either model, the brake pedal height should be checked. It will be noted that a bolt and locknut are set in the footrest mounting plate on RD125 LC models, and in a lug on the frame on DT125 LC models. These are provided so that the height of the brake pedal can be set to suit the rider's preference. The usually accepted position is that, with the rider seated in the normal riding position wearing the usual riding boots or shoes, the brake pedal is either underneath

the foot, ready for instant use, or immediately adjacent to it, as preferred. These positions will be achieved if the brake pedal is adjusted so that it is 30 mm (1.2 in) below the footrest on RD125 LC models, or 10 mm (0.4 in) below on DT125 LC models. It is, of course, entirely up the machine's rider to ensure that the controls suit his preference for maximum efficiency and safety on the road.

4   When the brake pedal height has been checked, and reset if necessary, the brake itself must be adjusted. Both machines described in this Manual are the same in the adjustment procedure required. The nut fitted to the extreme rear end of the brake operating rod must be rotated as necessary so that there is 20 - 30 mm (0.8 - 1.2 in) of free play, measured at the brake pedal tip, before the brake shoes are felt coming into firm contact with the brake drum. If the brake adjustment (or pedal height) has been altered, the stop lamp rear switch height must be altered to compensate. Hold the body of the switch and rotate the plastic sleeve nut as necessary so that the stop lamp lights just as the brake pedal has taken up its free play and is starting to engage the brake.

16.1a Do not omit wheel bearing central spacer

16.1b Bearings are fitted with sealed surface facing outwards

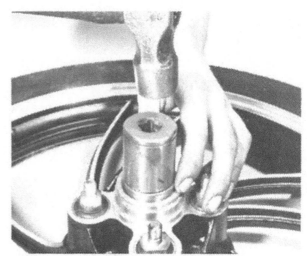

16.1c Using socket as a drift to tap bearings into place

16.1d Hub oil seals should be renewed as a matter of course ...

16.1e ... and are fitted by same method as bearings

17.1 Rear brake is identical on both machines

17.4a Rear brake adjustment is made at end of brake rod

17.4b Do not forget to check stop lamp switch setting

18 Rear sprocket and cush drive assembly: removal, examination and refitting

RD125 LC models

1   The transmission on this machine is fitted with a shock absorber to damp out shock loads in the drive train, thus producing a smoother ride and extending the life of the transmission components. While the components can be examined when in place on the machine, and the sprocket mounting nuts can be tightened in this position, if necessary, if any components are found to be in need of renewal, the rear wheel must be removed to gain access to them.

2   Examine the teeth of the rear sprocket. If these are hooked, chipped, or otherwise damaged, the sprocket must be renewed. Note that it is considered bad practice to renew just one sprocket or the chain alone; both front and rear sprockets and the chain should be renewed together at all times. When the sprocket has been checked, lock the wheel by applying hard the back brake, and attempt to rotate the sprocket backwards and forwards. While some movement should be evident due to the

damping action of the cush drive, excessive movement, which is normally revealed by a rough and jerky ride especially when changing gear or riding at low speed, will indicate that the cush drive rubbers are worn and in need of renewal.

3   With the wheel removed from the machine and the brake backplate withdrawn, lay the wheel on a convenient working surface with the sprocket uppermost. It is useful to place a sheet of cardboard or several layers of newspaper on the working surface to protect the wheel's finish. Using a suitable pair of circlip pliers, remove the circlip which retains the sprocket on the hub and lift away the spacer immediately below it. Bend back the locking tabs of the tab washers, then slacken and remove the four nuts which secure the cush drive rubbers to the sprocket. The sprocket can then be lifted clear and the new one fitted by reversing the above procedure. Tighten the four nuts to a torque setting of 2.3 kgf m (17 lb ft).

4   If the cush drive rubbers are to be renewed, some skill and patience will be required as they are of the bonded rubber type which are always difficult to remove, especially when subjected to the corrosion that will be inevitable due to the road dirt and salt that are always flying around the rear wheel. The first step is to apply a liberal dose of penetrating fluid to each cush drive bush, removing the nylon ring from each to permit this, and to leave the wheel for as long as possible to allow the fluid to work. When ready, invert the wheel and place it on the work surface on top of two wooden blocks placed as close around the hub as possible to give maximum support. The blocks must be thick enough for the wheel to be held at a height that will permit the removal of the bushes.

5   If the interior of the brake drum is examined, it will be seen that there are four holes in its rear face. These have been provided by Yamaha to give access to the rear of each bush, the rubber centre of which will be seen in each hole. Find the largest drift available which will fit through the access holes, and using a suitably heavy hammer, drive out the bush with a few healthy blows. If one is extremely lucky, the bush will come out in one piece. What is more likely to happen is that the rubber will distort and shear, causing the metal centre of the bush to be driven clear leaving the remains of the rubber and the bush metal outer still in place in the hub. In this case, turn the wheel over and use a smaller hammer and a finely-pointed centre punch or a pin punch to drive part of the metal outer inwards towards the centre. This should distort the outer and loosen it to the point where it can be gripped and drawn clear with a suitably heavy pair of pliers. Take the greatest of care not

to damage the hub casting. When the bushes have finally been removed, thoroughly clean the inside of each bush recess, removing all burrs, scratches, and any traces of corrosion.

6   While it is realised that the above procedure is not the most efficient or precise of removal methods, there would appear to be few alternatives. The normal solution would be to loosen the bushes by heating the casting; dry heat, in the form of a blowlamp or welding torch, would inevitably distort the entire wheel casting and there are not many ovens which are large enough to accept a complete motorcycle wheel. Heat in the form of boiling water would require a very large container and a considerable amount of boiling water to be effective. It should be pointed out that if the application of any form of heat is envisaged, great care must be taken to avoid personal injury when handling the wheel or the equipment, and the tyre, the tube, and all hub components such as wheel bearings and oil seals must be removed first. Remember that the entire wheel casting must be heated evenly to prevent distortion which would otherwise occur. The only other method of cush drive bush removal that is possible is to make up an extractor which can be attached to the protruding threaded length of the bush centre. This has the same disadvantage, however, as the method originally proposed; namely that it risks tearing the metal centre out of the bush. It is recommended that if the owner is at all doubtful about his ability, he takes the wheel to a local Yamaha dealer for the work to be carried out.

7   On reassembly, check that the hub recesses into which the new bushes are to be pressed, and the outside surface of the new bushes themselves are quite clean and free from burrs, scratches or other excrescences. Use fine emery cloth to polish away any that are found, then apply a thin smear of grease to each surface to aid assembly and to prevent corrosion. Tap each bush firmly into its recess in the hub using a hammer and a tubular drift (a socket spanner is ideal) which bears only on the metal outer of the bush. Cease tapping when the metal outer is flush with the surrounding surface of the hub. Be careful to remove all surplus grease, especially from inside the brake drum where a petrol-soaked rag must be wiped around to ensure that all traces of grease have been removed. Place the four nylon rings over the ends of the bush centres and refit the sprocket as described above, tightening the sprocket mounting nuts to a torque setting of 2.3 kgf m (17 lbf ft). Do not omit to lock each nut by bending up an unused portion of the tab washer against one of its flats.

DT125 LC models

8   The transmission of these machines has no shock absorber, the sprocket being mounted directly on the hub by four studs and retaining nuts. Again, while the condition and security of the sprocket can be checked, and the sprocket nuts tightened if necessary, with the wheel installed in the machine, if any components are found to be in need of attention the rear wheel must be removed as described in the relevant part of Section 15 of this Chapter. Check the condition of the sprocket teeth as described in paragraph 2 of this Section for the RD125LC models, then check that the sprocket is secure on its mountings. There must be no movement in any direction. While the nuts can be tightened with the wheel in place, if any damage has been done to the sprocket or the studs the wheel must be removed for repairs.

9   With the wheel removed from the machine and the brake backplate withdrawn, lay the wheel on a convenient working surface with the sprocket uppermost, using a sheet of cardboard or layers of newspaper to protect the wheel's finish. Bend back the locking tabs of the two tab washers, then slacken and remove the four sprocket retaining nuts and lift the sprocket away.

10   Carefully examine the studs. There should be no damage at any point along their exposed length, but especially at that point on which the sprocket bears when it is installed. If any damage is revealed the studs must be renewed. Lock two of the sprocket mounting nuts together on the exposed thread and, by applying a spanner to the lower nut, unscrew the stud from the hub. On refitting, apply a few drops of thread locking compound to that part of the stud which will screw into the hub, lock the two nuts together on that part which will be exposed, and screw the stud into the hub. By applying the torque wrench to the upper of the two nuts, tighten the stud to a torque setting of 3.9 kgf m (28 lbf ft). Release and remove the two nuts.

11   When the studs have been examined and renewed, if necessary, check that both faces of the sprocket are absolutely clean and dry, and check that the mating surface of the hub is flat and clean. Refit the sprocket over the studs, press it down to rest on the hub, and check that it cannot move backwards or forwards in the direction of rotation. Refit the two tab washers and the four retaining nuts, tightening the nuts securely to a torque setting of 4.5 kgf m (32.5 lbf ft) and locking each one by bending an unused portion of the tab washer against one of its flats. The wheel can then be refitted to the machine.

18.5a Rear of cush drive bushes can be seen in brake drum

18.5b Drifting cush drive bushes out of rear hub

18.7a Replace nylon damper rings around cush drive bushes

18.7b Refit sprocket ...

18.7c ... then the spacer ...

18.7d ... and the large circlip ...

18.7e Sprocket nuts are secured by two tab washers

18.7f Tighten nuts to correct torque setting ...

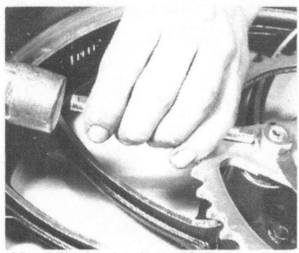

18.7g ... and bend up tab washer to secure them

19 Final drive chain: adjustment, examination, and lubrication

1   As the two models described in this Manual employ different methods of chain adjustment, this is discussed first, with each model being described under a separate heading. The Section then goes on to discuss the procedures of chain care and maintenance in paragraph 11, these applying equally to each of the two models.
2   Note that as the final drive chain is fully exposed on both models, its only protection being a light chainguard over the top run of the chain, and as no separate means of lubrication is provided, the chain willl lead a hard life unless properly cared for. Frequent and regular applications of lubricant will help reduce the rate of wear, and periodic adjustment will compensate for the wear that is inevitable and maintain the chain at its correct tension.

RD125 LC models

3   Chain tension is measured at a point midway between the two sprockets on the lower run of the chain. Due to the fact that chains never wear evenly along their length, tight spots will develop which must be taken into account when adjusting the chain by measuring the chain tension at the tightest point. As the machine must be supported by its wheels at the same time, push the machine off its stand and roll it backwards and forwards as necessary, testing the chain tension at points all along its length by using the fingers of one hand in the position indicated. When the tightest point in the chain has been found and is midway between the two sprockets on the chain lower run, support the machine on its prop stand and measure the free play. There should be 35 - 40 mm (1.4 - 1.6 in). If adjustment is necessary, straighten and remove the split-pin securing the rear wheel spindle nut, then slacken the nut by just enough to permit the spindle to be moved, and slacken the locknuts of the adjuster drawbolts on each fork end of the subframe.
4   If the chain is too tight, the adjuster drawbolts must be slackened and the spindle pushed forward; if the chain is too slack, the spindle must be drawn backwards by tightening the drawbolts as necessary. Whichever is the case, the spindle must be moved by exactly the same amount on each side so that the wheels remain accurately aligned. To assist this, a series of vertical index marks are stamped in each fork end immediately above the adjuster, and are matched by a single reference mark stamped in the adjuster itself. Ensure that the same index mark is aligned with the adjuster reference mark on each side. If desired, wheel alignment can be checked by running a plank of wood parallel to the machine, so that it touches the side of the rear tyre. If wheel alignment is correct, the plank will be equidistant from each side of the front wheel tyre, when tested on both sides of the rear wheel. It will not touch the front wheel tyre as this tyre is of smaller cross section. See the accompanying diagram.
5   When the chain tension is correctly adjusted, apply firmly the back brake to centralise the shoes on the drum, and tighten the wheel spindle nut to a torque setting of 8.5 kgf m (61.5 lbf ft), then secure the nut by fitting a new split-pin and spreading its ends correctly. Tighten securely the adjuster locknuts. Check and reset if necessary, the adjustment of the rear brake and the height of the stop lamp rear switch. Check that the rear wheel is free to rotate and that the rear brake works efficiently before taking the machine out on the road.

DT125 LC models

6   The chain tension of this model must be measured when with the sprung-loaded chain tensioner not acting on the chain and with the machine supported on its wheels, the tightest point along the chain's entire length is midway between the sprockets on the lower run of the chain. To check the tension, first press the chain tensioner down so that it is well clear of the chain and wedge it there with a suitable length of wood or similar placed against the bottom of the subframe. Raise the stand and find the tightest point in the chain exactly as described in paragraph 3 above for the RD125 LC models. When the chain is positioned correctly, measure the free play which should be in the range of 45 - 55 mm (1.77 - 2.17 in).
7   If adjustment is necessary, straighten and remove the split-pin securing the rear wheel spindle nut, then slacken the nut by just enough to permit the snail cams to be rotated. Rotate the two snail cams as necessary to permit the spindle to be pushed forwards or pulled backwards to achieve the correct chain free play. As mentioned above, it is essential to preserve accurate wheel alignment by rotating each cam exactly the same amount. To assist this, the cut-outs in each cam periphery which engage with the raised lug in each fork end are numbered, with two cut-outs between each numbered one. The larger the cut-out number, the further back the wheel spindle is positioned. Ensure that the same cut-out in each cam is positioned against its respective lug. Again the wheel alignment can be checked with a plank of wood as described above for the RD125 LC models.
8   When the chain tension is correctly adjusted, apply firmly the back brake and tighten the wheel spindle nut to a torque setting of 8.5 kgf m (61.5 lbf ft). Secure the nut with a new split-pin, ensuring that its ends are correctly spread and release the chain tensioner so that it springs up to press on the chain. Check and reset if necessary, the adjustment of the rear brake and the height of the stop lamp rear switch, then check that all nuts and bolts are securely fastened and that the rear wheel is free to rotate.
9   It is recommended that whenever the chain on a DT125 LC model is adjusted or lubricated, some time is devoted to examining those components which have been fitted to control the chain. Due to the long travel of the rear suspension on this model, the chain will undergo quite considerable variations in tension as the rear wheel moves up and down. These variations produce excessive slackness which will permit the chain to flail around and to damage both itself and any component that it touches. The following components have been fitted to reduce these variations to a minimum and to control the chain run: a sprung-loaded chain tensioner, a nylon-lined chain guide, and a nylon roller guide. A thick nylon pad has been fitted to protect both sides of the subframe left-hand pivot.
10 The subframe pivot protector and the fixed guide should be examined and renewed if they are badly worn or chewed away by the chain. The pivot protector is retained by a single bolt screwed into the subframe side, the chain guide by two bolts

Tyre changing sequence - tubed tyres

A — Deflate tyre. After pushing tyre beads away from rim flanges push tyre bead into well of rim at point opposite valve. Insert tyre lever adjacent to valve and work bead over edge of rim.

B — Use two levers to work bead over edge of rim. Note use of rim protectors

C — Remove inner tube from tyre

D — When first bead is clear, remove tyre as shown

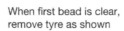

E — When fitting, partially inflate inner tube and insert in tyre

F — Work first bead over rim and feed valve through hole in rim. Partially screw on retaining nut to hold valve in place.

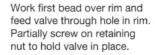

G — Check that inner tube is positioned correctly and work second bead over rim using tyre levers. Start at a point opposite valve.

H — Work final area of bead over rim whilst pushing valve inwards to ensure that inner tube is not trapped

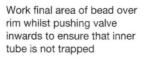

passing through brackets welded to the underside of the subframe. Neither of the above two components requires any further maintenance. The roller guide rotates freely around a shaft which is secured by a bolt and nut to a bracket welded to the frame immediately above the subframe pivot. Due to the fact that it can rotate freely, the roller is unlikely to wear to any great extent until a considerable mileage has been covered. Maintenance is restricted to ensuring that the roller is free to rotate easily and is not excessively worn. All the individual components are available from Yamaha dealers if repair work is necessary. The sprung-loaded chain tensioner is much the most important of all these chain guide assemblies, and will suffer the greatest degree of wear. Carefully clean and examine the nylon block which rubs on the chain. This will require renewal at frequent intervals and it is worth having a spare to carry around on the machine for use in emergencies. Remember that the two screws which pass through it to secure it to the end of the tensioner arm must not be allowed to touch the chain or they will wear through and the block and its mountings will be lost. The remainder of the tensioner assembly must be lubricated at the same time as the chain, and checked to ensure that it is securely fixed on its mountings but is free to pivot correctly. The spring is not likely to wear to any great extent but must be renewed if it does appear to have become fatigued, or if it actually breaks. Again all the individual components are available from Yamaha dealers if repair work is necessary.

All models

11 Never run the chain overtight to compensate for uneven wear. A tight chain will place excessive stresses on the gearbox and rear wheel bearings, leading to their early failure, and will absorb a surprising amount of power.

12 To assess the state of wear of the chain, a simple test can be conducted without removing the chain. With the chain tensioned correctly, try to pull the chain backwards off the rear sprocket. If the links can be pulled clear of the sprocket teeth, the chain must be considered worn out, and renewed with both sprockets.

13 A more accurate check will involve the removal of the chain from the machine. Disconnect the chain at its spring connecting link and pull it away. It is useful to acquire a scrap length of chain, perhaps from a local dealer's dustbin, and to connect this to one end of the chain to be removed. By pulling on the free end of the original chain, the scrap length will pass around both sprockets until the original length is clear. This process can be reversed to fit the cleaned and lubricated original chain, thus obviating the need to remove the crankcase left-hand cover which would otherwise be necessary to feed the chain around the front sprocket. When the chain is removed, clean it thoroughly in a petrol/paraffin mixture, swilling it around and using a brush to remove all traces of old lubricant or road dirt. When clean, dry the chain thoroughly so that the maximum amount of free play is present. Lay the chain out in a straight line and compress it endwise until all the free play is taken up. Measure the length of the chain, anchor one end and pull the chain out as far as possible. Measure the new length. If the chain has extended by more than $\frac{1}{4}$ inch per foot of original length it should be renewed.

14 Note that if the chain is to be renewed, this should only be done in conjunction with both sprockets. Running a new chain on worn sprockets will very rapidly wear the chain out.

15 When replacing the chain, make sure that the spring link is seated correctly, with the closed end facing the direction of travel.

16 An equivalent British-made chain of the correct size is available from Renold Limited. When ordering a new chain always quote the size (length and width at each pitch), the number of links and the machine to which it is fitted. For example, the standard chain fitted to the models described in this Manual is 428 ($\frac{1}{2}$ x $\frac{5}{16}$ in) size chain by 118 links. The length will obviously vary if non-standard sprockets are fitted.

17 Chain lubrication is most effectively accomplished by the use of a special chain grease such as Linklyfe or Chainguard. This is done by removing the chain as described above, cleaning it thoroughly in a petrol/paraffin mixture to remove all traces of old lubricant or road dirt, and immersing the chain in the grease which should be heated according to the manufacturer's instructions. This long and potentially messy process ensures that the innermost bearings in the chain are fully cleaned and lubricated. The grease itself is of a special type which stays on the chain more readily and is not flung off by centrifugal force as easily as thinner lubricants. A better solution for routine maintenance is the use of one of the many proprietary chain greases applied with an aerosol can. These are far easier to use, and very much cleaner, but should only be thought of as an addition to Linklyfe or Chainguard, and not a substitute for them. In spite of the manufacturer's claims, lubricant applied to the exterior of the chain by aerosol cannot penetrate to the inner bearing surfaces of the chain as effectively as molten grease. Ordinary engine oil can be used, but only if nothing better is available, as it is far too thin, and too easily flung off, to lubricate the chain effectively.

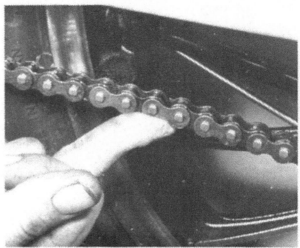

19.3 Chain tension is checked in position shown

19.4a Rotate drawbolts to move wheel spindle ...

19.4b ... using vertical stamp marks to preserve wheel alignment

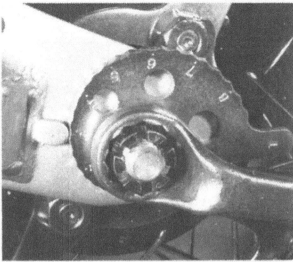

19.7 Rotate snail cams using numbered divisions to preserve wheel alignment

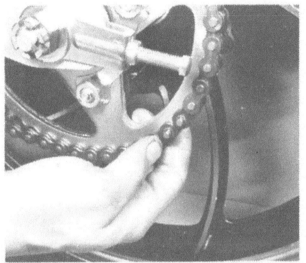

19.13 Disconnect chain at connecting link on removal

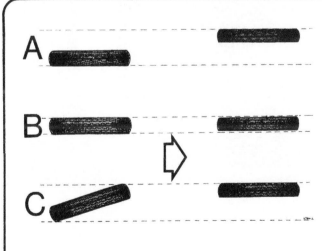

19.15 Always refit spring link as shown

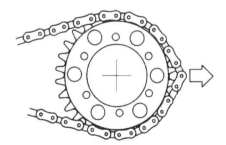

Fig. 6.9 Method of checking wheel alignment

A & C - Incorrect   *B - Correct*

Fig. 6.10 Checking for an excessively worn final drive chain

20 Tyres: removal, repair and refitting

1 At some time or other the need will arise to remove and replace the tyres, either as a result of a puncture or because replacements are necessary to offset wear. To the inexperienced, tyre changing represents a formidable task, yet if a few simple rules are observed and the technique learned, the whole operation is surprisingly simple.

2 To remove the tyre from either wheel, first detach the wheel from the machine. Deflate the tyre by removing the valve core, and when the tyre is fully deflated, push the bead away from the wheel rim on both sides so that the bead enters the centre well of the rim. Remove the locking ring and push the tyre valve into the tyre itself.

3 Insert a tyre lever close to the valve and lever the edge of the tyre over the outside of the rim. Very little force should be necessary; if resistance is encountered it is probably due to the fact that the tyre beads have not entered the well of the rim, all the way round. If aluminium rims are fitted, damage to the soft alloy by tyre levers can be prevented by the use of plastic rim protectors.

4 Once the tyre has been edged over the wheel rim, it is easy to work round the wheel rim, so that the tyre is completely free from one side. At this stage the inner tube can be removed.

5 Now working from the other side of the wheel, ease the other edge of the tyre over the outside of the wheel rim that is furthest away. Continue to work around the rim until the tyre is completely free from the rim.

6 If a puncture has necessitated the removal of the tyre, reinflate the inner tube and immerse it in a bowl of water to trace the source of the leak. Mark the position of the leak, and deflate the tube. Dry the tube, and clean the area around the puncture with a petrol soaked rag. When the surface has dried, apply rubber solution and allow this to dry before removing the backing from the patch, and applying the patch to the surface.

7 It is best to use a patch of self vulcanizing type, which will form a permanent repair. Note that it may be necessary to remove a protective covering from the top surface of the patch after it has sealed into position. Inner tubes made from a special synthetic rubber may require a special type of patch and adhesive, if a satisfactory bond is to be achieved.

8 Before replacing the tyre, check the inside to make sure that the article that caused the puncture is not still trapped inside the tyre. Check the outside of the tyre, particularly the tread area to make sure nothing is trapped that may cause a further puncture.

9 If the inner tube has been patched on a number of past occasions, or if there is a tear or large hole, it is preferable to discard it and fit a replacement. Sudden deflation may cause an accident, particularly if it occurs with the rear wheel.

10 To replace the tyre, inflate the inner tube for it just to assume a circular shape but only to that amount, and then push the tube into the tyre so that it is enclosed completely. Lay the tyre on the wheel at an angle, and insert the valve through the rim tape and the hole in the wheel rim. Attach the locking ring on the first few threads, sufficient to hold the valve captive in its correct location.

11 Starting at the point furthest from the valve, push the tyre bead over the edge of the wheel rim until it is located in the central well. Continue to work around the tyre in this fashion until the whole of one side of the tyre is on the rim. It may be necessary to use a tyre lever during the final stages.

12 Make sure there is no pull on the tyre valve and again commencing with the area furthest from the valve, ease the other bead of the tyre over the edge of the rim. Finish with the area close to the valve, pushing the valve up into the tyre until the locking ring touches the rim. This will ensure that the inner tube is not trapped when the last section of bead is edged over the rim with a tyre lever.

13 Check that the inner tube is not trapped at any point. Reinflate the inner tube, and check that the tyre is seating correctly around the wheel rim. There should be a thin rib moulded around the wall of the tyre on both sides, which should be an equal distance from the wheel rim at all points. If the tyre is unevenly located on the rim, try bouncing the wheel when the tyre is at the recommended pressure. It is probable that one of the beads has not pulled clear of the centre well.

14 Always run the tyres at the recommended pressures and never under or over inflate. The correct pressures are given in the Specifications Section of this Chapter.

15 Tyre replacement is aided by dusting the side walls, particularly in the vicinity of the beads, with a liberal coating of french chalk. Washing up liquid can also be used to good effect, but this has the disadvantage, where steel rims are used, of causing the inner surface of the wheel rim to rust.

16 Never replace the inner tube and tyre without the rim tape in position. If this precaution is overlooked there is a good chance of the ends of the spoke nipples chafing the inner tube and causing a crop of punctures.

17 Never fit a tyre that has a damaged tread or sidewalls. Apart from legal aspects, there is a very great risk of a blowout, which can have very serious consequences on a two wheeled vehicle.

18 Tyre valves rarely give trouble, but it always advisable to check whether the valve itself is leaking before removing the tyre. Do not forget to fit the dust cap, which forms an effective extra seal.

21 Valve cores and caps

1 Valve cores seldom give trouble, but do not last indefinitely. Dirt under the seating will cause a puzzling 'slow-puncture'. Check that they are not leaking by applying spittle to the end of the valve and watching for air bubbles.

2 A valve cap is a safety device, and should always be fitted. Apart from keeping dirt out of the valve, it provides a second seal in case of valve failure, and may prevent an accident resulting from sudden deflation.

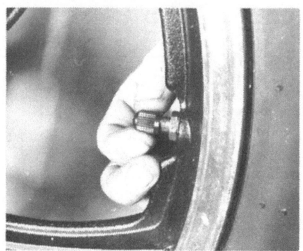

21.2 Always refit dust caps on tyre valves

22 Wheel balancing

1   It is customary on all high performance machines to balance the wheels complete with tyre and tube. The out of balance forces which exist are eliminated and the handling of the machine is improved in consequence. A wheel which is badly out of balance produces through the steering a most unpleasant hammering effect at high speeds.

2   Some tyres have a balance mark on the sidewall, usually in the form of a coloured spot. This mark must be in line with the tyre valve, when the tyre is fitted to the inner tube. Even then the wheel may require the addition of balance weights, to offset the weight of the tyre valve itself.

3   If the wheel is raised clear of the ground and is spun, it will probably come to rest with the tyre valve or the heaviest part downward and will always come to rest in the same position. Balance weights must be added to a point diametrically opposite this heavy spot until the wheel will come to rest in ANY position after it is spun.

4   When working on RD125 LC models, note that it may be necessary to remove the caliper so that the wheel is completely free to rotate. Special weights are available from Yamaha dealers in 10, 20 and 30 gram sizes; these weights being constructed to clamp on to each side of the raised rib cast on the wheel rim. While balance weights are not specifically listed by Yamaha for the DT125 LC model, weights of a suitable type will be available at any good motorcycle dealer or tyre fitting agency. The usual type is designed to be clamped around a spoke, next to the rim, and will be available in many different sizes. Finally, when working on either of the machines described in this Manual, note that it will be rarely necessary to balance the rear wheel but that the procedure is exactly the same as for the front wheel. The chain should be disconnected and the rear brake adjustment slackened off to allow the wheel to rotate easily.

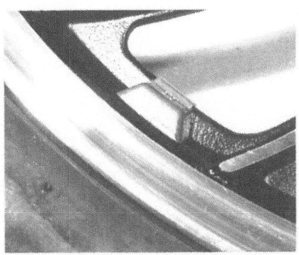

22.4 Note type of weight available for RD125 machines

Chapter 7 Electrical system

For modifications, and information relating to later models, see Chapter 8

Contents

Specifications

Electrical system	**RD125 LC**	**DT125 LC**
Voltage	12	6
Earth	Negative	Negative
Battery		
Make	FB	GS
Type	12N5-3B	6N6-3B-1
Capacity	5Ah	6Ah
Voltage regulator		
Manufacturer	Shindengen	Stanley
Model	5H0	3H6
Regulated voltage	14 – 15 volts	6.9 – 7.5 volts
Rectifier		
Manufacturer	Shindengen	Toshiba
Model	5H0	S5280

Fuse rating	20A	10A
Horn		
Manufacturer	Nikko	Nikko
Model	SF4-12	YF-6
Maximum amperage	3.0A	3.0A
Flashing indicator relay		
Manufacturer	Nippondenso	Nippondenso
Model	5Y3	10V
Wattage	21W x 2 + 3.4W	21W x 2 + 3W
Oil level switch		
Manufacturer	Stanley	Stanley
Model	4L0	2E7-02
Bulbs		
Headlamp	12V, 45/40W	6V, 45/40W
Stop/tail lamp	12V, 21/5W	6V, 21/5W
Flashing indicators	12V, 21W	6V, 21W
Parking lamp	12V, 3.4W	6V, 3W
All instrument/warning lamps	12V, 3.4W	6V, 3W

1 General description

The 12 volt electrical system fitted to the RD 125 LC model is powered by a crankshaft-mounted alternator located behind the crankcase left-hand outer cover. Output from the alternator is fed to a combined rectifier/regulator unit where it is converted from alternating current (ac) to direct current (dc) by the full-wave rectifier section, and the system voltage is regulated to 14 – 15 volts by the electronic voltage regulator.

In the case of the DT 125 LC model a simpler and less sophisticated 6 volt system is powered by a flywheel generator mounted on the crankshaft left-hand end, the output generated being fed largely to the lights. A voltage regulator is fitted to control the current to prevent the bulbs from blowing through voltage surges, and to soak up the excess current generated when the lights are switched to the 'Off' or 'P' positions. The remainder of the generator output is converted to direct current (dc) by a silicon rectifier and is then used to charge the battery which powers the ancillary electrical equipment such as horn, stop lamp, and flashing indicator lamps.

2 Testing the electrical system

1 Simple continuity checks, for instance when testing switch units, wiring and connections, can be carried out using a battery and bulb arrangement to provide a test circuit. For most tests described in this chapter, however, a pocket multimeter should be considered essential. A basic multimeter capable of measuring volts and ohms can be bought for a very reasonable sum and will prove an invaluable tool. Note that separate volt and ohm meters may be used in place of the multimeter, provided those with the correct operating ranges are available. In addition, if the generator output is to be checked, an ammeter of 0-5 amperes range will be required.

2 Care must be taken when performing any electrical test, because some of the electrical components can be damaged if they are incorrectly connected or inadvertently shorted to earth. This is particularly so in the case of electronic components. Instructions regarding meter probe connections are given for each test, and these should be read carefully to preclude accidental damage occurring.

3 Where test equipment is not available, or the owner feels unsure of the procedure described, it is strongly recommended that professional assistance is sought. Errors made through

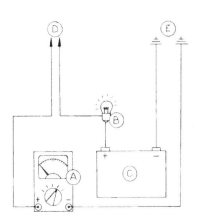

Fig. 7.1 Simple circuit testing arrangement

A Multimeter
B Bulb
C Battery
D Positive probe
E Negative probe

carelessness or lack of experience can so easily lead to damage and need for expensive replacement part.

4 A certain amount of preliminary dismantling will be necessary to gain access to the components to be tested. Normally, removal of the seat and side panels will be required, with the possible addition of the fuel tank and headlamp unit to expose the remaining components.

3 Wiring: layout and examination

1 The wiring harness is colour-coded and will correspond with the accompanying wiring diagram. When socket connectors are used, they are designed so that reconnection can be made in the correct position only.

2 Visual inspection will usually show whether there are any breaks or frayed outer coverings which will give rise to short circuits. Occasionally a wire may become trapped between two

components, breaking the inner core but leaving the more resilient outer cover intact. This can give rise to mysterious intermittent or total circuit failure. Another source of trouble may be the snap connectors and sockets, where the connector has not been pushed fully home in the outer housing, or where corrosion has occurred.

3   Intermittent short circuits can often be traced to a chafed wire that passes through or is close to a metal component such as a frame member. Avoid tight bends in the lead or situations where a lead can become trapped between casings.

4   Charging system: checking the output

1   In the event that the charging system appears to be over- or under-charging the battery, the output should be checked first, a task which will require the use of a multimeter or other testing equipment, as appropriate.

RD125 LC models

2   With the meter set on the appropriate ac voltage scale, remove the radiator cover and disconnect the three-pin block connector which connects the three white wires from the alternator into the main loom, and connect the meter probes to any two of the wire terminals leading up from the engine. Start the engine but **do not** increase engine revs above idle speed. The meter should give a reading of 12 – 20 volts or more at idle speed. Remove one of the meter leads and connect it to the remaining white wire terminal. The reading obtained should be the same.

3   If the readings obtained by the above check are found to be correct, then the alternator is functioning correctly and the fault must be in the rectifier/regulator unit, the battery, the switches, or in the wiring between them. Check these in turn as described in the subsequent Sections of this Chapter. If the readings are incorrect, the condition of the alternator coils must be checked as described in the next Section.

DT125 LC models – charging system

4   If symptoms arise which indicate a fault in the charging or lighting systems, the two must be checked in separate tests. In both cases, the battery must be in good condition and fully charged, as described in the relevant Sections of this Chapter, and the engine must be warmed up to normal operating temperature. To check the charging system, remove the right-hand sidepanel and disconnect the battery red wire at the snap connector between the battery positive (+) terminal and the fuse. Connect a dc ammeter, or multimeter set to the appropriate scale, between the two connector terminals, the meter positive (+) lead to the fuse terminal and its negative (–) lead to the battery terminal. Set the meter to the 0 – 5 amp scale, then check that the lights are switched off. Start the engine, then check the output as shown below.

5   The readings obtained at the engine speeds specified should be as follows:

Charging rate – daytime
    0.8 – 1.5A minimum @ 3000 rpm
    2.0A maximum @ 8000 rpm

Switch the lights to the 'On' position and the dipswitch to the 'Hi' position and repeat the test. The readings obtained should be as follows:

Charging rate – night time
    0.7 – 1.5A minimum @ 3000 rpm
    2.0A maximum @ 8000 rpm

As soon as the test is finished, stop the engine, then disconnect the meter and re-connect the battery red wire. If the results obtained were satisfactory the generator is functioning correctly and the fault lies in the rectifier, the battery, or in the wiring. If the test indicates a fault in the generator; the condition of its coils must be checked, as described in the next Section.

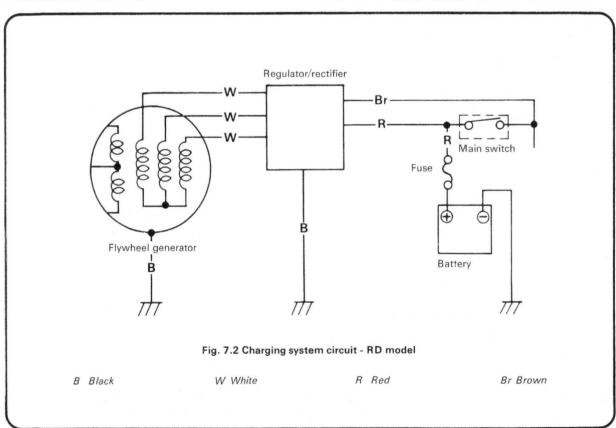

Fig. 7.2 Charging system circuit - RD model

B Black   W White   R Red   Br Brown

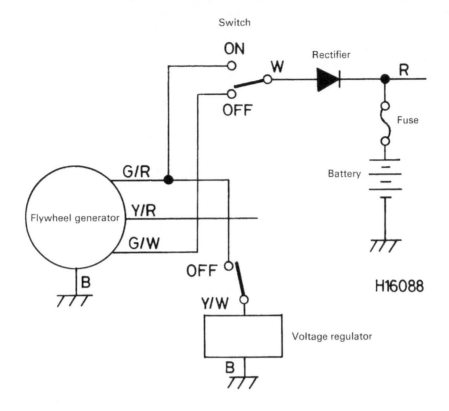

Fig. 7.3 Charging system circuit - DT model

B	Black	G/R	Green/red	G/W	Green/white	W	White
R	Red	Y/R	Yellow/red	Y/W	Yellow/white		

DT125 LC models – lighting system

6   If the lights appear to be dim, or if constant bulb-blowing is experienced, and the bulb filaments appear to have been melted, the lighting system must be checked to ensure that the correct voltage is being applied. Remove the headlamp unit as described in Section 11 of this Chapter and disconnect the bulb wires. Switch the lights to the 'On' position and the dipswitch to the 'Hi' position. Using an ac voltmeter or a multimeter set to the 0 – 10 volts ac scale, connect the meter positive (+) lead to the yellow headlamp wire terminal and the meter negative (–) lead to the black headlamp wire terminal Start the engine and increase engine speed as specified. The readings obtained should be as follows:

Lighting voltage
    5.7V minimum @ 2000 rpm
    8.6V maximum @ 8000 rpm

7   If the readings obtained are satisfactory, the system is functioning correctly and the bulb wattage should be checked to ensure that the correct items are fitted. Also check that there is no intermittent fault such as a dirty or loose conenction, or a faulty switch. If the readings are not satisfactory the condition of the stator coils should be checked, as described in the next Section, and then the voltage regulator, the switches, and the wiring between them.

---

5   Generator coils: resistance tests

---

1   On both models, if the tests described in Section 4 have revealed a fault in the alternator or generator itself, the condition of the individual coils can be checked by measuring their respective resistances.

RD125 LC models

2   Using an ohmmeter or a multimeter set to the x1 ohms scale, measure the resistance between each of the three white wires leading from the alternator. The readings obtained should be 0.31 – 0.42 ohm ± 20% at 20°C (68°F) in each case.

DT125 LC models

3   Remove the seat, the sidepanels, and the petrol tank, then identify and disconnect, as appropriate, the wires mentioned in the tests below. The connectors are in the form of a multi-pin block connector and two snap connectors and are to be found in the vicinity of the rear suspension unit top mounting. Using an ohmmeter or a multimeter set to the X1 ohms scale, measure the resistances between the pairs of wire shown. The readings obtained should be as follows:
Charging coil resistance:
    Day – Green/white to Black 0.21 ohm ± 20% at 20°C (68°F)
    Night – Green/red to Black 0.28 – 0.40 ohm ± 20% at 20°C (68°F)
Lighting coil resistance:
    Yellow/red to Black 0.24 ohm ± 10% at 20°C (68°F)

All models

4   With both machines, if any of the tests described above show one or more stator coils to be faulty, the complete machine should be taken to an authorised Yamaha dealer for

accurate testing as the only solution that is likely to be practicable is to renew the coil concerned. While an experienced auto-electrician may be able to repair or rewind a faulty coil, this is a task for the expert alone.

5   It should be noted that while the coils on the DT125 LC model are available as separate components, on the RD125 LC model a fault in any one of the coils will require the purchase of a new alternator assembly comprising rotor and stator. No individual components are available for this model. It is recommended, therefore, that an expert second opinion is sought before any component is finally condemned, to save unnecessary expense. Close examination may reveal a broken or trapped wire which can be repaired. Such a task is easy for the expert, and it is infinitely preferable to pay a small sum for the time involved in checking your findings and to make any small repairs than to pay a large amount for unnecessary new components.

6   Regulator/rectifier unit – location and testing: RD125 LC

1   The combined regulator/rectifier unit fitted to the RD125 LC models is a heavily finned, sealed metal unit secured to the battery tray by two screws. It should be noted at the outset that the unit cannot be repaired if found faulty, renewal is the only possible solution.

2   A quick test of the regulator is as follows. Remove the left-hand sidepanel and, using a dc voltage or a multimeter set to the 0 – 20 volts dc scale, measure the battery voltage. A reading of 11 – 12 volts should be obtained when the ignition switch is in the 'On' position. If not, the battery must be examined and re-charged or renewed as necessary, as described in the relevant Sections of this Chapter. It must be stressed that the battery must be in peak condition if the test is to be at all accurate. When the condition of the battery is known

to be satisfactory, the test may proceed, but note that the battery leads must not be disconnected during the test. **Note**: If the machine is run with the battery disconnected the increased voltage across the alternator terminals will rise, causing damage to the regulator/rectifier unit or to the alternator windings.

3   Connect the positive (red) probe lead to the positive (+) battery terminal and the nagative (black) probe lead to the negative (–) battery terminal. Start the engine and note the voltage reading at 2000 rpm. This should be 14.5 ± 0.5 volts if the system is operating correctly. If the voltage is outside this range it will be necessary to check the regulator/rectifier unit further by isolating it.

4   Unplug the multi-pin connector which connects the brown, the black, and the red wires from the regulator/rectifier to the main loom. Connect the position (+) lead of the dc voltmeter or multimeter to the brown wire terminal of that half of the multi-pin connector which leads into the main loom, and the meter negative (–) lead to a good earth point. Switch on the ignition and check that a reading of 11 – 12 volts is obtained. If so, having conducted the tests described in Section 4 and in this Section, the alternator, the battery, the ignition switch, and the wiring will have been eliminated and the fault must lie in the regulator/rectifier unit. If full battery voltage is not measured in the brown wire, the ignition switch and wiring should be checked for faults, particularly the continuity of the brown wire itself.

5   As a final check, connect the regulator/rectifier back into the main loom and measure the voltage across the brown wire by inserting a meter probe into each side of the connector block to contact each terminal of the brown wire. Start the engine and increase speed to 2000 rpm. Once again a voltage reading of 14 – 15 volts should be obtained if all is well, but check also that the reading drops to exactly 12 volts when the lights are switched to the 'On' position and the dipswitch is in the 'Hi' position. If the readings obtained are not satisfactory, the regulator/rectifier unit should be considered faulty, but the

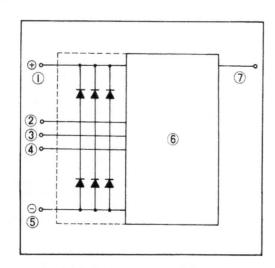

1. B (Red)
2. U (White)
3. V (White)
4. W (White)
5. E (Black)
6. I.C. regulator
7. L (Brown)

Checking element	Pocket test connecting point		Good	Replace (element shorted)	Replace (element opened)
	(+) (red)	(–) (black)			
D_1	B	U	O	O	×
	U	B	×	O	×
D_2	B	V	O	O	×
	V	B	×	O	×
D_3	B	W	O	O	×
	W	B	×	O	×
D_4	U	E	O	O	×
	E	U	×	O	×
D_5	V	E	O	O	×
	E	V	×	O	×
D_6	W	E	O	O	×
	E	W	×	O	×

O ..... Continuity
× ..... Discontinuity

Fig. 7.4 Rectifier resistance test - RD125 LC

machine should be taken to a Yamaha dealer for accurate checking before the component is finally condemned and a new one purchased.

6   The rectifier is an arrangement of six diodes, connected in a bridge pattern to provide full-wave rectification. This means that the full output of the alternator is converted to dc, rather then half of it, as is the case with simple half-wave rectifiers as used on the DT125 LC model.

7   The condition of the rectifier can be checked using a multimeter, set on its resistance scale, as a continuity tester. Each of the diodes acts as a one-way valve, allowing current to flow in one direction, but blocking it if the polarity is reversed. Perform the resistance check by following the table accompanying Fig. 7.4. If any one test produces the wrong reading the rectifier will have to be renewed.

---

7   Regulator and rectifier units – location and testing: DT125 LC

1   The voltage regulator unit fitted to the DT125 LC models is a heavily-finned, silver, sealed metal unit which is fastened to the right-hand side of the frame top tubes immediately behind the steering head. It requires no maintenance except for a periodic check, whenever the petrol tank is removed, that it is clean and free from dirt and corrosion and that its two retaining screws are securely fastened.

2   The regulator unit operates on the same principle as a Zener diode, diverting excess lighting voltage away from the bulbs and disposing of the surplus in the form of heat through its finned cover. In this way the varying demands of the bulbs are met, and the risk of bulbs blowing due to voltage surges is reduced greatly.

3   Unfortunately the unit can be tested only with a special test meter, no other data being provided. If the condition of the unit is suspect after the test of the lighting voltage described in Section 4, and other possibilities such as the bulbs, switches, and wiring have been eliminated, it must be returned to a Yamaha dealer to be checked. If found to be faulty, it must be renewed as repairs are not possible.

4   The silicon diode rectifier fitted to these machines is a small rectangular black plastic block with two male spade terminals projecting from its underside, which is retained by a single

screw to the left-hand side of the frame top tubes, immediately behind the steering head. It requires no maintenance at all, save a periodic check that it is clean, and that both it and its connections are securely fastened.

5   The rectifier consists of a small diode and serves to convert the ac output of the flywheel generator into dc to charge the battery. It should be thought of as a one-way valve, in that it will allow the current to flow in one direction only, thus blocking half of the output wave from the generator.

6   Before removing the unit, identify the polarity of the two terminals by the colour of the wire leading to each one. The red wire leads to the positive (+) terminal and the white wire to the negative (−) or ac (∼) terminal.

7   Using a multimeter set to the resistance mode, check for continuity between the two terminals. There should only be continuity from the negative (−) to the positive (+) terminal. This direction of flow may be shown by an arrow on the surface of the unit. If there is continuity in the reverse direction, or if resistance is measured in both directions, the rectifier is faulty and must be renewed. No repair is possible.

6.1 Location of regulator/rectifier unit –RD125

7.1 Location of voltage regulator and CDI unit – DT125

7.4 Location of rectifier and flashing indicator relay – DT125

8 Battery: examination and maintenance

1   The battery fitted on the RD125 LC model is housed in a tray located behind the left-hand sidepanel and is retained in position by a bracket which is hinged from the top of the tray and locked in position with a single screw. On the DT125 LC model, the battery is housed in a tray behind the right-hand sidepanel and is secured by a rubber strap.

2   The transparent plastic case of the battery permits the upper and lower levels of the electrolyte to be observed without disturbing the battery by removing the side cover. Maintenance is normally limited to keeping the electrolyte level between the prescribed upper and lower limits and making sure that the vent tube is not blocked. The lead plates and their separates are also visible through the transparent case, a further guide to the general condition of the battery. If electrolyte level drops rapidly, suspect over-charging and check the system.

3   Unless acid is spilt, as may occur if the machine falls over, the electrolyte should always be topped up with distilled water to restore the correct level. If acid is spilt onto any part of the machine. It should be neutralised with an alkali such as washing soda or baking powder and washed away with plenty of water, otherwise serious corrosion will occur. Top up with sulphuric acid of the correct specific gravity (1.260 to 1.280) only when spillage has occurred. Check that the vent pipe is well clear of the frame of any of the other cycle parts.

4   It is seldom practicable to repair a cracked battery case because the acid present in the joint will prevent the formation of an effective seal. It is always best to renew a cracked battery especially in view of the corrosion which will be caused if the acid continues to leak.

5   If the machine is not used for a period of time, it is advisable to remove the battery and give it a 'refresher' charge every six weeks or so from a battery charger. The battery will require recharging when the specific gravity falls below 1.260 (at 20°C − 68°F). The hydrometer reading should be taken at the top of the meniscus with the hydrometer vertical. If the battery is left discharged for too long, the plates will sulphate. This is a grey deposit which will appear on the surface of the plates, and will inhibit recharging. If there is sediment on the bottom of the battery case, which touches the plates, the battery needs to be renewed. Prior to charging the battery refer to the following Section for correct charging rate and procedure. If charging from an external source with the battery on the machine, disconnect the leads, or the rectifier will be damaged.

6   Note that when moving or charging the battery, it is essential that the following basic safety precautions are taken:

(a)   Before charging check that the battery vent is clear or where no vent is fitted, remove the combined vent/filler caps. If this precaution is not taken the gas pressure generated during charging may be sufficient to burst the battery case, with disastrous consequences.

(b)   Never expose a battery on charge to naked flames or sparks. The gas given off by the battery is highly explosive.

(c)   If charging the battery in an enclosed area, ensure that the area is well ventilated.

(d)   Always take great care to protect yourself against accidental spillage of the sulphuric acid contained within the battery. Eyeshields should be worn at all times. If the eyes become contaminated with acid they must be flushed with fresh water immediately and examined by a doctor as soon as possible. Similar attention should be given to a spillage of acid on the skin.

Note also that although, should an emergency arise, it is possible to charge the battery at a more rapid rate than that stated in the following Section, this will shorten the life of the battery and should therefore be avoided if at all possible.

7   Occasionally, check the condition of the battery terminals to ensure that corrosion is not taking place, and that the electrical connections are tight. If corrosion has occurred, it should be cleaned away by scraping with a knife and then using emery cloth to remove the final traces. Remake the electrical connections whilst the joint is still clean, then smear the assembly with petroleum jelly (NOT grease) to prevent recurrence of the corrosion. Badly corroded connections can have a high electrical resistance and may give the impression of complete battery failure.

8.1a Slacken and remove screw to release battery

8.1b DT125 battery is released by unhooking retaining strap

9   Battery: charging procedure

1   Whilst the machine is used on the road it is unlikely that the battery will require attention other than routine maintenance because the generator will keep it fully charged. However, if the machine is used for a succession of short journeys only, mainly during the hours of darkness when the lights are in full use, it is possible that the output from the generator may fail to keep pace with the heavy electrical demand, especially if the machine is parked with the lights switched on. Under these circumstances it will be necessary to remove the battery from time to time to have it charged independently.

2   The normal maximum charging rate for any battery is 1/10 the rated capacity. Hence the charging rate for the battery fitted to the RD125 LC model is 0.5 amp, while that of the 6 Ah battery fitted to the DT125 LC model is 0.6 amp. A slightly higher charge rate may be used in emergencies only, but this should never exceed 1 amp.

3   Ensure that the battery/charger connections are properly made, ie the charger positive (usually coloured red) lead to the battery positive (the red wire) lead, and the charger negative (usually coloured black or blue) lead to the battery negative (the blue wire) lead. Refer to the previous Section for precautions to be taken during charging. It is especially important that the battery cell cover plugs are removed to eliminate any possibility of pressure building up in the battery and cracking its casing. Switch off the charger if the cells become overheated, ie over 45°C (117°F).

4   Charging is complete when the specific gravity of the electrolyte rises to 1.260 – 1.280 at 20°C (68°F). A rough guide to this state is when all cells are gassing freely. At the normal (slow) rate of charge this will take between 3 – 15 hours, depending on the original state of charge of the battery.

5   If the higher rate of charge is used, never leave the battery charging for more than 1 hour as overheating and buckling of the plates will inevitably occur.

10   Fuse: location and renewal

1   The electrical system is protected by a single fuse of 10 or 20 amp rating as appropriate. It is retained in a plastic casing set in the battery positive (+) terminal lead, and is clipped to a holder immediately in front of the battery. If the spare fuse is ever used, replace it with one of the correct rating as soon as possible.

2   Before renewing a fuse that has blown, check that no obvious short circuit has occurred, otherwise the replacement fuse will blow immediately it is inserted. It is always wise to check the electrical circuit thoroughly, to trace the fault and eliminate it.

3   When a fuse blows while the machine is running and no spare is available, a 'get you home' remedy is to remove the blown fuse and wrap it in silver paper before replacing it in the fuse holder. The silver paper will restore the electrical continuity by bridging the broken fuse wire. This expedient should never be used if there is evidence of short circuit or other major electrical fault, otherwise more serious damage will be caused. Replace the 'doctored' fuse at the earliest possible opportunity, to restore full circuit protection.

11   Headlamp: bulb renewal and beam adjustment

1   To gain access to the headlamp bulb on the RD125 LC model it will be necessary to remove the headlamp fairing. Slacken and remove the two bolts which secure the upper part

10.1a Always keep a spare fuse in case of emergency

10.1b Fuse holder is clipped to front of battery

of the fairing to brackets bolted to the headlamp bracket, then pull the bottom edge of the fairing carefully forwards to disengage it from its lower mounting grommets. Slacken and remove the two screws which secure the headlamp rim in the headlamp shell, then withdraw the rim and reflector unit far enough for the block connector to be released from the back of the bulb.

2   On DT125 LC models, the headlamp cowling is retained by three bolts, two set horizontally in the upper half of the cowling and the third at the front left-hand side, below the light unit. Pull the cowling forwards far enough for the block connector to be released from the back of the bulb.

3   The headlamp bulbs are of a similar type on both machines, removal and refitting is, therefore, exactly the same. Peel away the moulded rubber seal from around the bulb. Depress the large bulb holder ring against spring pressure and rotate it anticlockwise to release it. The bulb can be lifted away. On refitting, note that there is a projection from the periphery of the bulb which must engage in the cut-out provided for it in the rear of the reflector; the remainder of the operation being a straightforward reversal of the removal procedure.

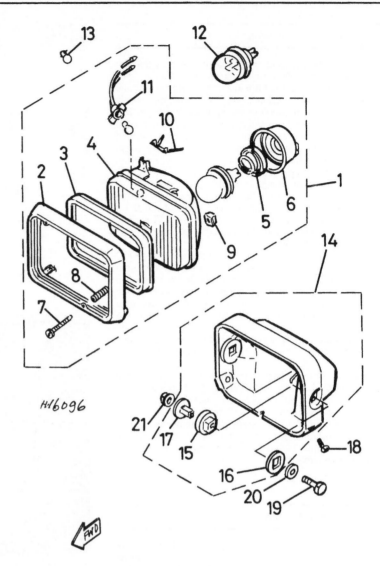

Fig. 7.5 Headlamp – RD model

1 Headlamp assembly
2 Rim
3 Inner rim
4 Reflector unit
5 Bulb holder ring
6 Rubber seal
7 Adjusting screw
8 Spring
9 Nut
10 Spring clip
11 Parking lamp bulb holder
12 Bulb
13 Parking lamp bulb
14 Headlamp shell
15 Damping rubber - 2 off
16 Damping rubber - 2 off
17 Collar - 2 off
18 Screw - 2 off
19 Bolt - 2 off
20 Washer - 2 off
21 Nut - 2 off

H16096

4 The parking lamp bulb holder can be removed by depressing it and rotating it anticlockwise, whereupon the bulb itself is released from the holder in the same way.

5 Headlamp beam vertical alignment is adjusted on the RD125 LC model by slackening the two bolts which secure the headlamp shell to the headlamp brackets and by tilting the unit as required to align it with the fairing. If sufficient care is exercised, there is no need to remove the headlamp fairing to reach these bolts. A spring-loaded screw is set in the headlamp bottom edge for fine adjustment. On the DT125 LC model the headlamp cowling is fixed and so a sprung-loaded screw is set in the bottom edge of the light unit to be rotated as necessary to tilt the headlamp unit to the required setting.

6 In the UK, regulations stipulate that the headlamp must be arranged so that the light will not dazzle a person standing at a distance greater than 25 feet from the lamp, whose eye level is not less than 3 feet 6 inches above that plane. It is easy to approximate this setting by placing the machine 25 feet away from a wall, on a level road, and setting the dipped beam height so that it is concentrated at the same height as the distance of the centre of the headlamp from the ground. The rider must be seated normally during this operation and also the pillion passenger, if one is carried regularly.

7 Note that horizontal beam alignment is also adjustable, by means of a sprung-loaded screw in both cases, this adjusting screw being set in the headlamp top edge on both machines.

11.1 Headlamp rim is secured by two screws

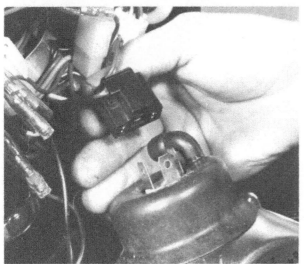

11.2 Disconnect wires by pulling off connector block

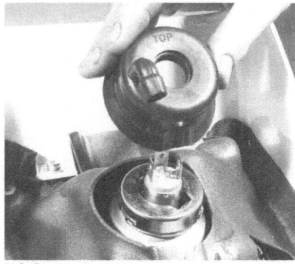

11.3a Remove rubber seal ...

11.3b ... then bulb holder ring ...

11.3c ... and withdraw bulb

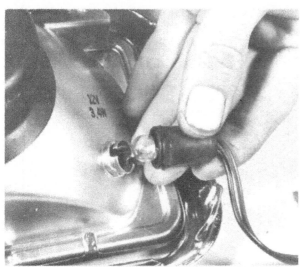

11.4a Parking lamp bulb holder is a bayonet fit in reflector ...

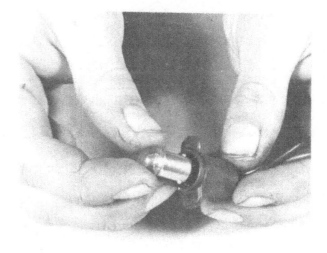

11.4b ... bulb fits the holder in the same way

11.5 Headlamp beam vertical adjusting screw

11.7 Headlamp beam horizontal adjusting screw

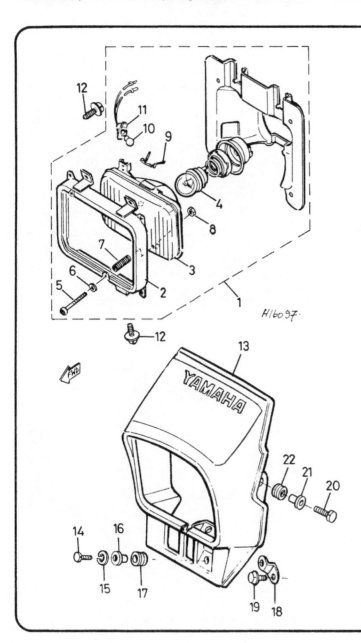

Fig. 7.6 Headlamp and cowling – DT model

1   Headlamp assembly
2   Rim
3   Reflector unit
4   Bulb
5   Adjusting screw
6   Washer
7   Spring
8   Nut
9   Spring clip
10 Parking lamp bulb
11 Parking lamp bulb holder
12 Bolt and washer - 3 off
13 Cowling
14 Bolt
15 Spring washer
16 Collar
17 Grommet
18 Bracket
19 Bolt
20 Bolt - 2 off
21 Collar - 2 off
22 Damping rubber - 2 off

12 Stop and tail lamp: bulb renewal

1   The combined stop and tail lamp bulb contains two filaments, one for the stop lamp and one for the tail lamp.
2   The offset pin bayonet fixing bulb can be renewed after the plastic lens cover and screws have been removed.

13 Flashing indicator lamps: bulb renewal

1   The flashing indicator lamp assemblies are plastic mouldings bolted to metal stalks. The earth connections should be checked at convenient intervals to ensure that they are tight and free from dirt or corrosion. If a bulb or wiring connection fails, the affected lamp will cease operation, the failure being indicated by rapid flashing of the remaining bulb.
2   The lamp assemblies of the two machines are slightly different. On RD125 LC models, the lens is clipped to the body of the lamp and can be removed by using a coin or broad-bladed screwdriver in the slot provided to lever the lens off. Both the lens and the body are of plastic construction, so care must be taken to avoid damage during removal. The bulb is of the conventional bayonet type and can be removed by pushing inwards, twisting gently anti-clockwise and releasing. On DT125 LC models, each lens is retained by two screws. Slacken and remove these and lift the lens away. Bulb removal and refitting is as described for RD125 LC models.
3   On RD125 LC models it should be noted that no reflectors are fitted to the indicator lamps, and owners may wish to improve their visibility in bright sunlight by lining the inside of the lamp body with aluminium cooking foil. The lamps are, however, adequate in most normal conditions.

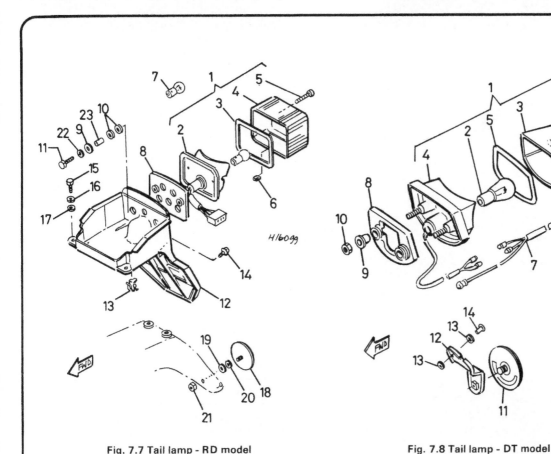

Fig. 7.7 Tail lamp - RD model

Fig. 7.8 Tail lamp - DT model

1	Tail lamp assembly	
2	Reflector	13 Clip
3	Sealing ring	14 Bolt and washer - 2 off
4	Lens	15 Bolt - 2 off
5	Screw – 2 off	16 Spring washer - 2 off
6	Grommet	17 Washer - 2 off
7	Bulb	18 Reflector
8	Damping rubber	19 Washer
9	Washer - 2 off	20 Spring washer
10	Washer - 4 off	21 Nut
11	Bolt - 2 off	22 Spring washer - 2 off
12	Mounting bracket	23 Spacer - 2 off

1	Tail lamp assembly	8	Damping rubber
2	Bulb	9	Collar - 2 off
3	Lens	10	Nut - 2 off
4	Reflector	11	Reflector
5	Sealing ring	12	Mounting bracket
6	Screw - 2 off	13	Washer - 4 off
7	Wiring harness	14	Rivet - 2 off

12.2a Slacken and remove lens fixing screws

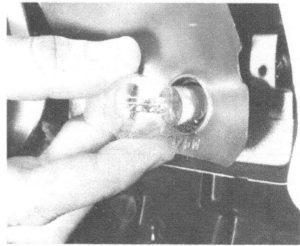

12.2b Bulb is a bayonet fitting in taillamp

13.2a Flashing indicator lens removal – RD125

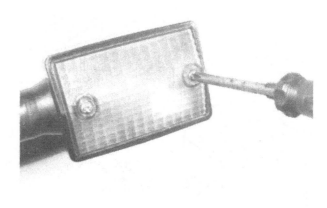

13.2b Flashing indicator lens removal – DT125

13.2c Bulb is a bayonet fitting on both models

14 Instrument panel and warning lamps: bulb renewal

1   Access is gained to the various bulbs in the instrument panel of the RD125 LC model by removing the headlamp fairing, and unscrewing the knurled rings which secure the speedometer and tachometer drive cables to the instrument bases. If the underside of the panel is studied, it will be seen that the bulbs are retained by the panel mounting bracket. Slacken and remove the two nuts which secure the instrument panel assembly to the mounting bracket, then lift the panel carefully up until the bulb holders can be pulled out. Each bulb is rated at 12 volt 3.4 W and is of the capless type, pressed into a bulb holder whih is a push fit in the instrument panel base.

2   On the DT125 LC model, release the headlamp cowling by slackening and removing the three retaining bolts, and pull the cowling forwards far enough to expose the instrument panel base. Using a suitable pair of pliers, unscrew the knurled ring which secures the speedometer drive cable to the speedometer, then pull out the three spring clips which secure the instruments to the panel mounting bracket, the speedometer being secured by the two left-hand clips and the temperature gauge assembly being secured by the single right-hand clip. Catch the plain

metal washers as they fall clear. Lift the speedometer or temperature gauge assembly as appropriate until the bulb holders are exposed. Again, the bulb holders are a push fit in the instrument base, but in this case the bulbs are all of 6 volt 3 W rating.

15 Horn: location and testing

1   The horn is mounted on a flexible steel bracket between the fork legs, that of the RD125 LC model being bolted to the headlamp bracket, and that of the DT125 LC model being bolted to the bottom yoke. No maintenance is required other than regular cleaning to remove road dirt and occasional spraying with WD40 or a similar water dispersant spray to minimise internal corrosion.

2   If the horn note becomes weak or intermittent, usually after a considerable mileage use, and the switches, battery, and wiring are known to be in good order, the fault may be due to wear in the internal contacts which can be adjusted by means of the screw and locknut on the outside of the horn. Slacken the locknut and rotate the screw very carefully until the loudest and clearest note is obtained. Tighten the locknut to retain the setting.

3   If the horn fails to work, first test that power is reaching the instrument by disconnecting the wires and substituting a 6 or 12 volt bulb as appropriate. Switch on the ignition and press the horn button. If the bulb fails to light, check the horn button and wiring as described below. If the bulb does light, the horn circuit is proved good and the horn must be checked. Connect a fully charged 6 or 12 volt battery, as appropriate, directly to the horn. If it does not sound, a gentle tap on the outside may serve to free the internal contacts. If this fails the renewal of the horn is the only solution as repair is not possible.

4   If a multimeter is available, the wiring can be checked as follows:

   a)   Check for full battery voltage at the horn brown wire, connecting one meter probe to the wire terminal and the other to a good earth.

   b)   Switch the meter to the resistance function and transfer the probe from the brown wire terminal to the pink wire terminal. Check for continuity when the horn button is depressed.

5   If the first of the above tests is unsatisfactory, check the main feed back to the battery via the brown wire, the ignition switch, the red wire and the fuse. If the second test is unsatisfactory, check the condition of the pink wire and the horn button itself.

14.1a Bulb holders are a push fit in instrument base

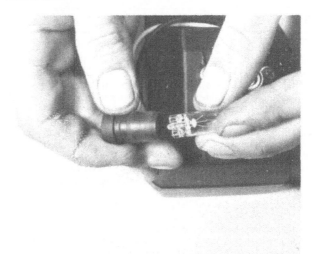

14.1b Bulbs of capless type are employed – RD125

14.2 Release spring clips to lift instruments for bulb renewal

15.1a Location of horn – DT125

15.1b RD125 horn is fitted similarly. Note adjustment screw and locknut

16 Flashing indicator relay : location and testing

1   The flashing indicator relay fitted to both machines described in this Manual is a rectangular black plastic sealed unit mounted on the frame immediately behind the steering head. It is fitted on the right-hand side on RD125 LC models and on the left-hand side of DT125 LC models and is identified readily by its rubber mounting and the two-pin plastic connector which joins it to the wiring loom. Note that the relay is a delicate unit and must be handled carefully at all times. If it is found to be faulty at any time it must be renewed as repairs are not possible. It will be necessary to remove the petrol tank as described in Section 2 of Chapter 3 to gain access to the relay.
2   If the flashing indicator lamps cease to function correctly, there may be any one of several possible faults responsible which should be checked before the relay is suspected. First check that the flashing indicator lamps are correctly mounted and that all the earth connections are clean and tight. Check that the bulbs are of the correct wattage and that corrosion has not developed on the bulbs or in their holders. Any such corrosion must be thoroughly cleaned off to ensure proper bulb contact. Also check that the flashing indicator switch is functioning correctly and that the wiring is in good order. Finally ensure that the battery is fully charged. The following check list will be of assistance in tracing faults in the flashing indicator system:

  a) Check bulbs
  b) Right circuit:
      1 Check for full battery voltage on dark green wire to light.
      2 Check for ground on black wire to light assembly.
  c) Left circuit:
      1 Check for battery voltage on dark brown wire to light
      2 Check for ground on black wire to light assembly.
  d) Right and left circuits do not work:
      1 Check for battery voltage on brown/white wire to flasher switch on left handlebar.
      2 Check for battery voltage on brown wire to flasher relay.
      3 Replace flasher relay.
      4 Replace flasher switch.

3   Faults in any one or more of the above items will produce symptoms for which the flashing indicator relay may be blamed unfairly. If the fault persists even after the preliminary checks

have been made, the relay must be at fault. Unfortunately the only practical method of testing the relay is to substitute a known good one. If the fault is then cured, the relay is proven faulty and must be renewed. Fortunately relay failure is a rare occurrence.

17 Stop lamp switches: location and testing

1   Both machines described in this Manual are fitted with two switches to activate the stop lamp when the brakes are applied. The front switch is fitted in the handlebar clamp and non-adjustable but the rear switch is mounted on a frame bracket above the rear brake pedal and is adjustable as described in Section 17 of Chapter 6. Neither switch can be repaired; renewal is the only answer if a fault is found.
2   If a fault develops, use the test sequence below to find the problem:

  a) Check bulb and connections.
  b) Check for full battery voltage on yellow lead to brake lamp.
  c) Check for full battery voltage on brown lead to front and rear brake switches.
  d) Check black earth lead from lamp unit to frame (continuity test).

16.1 Location of flashing indicator relay – RD125 model

18 Handlebar switches: general

1   While in general the switches fitted to the machines described in this Manual should give little trouble, they can be tested using a multimeter set to the resistance function or a battery and bulb test circuit. Using the information given in the wiring diagram at the end of this Manual, check that full continuity exists in all switch positions and between the relevant pairs of wires. When checking a particular circuit, follow a logical sequence to eliminate the switch concerned. Note for example the test sequences given above to check for faults in the horn, flashing indicator lamp, and stop lamp circuits.
2   As a simple precaution always disconnect the battery before removing any of the switches, to prevent the possibility of a short circuit. Most troubles are caused by dirty contacts, but in the event of the breakage of some internal part, it will be necessary to renew the complete switch.

3 It should, however, be noted that if a switch is tested and found to be faulty, there is nothing to be lost by attempting a repair. It may be that worn contacts can be built up with solder, or that a broken wire terminal can be repaired, again using a soldering iron. The handlebar switches can all be dismantled to a greater or lesser extent. It is however, up to the owner to decide if he has the skill to carry out this sort of work.

4 While none of the switches require routine maintenance of any sort, some regular attention will prolong their life a great deal. In the author's experience, the regular and constant application of WD40 or a similar water-dispersant spray not only prevents problems occurring due to waterlogged switches and the resulting corrosion, but also makes the switches much easier and more positive to use. Alternatively, the switch may be packed with a silicone-based grease to achieve the same result.

19 Ignition switch: removal and refitting

1 The ignition switch fitted to the RD125 LC model is secured to the front fork top yoke by two bolts which thread into the yoke from underneath. It will be necessary, therefore, to remove the headlamp fairing, the complete headlamp assembly, and the instrument panel in order to withdraw the switch.

2 On the DT125 LC model, the switch is mounted on the instrument panel and is secured by two screws from underneath. Remove the headlamp cowling, disconnect the switch wires at the multi-pin connector, then slacken and remove the two screws. Pull the switch upwards through the panel to remove it.

3 While repairs are rarely a practicable proposition, if the switch is known to be faulty there is nothing to be lost in attempting to dismantle it and cure the problem, as the only alternative is to renew the switch. In the long term it would be preferable to renew the switch anyway to ensure reliability.

20 Neutral indicator switch: location and testing

1 The neutral indicator switch is a black plastic unit screwed into the crankcase to contact the left-hand end of the selector drum. The crankcase left-hand cover must be removed first, so that the switch wire can be disconnected and the switch unscrewed.

2 To test the switch, first check that the bulb and its connections are in good order, then use a multimeter to check that full battery voltage is available between the terminal of the light blue switch wire and the crankcase. If the above checks are satisfactory the switch is almost certainly at fault. As a final check, switch the meter to the resistance function, apply one meter probe to the switch terminal and the other to the crankcase wall, then check that continuity exists only when neutral is selected.

3 If the switch is found to be faulty it must be renewed as repairs are not possible.

21 Oil level warning lamp: description and testing

1 The oil level warning lamp is operated by a float-type switch mounted in the oil tank. The circuit is wired through the neutral switch so that when the ignition is switched on and the machine is in neutral, the lamp comes on as a means of checking its operation. As soon as a gear is selected the lamp should go out unless the oil level is low.

2 In the event of a fault the bulb can be checked by switching the ignition on and selecting neutral. If this proves sound, check for full battery voltage on the black/red lead to the switch. If the switch proves to be defective it can be unclipped from the tank and withdrawn.

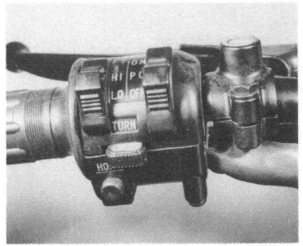
18.1 Handlebar switches give little trouble but need attention

19.3 Repairs to ignition switch are rarely successful – it is better to renew

20.1 Neutral indicator switch rarely goes wrong

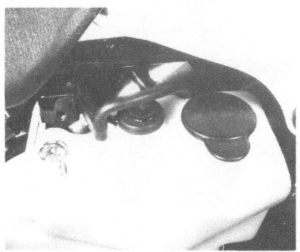

21.1 Oil level switch is clipped to top of oil tank

22 Water temperature gauge and sender unit: location and testing

1   A water temperature gauge is fitted to the LC models to monitor the coolant temperature. The system comprises a gauge built into the instrument panel and a sender unit mounted on the cylinder head. The sender unit is a thermocouple, the resistance of which reduces as it gets hotter. It is connected between the gauge and earth and thus controls the needle position according to engine temperature.

2   In the event of a fault occurring in the temperature gauge circuit, a simple test sequence has been devised to trace the fault. Note that if the sender unit or the gauge is found to be faulty, they must be renewed as repairs are not possible. To remove the sender unit, disconnect the wire leading to it, then slacken it off using a suitable spanner. Unscrew the unit as fast as possible and withdraw it, placing a finger to stop the coolant escaping. Check the condition of the sealing washer, renewing it if necessary, and fit the new unit and washer. If sufficient care is taken the minimum of coolant will escape and there will be no need to drain and refill the system, but remember to top up the radiator or expansion tank, and to wash off all surplus coolant. Tighten the sender unit to a torque setting of 1.0 kgf m (7 lbf ft) and connect its lead wire again.

3   The temperature gauge unit fitted to the RD125 LC model is bolted to the bottom half of the instrument panel by three nuts; these nuts also serve as connectors for the gauge wires. The instrument panel must be removed from the machine and dismantled, as described in Section 16 of Chapter 5. Remove also the panel mounting bracket to expose the gauge connections, the mounting bracket being retained by two nuts, then slacken and remove the three nuts to disconnect the gauge wires and to release the gauge. Make a careful note of the relative positions of the various washers, mounting plates, mounting rubbers, and metal spacers. On reassembly, note that the positions of the gauge wires are indicated by letters denoting their respective colours that are moulded into the lower half of the panel.

4   On DT125 LC models the temperature gauge is only available as part of the warning lamp console fitted next to the speedometer; this is easily removed after the warning lamp bulbs and gauge wiring have been disconnected as described in Section 13 of this Chapter.

5   To test the complete temperature gauge circuit, first check that the battery is in good condition, then proceed as follows:

   a)   Switch on the ignition and disconnect the sender unit lead wire from the sender unit. The gauge needle

should point at 'C'. Earth the sender unit wire on the cylinder head, whereupon the needle should immediately swing over to 'H'. If the needle reacts as described the gauge is shown to be serviceable and the sender unit is proven faulty and must be renewed. A second test is given below if any doubts exist and confirmation is needed. If the needle does not move, proceed to the next test to eliminate the wiring.

   b)   On RD125 LC models, remove the headlamp fairing and headlamp rim and reflector unit, and on DT125 LC models, remove the headlamp cowling. Disconnect the green/red wire from the temperature gauge at its single snap connector and use a multimeter set to the relevant dc volts scale to measure the voltage between the gauge and a good earth point when the ignition is switched on. Full battery voltage should be measured. If this is the case, then the fault must be in the green/red wire between the connection just tested and the sender unit terminal. Establish this by checking for continuity between the two ends of the wire with the multimeter set to the resistance function. If heavy resistance is measured, the wire is broken or damaged and must be repaired. If, on the other hand, full battery voltage is not found at the gauge end of the green/red wire, then the gauge is proven faulty and must be renewed.

6   If the water temperature sender appears to be faulty it can be tested by measuring its resistance at various temperatures. To accomplish this it will be necessary to gather together a heatproof container into which the sender can be placed, a burner of some description (a small gas-powered camping burner would be ideal), a thermometer capable of measuring between 40°C and 100°C (104°F – 212°F) and an ohmmeter or multimeter capable of measuring 0 – 240 ohms with a reasonable degree of accuracy.

7   Fill the container with cold water and arrange the sender unit on some wire so that the probe end is immersed in it. Connect one of the meter leads to the sender body and the other to the terminal. Suspend the thermometer so that the bulb is close to the sender probe.

8   Start to heat the water, and make a note of the resistance reading at the temperature shown in the table below. If the unit does not give readings which approximate quite closely to those shown it must be renewed.

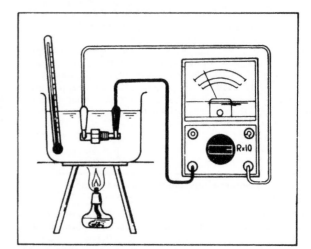

Water Temperature	40°C	60°C	80°C	100°C
Resistance	240Ω	94 ~ 114Ω	52Ω	16 ~ 30Ω

Fig. 7.9 Testing the water temperature gauge sender

22.2a Temperature gauge must be renewed if faulty

22.2b Sender unit mounted in cylinder head

22.3 Note colours of gauge wires indicated by stamped letters in panel base

22.4 DT125 temperature gauge incorporates warning lamp console next to speedometer

The DT125 LC II model

The DT125 LC III model

Chapter 8 The RD/DT125 LC II and III models

Contents

Specifications

Note: *The specifications shown below relate to the RD125 LC II and III, and to the DT125 LC II and III models covered in this update Chapter. Unless shown otherwise, these specifications relate to all models.*

Specifications relating to Chapter 1

Engine

Type	Water-cooled single cylinder two-stroke
Bore	56 mm (2.20 in)
Stroke	50 mm (1.97 in)
Capacity	123 cc (7.5 cu in)
Compression ratio:	
RD	6.4:1
DT	6.8:1

Cylinder head

Type	Cast aluminium, with integral water passages
Maximum warpage	0.03 mm (0.0012 in)

Cylinder barrel

Type	Cast aluminium, with cast iron liner
Standard bore size	56.000 – 56.020 mm (2.205 – 2.206 in)
Service limit	Not available
Maximum taper:	
RD	0.05 mm (0.002 in)
DT	0.08 mm (0.003 in)
Maximum ovality:	
RD	0.01 mm (0.0004 in)
DT	0.05 mm (0.0020 in)

Piston rings

Top:	
Type	Keystone
Thickness	1.2 mm (0.047 in)
Width	2.2 mm (0.087 in)
End gap (installed)	0.30 – 0.45 mm (0.012 – 0.018 in)
Ring to groove clearance, RD	0.02 – 0.06 mm (0.0008 – 0.0024 in)
Ring to groove clearance, DT	Not available
Second:	
Type	Plain
Thickness	1.2 mm (0.047 in)
Width	1.85 mm (0.073 in)
End gap – installed, RD	0.30 – 0.50 mm (0.012 – 0.020 in)
End gap – installed, DT	0.30 – 0.45 mm (0.012 – 0.018 in)
Ring to groove clearance, RD	0.03 – 0.07 mm (0.0012 – 0.0028 in)
Ring to groove clearance, DT	Not available

Piston

Standard OD (measured at 10 mm from base of skirt)	55.94 – 56.00 mm (2.202 – 2.205 in)
Piston to cylinder clearance:	
RD II	0.050 – 0.060 mm (0.0020 – 0.0024 in)
RD III	0.055 – 0.060 mm (0.0022 – 0.0024 in)
DT II and III	0.050 – 0.055 mm (0.0020 – 0.0022 in)
Oversizes:	
1st	+0.25 mm (+0.010 in)
2nd	+0.50 mm (+0.020 in)

Crankshaft

Big-end bearing deflection (at small-end)	2.0 mm (0.08 in)
Big-end side clearance:	
RD	0.2 – 0.7 mm (0.008 – 0.028 in)
DT	0.4 – 0.7 mm (0.016 – 0.028 in)
Maximum runout (at mainshafts)	0.03 mm (0.0012 in)
Width across flywheel outer faces	55.90 – 55.95 mm (2.201 – 2.203 in)

Primary drive

Type	Helical gear
Reduction ratio	3.227:1 (71/22T)

Clutch

Type	Wet, multiplate
No. of friction plates	6
No. of plain plates	5
No. of springs	4
Friction plate thickness	3.0 mm (0.12 in)
Service limit	2.7 mm (0.11 in)
Plain plate thickness	1.2 mm (0.05 in)
Maximum warpage	0.05 mm (0.002 in)
Spring free length	34.5 mm (1.36 in)
Service limit	32.5 mm (1.28 in)
Pushrod maximum runout	0.15 mm (0.006 in)
Outer drum thrust clearance	0 mm (0 in)
Outer drum radial clearance	0.015 – 0.049 mm (0.0006 – 0.0020 in)

Gearbox

Type	6-speed, constant mesh	
Gear ratios:	**RD125 LC**	**DT125 LC**
1st	2.833:1 (34/12T)	3.500:1 (35/10T)
2nd	1.812:1 (29/16T)	2.214:1 (31/14T)
3rd	1.368:1 (26/19T)	1.555:1 (28/18T)
4th	1.142:1 (24/21T)	1.190:1 (25/21T)
5th	1.000:1 (23/23T)	1.000:1 (23/23T)
Top	0.916:1 (22/24T)	0.084:1 (21/25T)

Final drive

Type	Chain and sprocket
Reduction ratio:	
RD	2.812:1 (45/16T)
DT	3.227:1 (53/16T)

Torque wrench settings

Component	RD125 LC kgf m	RD125 LC lbf ft	DT125 LC kgf m	DT125 LC lbf ft
Spark plug	2.0	14.5	2.0	14.5
Cylinder head nuts	2.1	15.0	2.2	16.0
Cylinder head studs	Not applicable	1.0	7.2	
Cylinder barrel nuts	2.8	20.0	2.5	18.0
Thermostat cover	0.8	5.8	0.8	5.8
YPVS components (not applicable to RD125 LC II):				
Valve holder (left)	0.5	3.6	0.5	3.6
Power valve	0.6	4.3	0.6	4.3
Pulley	1.0	7.2	1.0	7.2
Power valve cover	0.5	3.6	0.5	3.6
Power valve seal cap	0.5	3.6	0.5	3.6
Reed valve bolts	0.8	5.8	0.8	5.8
Exhaust pipe nuts	1.8	13.0	1.8	13.0
Exhaust pipe studs	1.0	7.2	1.0	7.2
Crankcase screws	0.8	5.8	0.8	5.8
Crankcase drain bolt	2.0	14.0	2.0	14.0
Neutral switch	0.4	2.9	0.4	2.9
Crankcase cover screws	1.0	7.2	1.0	7.2
Water pump cover screws	0.8	5.8	Not applicable	
Water pump cover drain bolt	1.0	7.2	Not applicable	
Oil pump cover screws	1.0	7.2	1.0	7.2
Oil pump mounting screws	0.5	3.6	0.5	3.6
Primary drive gear	6.5	47.0	8.0	58.0
Clutch centre nut	6.5	47.0	5.5	40.0
Clutch spring bolts	1.0	7.2	0.6	4.3
Balancer gear nut	5.5	40.0	5.5	40.0
Oil seal retainer	1.6	11.0	1.6	11.0
Bearing retainer	1.0	7.2	1.0	7.2
Kickstart lever	6.5	47.0	6.5	47.0
Stopper arm bolt	1.4	10.0	1.4	10.0
Gearbox sprocket bolt	1.0	7.2	1.0	7.2
Tachometer housing	0.5	3.6	0.5	3.6
Gearchange pedal	1.5	11.0	1.5	11.0
CDI rotor	8.3	60.0	8.3	60.0
CDI stator screws	1.0	7.2	1.0	7.2

Specifications relating to Chapter 2

	RD125 LC	**DT125 LC**

Coolant

Mixture type	50% distilled water, 50% corrosion-inhibited ethylene glycol antifreeze (aluminium engine type)	
Overall capacity	1.00 litre (1.76 Imp pint)	0.64 litre (1.13 Imp pint)
Reservoir tank capacity	0.25 litre (0.44 Imp pint)	0.13 litre (0.2 Imp pint)

Water pump

Type	Centrifugal impeller
Drive	Gear, from crankshaft via idler pinion

Radiator

Core width	160.0 mm (6.30 in)	122.5 mm (4.83 in)
Core height	160.0 mm (6.30 in)	240.0 mm (9.45 in)
Core thickness	32.0 mm (1.26 in)	32.0 mm (1.26 in)
Cap opening pressure	0.9 ± 0.15 kg/cm² (13 ± 2 psi)	

Torque wrench settings

Component	kgf m	lbf ft
Thermostat cover	0.8	5.8
Coolant drain plugs	1.0	7.2
Pump cover screws	1.0	7.2
Temperature gauge sender unit	1.0	7.2

Specifications relating to Chapter 3

Fuel tank capacity

	RD125 LC	DT125 LC
Overall	13 litre (2.86 Imp gal)	10 litre (2.20 Imp gal)
Reserve	1.9 litre (0.4 Imp gal)	1.5 litre (0.3 Imp gal)

Carburettor

	RD125 LC II	RD125 LC III
Make	Mikuni	Mikuni
Type	VM24SS	VM26SS
ID number	12A-01	IGU-00
Main jet	195	190
Main air jet	02.5	01.2
Jet needle	5GN36	406
Clip position	3rd from top	4th from top
Needle jet	0-2	P-2
Throttle valve cutaway	1.5	2.0
Pilot jet	20	25
Pilot mixture screw, turns out	$1\frac{1}{2}$	$1\frac{3}{4}$
Float valve seat	02.5	02.5
Starter jet	Not available	25
Float height	21 ± 1.0 mm (0.83 ± 0.04 in)	21 ± 1.0 mm (0.83 ± 0.04 in)
Fuel level	0.5 ± 1.0 mm (0.02 ± 0.04 in)	0 – 0.5 mm (0 – 0.02 in)
Idle speed	1300 ± 50 rpm	1300 – 1400 rpm

	DT125 LC II	DT125 LC III
Make	Mikuni	Mikuni
Type	VM26SS	VM26SS
ID number	34X00	34X00
Main jet	200	200
Main air jet	01.2	01.2
Jet needle	406	406
Clip position	4th from top	4th from top
Needle jet	P-4	P-2
Throttle valve cutaway	2.0	2.0
Pilot jet	27.5	25
Pilot mixture screw, turns out	$1\frac{3}{4}$	$1\frac{3}{4}$
Float valve seat	02.0	02.0
Starter jet	Not available	25
Float height	22 ± 1.0 mm (0.87 ± 0.04 in)	22 ± 1.0 mm (0.87 ± 0.04 in)
Fuel level	0.5 ± 1.0 mm (0.02 ± 0.04 in)	0.5 ± 1.0 mm (0.02 ± 0.04 in)
Idle speed	1300 ± 50 rpm	1300 ± 50 rpm

Reed valve assembly

Stopper plate height	10.3 mm (0.41 in)
Reed maximum warpage	0.5 mm (0.020 in)
Reed thickness	0.2 mm (0.008 in)

Air filter

Type	Oiled foam element

Engine lubrication system

	RD125 LC	DT125 LC
Type	Pump-fed total loss system (Yamaha Autolube)	
Oil tank capacity	1.1 litre (1.9 Imp pint)	1.2 litre (2.2 Imp pint)
Pump colour code:		
RD125 LC II	Red	
RD125 LC III	Orange	
DT125 LC II and III	Yellow	
Pump output per 200 strokes:		
Minimum	0.38 – 0.48 cc	0.50 – 0.63 cc
	(0.0134 – 0.0169 Imp fl oz)	(0.018 – 0.022 Imp fl oz)
Maximum	3.56 – 3.94 cc	4.65 – 5.15 cc
	(0.126 – 0.139 Imp fl oz)	(0.164 – 0.181 Imp fl oz)
Minimum stroke	0.20 – 0.25 mm (0.008 – 0.010 in)	
Maximum stroke	1.85 – 2.05 mm (0.073 – 0.081 in)	

Gearbox lubrication

Capacity:		
At oil change	0.55 litre (0.97 Imp pint)	
Dry	0.63 litre (1.11 Imp pint)	

Torque wrench settings

Component	RD125 LC		DT125 LC	
	kgf m	lbf ft	kgf m	lbf ft
Oil pump cover screws	1.0	7.2	1.0	7.2
Oil pump mounting screws	0.5	3.6	0.5	3.6
Reed valve mounting bolts	0.8	5.8	0.8	5.8
Exhaust mounting:				
Nut	1.8	13.0	1.8	13.0
Stud bolt	1.0	7.2	1.0	7.2
Transmission oil drain plug	2.0	14.0	2.0	14.0

Specifications relating to Chapter 4

Ignition system

	RD125 LC	DT125 LC
Type	Capacitor discharge ignition (CDI)	
Voltage	12 volt	
Ignition timing (BTDC):		
At 1300 rpm	16°	Not applicable
At 1350 rpm	Not applicable	8°
At 3000 rpm	30°	Not applicable
At 4000 rpm	Not applicable	30°

Flywheel generator

	RD125 LC	DT125 LC
Make	Yamaha	Yamaha
Type	F1GM	F34Y
Pickup coil resistance	350 $\Omega \pm$ 20% @ 20°C (68°F)	
Source coil resistances:		
Source coil 1	650 $\Omega \pm$ 20% @ 20°C (68°F)	355$\Omega \pm$ 20% @ 20°C (68°F)
Source coil 2	63 $\Omega \pm$ 20% @ 20°C (68°F)	Not applicable

CDI unit

Make	Yamaha	Hitachi
Type	F1GM	37F-MO

Ignition HT coil

Make	Yamaha	
Type	C2T4	
Output	13kV or more at 500 rpm, 23kV or less at 8000 rpm	
Resistances:		
Primary windings	1.6 $\Omega \pm$ 10% @ 20° (68°F)	
Secondary windings	6.6k $\Omega \pm$ 20% @ 20°C (68°F)	

Spark plug

Make	NGK	
Type	BR8ES or BR9ES	
Electrode gap	0.7 – 0.8 mm (0.028 – 0.031 in)	

Specifications relating to Chapter 5

Frame

	RD125 LC	DT125 LC
Type	Welded tubular steel	

Front forks

Type	Oil-damped coil spring telescopic	
Travel	140 mm (5.51 in)	240 mm (9.45 in)
Spring free length:		
Main	515.4 mm (20.28 in)	576.4 mm (22.69 in)
Service limit	510 mm (20.08 in)	570.6 mm (22.46 in)
Secondary	Not applicable	53.7 mm (2.11 in)
Service limit	Not applicable	53.2 mm (2.09 in)
Oil capacity (per leg)	238 cc (8.39 Imp fl oz)	366 cc (12.9 Imp fl oz)
Oil level (fork fully compressed, spring removed)	149 mm (5.87 in)	139.5 mm (5.5 in)
Fork oil grade	SAE 10W fork oil	SAE 10W/30 motor oil

Rear suspension

Type	Cantilever	Rising-rate (Yamaha Mono-cross)
Rear wheel travel	110 mm (4.33 in)	210 mm (8.3 in)
Shock absorber travel	55 mm (2.17 in)	74 mm (2.91 in)
Spring free length	234.5 mm (9.23 in)	238.0 mm (9.37 in)
Service limit	232 mm (9.13 in)	236 mm (9.29 in)
Gas pressure	15 kg/cm^2 (213 psi)	15 kg/cm^2 (213 psi)

Rear subframe

Free play service limits:		
End	1.0 mm (0.04 in)	1.0 mm (0.04 in)
Side	1.0 mm (0.04 in)	0.4 – 0.7 mm (0.016 – 0.028 in)

Torque wrench settings

Component	RD125 LC		DT125 LC	
	kgf m	lbf ft	kgf m	lbf ft
Engine mounting bolts:				
Front	2.5	18.0	3.3	24.0
Rear	2.5	18.0	3.8	27.0
Top yoke pinch bolts	2.3	17.0	2.3	17.0
Steering stem top bolt	4.0	28.0	7.0	50.0
Handlebar mounting bolts	1.0	7.2	1.5	11.0
Handlebar pinch bolts	2.0	14.0	Not applicable	
Lower yoke pinch bolts	3.0	22.0	2.0	14.0
Fork damper rod bolt	2.3	17.0	2.0	14.0
Front wheel spindle	7.5	54.0	8.5	61.0
Front wheel spindle pinch bolt	2.0	14.0	Not applicable	
Swinging arm pivot shaft	4.3	31.0	8.0	58.0
Rear wheel spindle	8.5	61.0	8.5	61.0
Rear suspension unit mounting bolt	3.2	23.0	4.2	30.0
Rear suspension unit connecting rod to relay arm	Not applicable		4.2	30.0
Relay arm to swinging arm	Not applicable		5.9	43.0
Connecting rod to frame	Not applicable		4.2	30.0
Rear brake torque arm:				
Brake end	1.9	13.0	Not applicable	
Frame end	2.3	17.0	Not applicable	

Specifications relating to Chapter 6

Wheels

Type	Cast alloy	Wire spoked
Sizes:		
Front	MT2. 15x16	1.60x21
Rear	MT2. 15x18	1.85x18
Rim runout (maximum):		
Radial/axial	2.0 mm (0.08 in)	2.0 mm (0.08 in)

Front brake

	RD125 LC	DT125 LC III
Front brake type	Single hydraulic disc brake	
Disc diameter	245 mm (9.64 in)	189 mm (7.44 in)
Disc thickness	4.0 mm (0.16 in)	3.5 mm (0.14 in)
Service limit	3.0 mm (0.12 in)	3.0 mm (0.12 in)
Pad thickness	5.5 mm (0.23 in)	6.0 mm (0.24 in)
Service limit	0.5 mm (0.02 in)	0.8 mm (0.03 in)
Master cylinder bore ID	12.7 mm (0.50 in)	11.0 mm (0.43 in)
Caliper bore ID	38.1 mm (1.50 in)	34.9 mm (1.37 in)
Hydraulic fluid type	DOT 3 or SAE J1703	DOT 3 or SAE J1703
Brake lever free play	5 – 8 mm (0.2 – 0.3 in)	5 – 8 mm (0.2 – 0.3 in)

Front brake

DT125 LC II

Type	Single leading shoe (sls) drum
Drum ID	130 mm (5.12 in)
Service limit	131 mm (5.16 in)
Lining thickness	4.0 mm (0.16 in)
Service limit	2.0 mm (0.08 in)
Return spring free length	23.0 mm (1.30 in)

Rear brake

	RD125 LC	**DT125 LC**
Type	Single leading shoe (sls) drum	
Drum ID	130 mm (5.12 in)	130 mm (5.12 in)
Service limit	131 mm (5.16 in)	131 mm (5.16 in)
Lining thickness	4.0 mm (0.16 in)	4.0 mm (0.16 in)
Service limit	2.0 mm (0.08 in)	2.0 mm (0.08 in)
Return spring free length	36.5 mm (1.44 in)	36.5 mm (1.44 in)
Brake pedal position	10 mm (0.4 in) below top surface of footrest	

Tyres

Size:		
Front	80/100-16 45P	2.75-21-4PR
Rear	90/90-18 51P	4.10-18-4PR
Pressures (cold):		
Up to 90 kg (198 lb) load:		
Front	25 psi (1.8 kg/cm^2)	18 psi (1.3 kg/cm^2)
Rear	28 psi (2.0 kg/cm^2)	22 psi (1.5 kg/cm^2)
Over 90 kg (198 lb) load:		
Front	25 psi (1.8 kg/cm^2)	22 psi (1.5 kg/cm^2)
Rear	32 psi (2.3 kg/cm^2)	26 psi (1.8 kg/cm^2)
High speed riding:		
Front	25 psi (1.8 kg/cm^2)	Not applicable
Rear	32 psi (2.3 kg/cm^2)	Not applicable

Torque wrench settings

Component	kgf m	lbf ft	kgf m	lbf ft
Front wheel spindle	7.5	54.0	8.5	61.0
Front wheel spindle pinch bolt	2.0	14.0	Not applicable	
Rear wheel spindle	8.5	61.0	8.5	61.0
Rear brake torque arm:				
Brake end	1.9	13.0	Not applicable	
Frame end	2.3	17.0	Not applicable	
Disc brake components (not applicable to DT125 LC II):				
Brake disc	2.0	14.0	Not available	
Caliper to fork leg	3.5	25.0	3.5	25.0
Caliper support pin bolt	Not applicable		1.8	13.0
Brake hose union bolts	2.7	19.0	2.7	19.0
Rear wheel sprocket bolts	4.5	32.0	6.2	45.0
Rear hub stud bolts	3.9	28.0	3.9	28.0

Specifications relating to Chapter 7

	RD125 LC	**DT125 LC**

Electrical system

Voltage	12 volt	12 volt
Earth (ground)	Negative (–)	Negative (–)

Battery

Type	12N5-3B	FB3L-B or GM3-3B
Capacity	5 Ah	3 Ah
Electrolyte specific gravity	1.280	1.260

Charging system

Type	AC generator	AC generator
Make	Yamaha	Yamaha
Model	F1GM	F34Y
Charging coil resistance	0.35 Ω +20% at 20°C (68°F)	0.43 Ω ± 20% at 20°C (68°F)
Lighting coil resistance	Not applicable	0.35 Ω ± 20% at 20°C (68°F)

Voltage regulator/rectifier

Type	Short circuit	Short circuit
Make	Shindengen	Matsushita
Model	5HO	EHU-01TR07
No-load regulated voltage	14.5	14.5

Horn
Make	Nikko	Nikko
Model	CF-12	MF-12
Maximum amperage	2.5	1.5

Turn signal relay
Type	Condenser	Condenser
Make	Nippondenso	Nippondenso
Model	34L	061300-7110
Frequency	60 – 120 cpm	60 – 120 cpm
Wattage	21W x 2 + 3.4W	21W x 2 + 3.4W

Oil level switch
Make	Stanley	Taiheiyo
Model	4LO	AST1

Fuse
Rating	20A	10A

Bulbs – all models
Headlamp	12V 45/40W
Parking (city) lamp	12V 3.4W
Tail/brake lamp	12V 5/21W
Turn signal lamp	12V 21W
Instrument panel lamps	12V 3W
Warning lamps	12V 3W

1 Introduction

The preceding Chapters of this manual cover the RD and DT125 models sold in the UK between 1982 and 1984. In 1985 two new models, the RD125 LC II and the DT125 LC II, were introduced, and these embodied a number of major modifications, as well as numerous detail and cosmetic changes. Both machines were aimed at the learner market in the UK, and as such were sold with engines restricted to comply with the 9kW (12.2 bhp) power limit. The normal unrestricted versions were not imported into the UK.

In 1986, two further revised models were imported. The DT125 LC III appeared first, in March of that year, and was followed in November by the RD125 LC III. These machines included further refinements and modifications to the Mk II models.

As is the case with the machines covered in Chapters 1 to 6 of this manual, it is useful to be able to identify the specific model from its frame number, particularly when ordering spare parts. Given below are the initial frame numbers of the restricted models available in the UK.

RD125 LC II	1GM-000101 onwards
DT125 LC II	35A-000101 onwards
RD125 LC III	2HK-000101 onwards
DT125 LC III	35A-020101 onwards

As a guide to owners, the major changes from the previous models are described below, and will be covered in detail in the remainder of this Chapter. This information, together with the accompanying specifications, line drawings and wiring diagrams should be used in conjunction with the main text.

RD125 LC II

The RD125 received a thorough cosmetic update with the introduction of the Mk II model, but retained the same basic engine/transmission unit of its predecessors with only minor alterations. The chassis was subject to more extensive changes, particularly at the front end. Most noticeable was the adoption of a 16 inch front wheel in place of the 18 inch wheel fitted previously.

New, larger diameter forks were fitted, these having 33 mm diameter stanchions in place of the 30 mm type used on earlier versions. The newer forks also featured a fork brace fitted between the two lower legs. These changes, in conjunction with revised castor and trail angles, were designed to give quicker and more responsive steering. Other chassis revisions included a new front disc brake which utlised an opposed piston caliper, and a revised instrument panel.

DT125 LC II

The DT125 was rather more extensively revised than was the RD model, and employed a specially restricted version of the power valve (YPVS) engine sold in other markets. The power valve was fitted to the engine, but pegged in one position to comply with the UK learner power limits. Other than the revised cylinder barrel with its integral power valve housing, the engine/transmission unit remained largely unchanged.

Like the RD model, the DT was the subject of significant chassis and suspension modifications. New 36 mm fork stanchions were adopted, as was a completely revised rear suspension arrangement. The simple cantilever system employed on the earlier models was replaced by a full rising-rate arrangement, giving much improved rear wheel control. Unlike the DT model supplied to other markets, however, the front drum brake was retained.

RD125 LC III

With the introduction of the Mk III model, the engine arrangement was brought in line with that of the DT models, and the same pegged power valve was fitted.

Another obvious change was the adoption of a new exhaust system as part of the cosmetic update. The side stand was deleted on the Mk III model.

DT125 LC III

The DT125 LC continued in Mk III form with the usual cosmetic and detail modifications, the most significant of these being the adoption of the semi-shrouded front disc brake in line with models sold in other markets. In 1987, the Mk III received plastic covers over the fork lower legs. These, and other cosmetic modifications, are discussed later in this Chapter.

2 Cylinder barrel and YPVS components: DT125 LC II, III and RD125 LC III models

1   In most countries, the above models are sold with a servo-controlled power valve (Yamaha Power Valve System, or YPVS). A cylindrical valve is incorporated in the exhaust port, and this can be rotated to control the exhaust port opening and position. The valve is linked by two cables to a frame-mounted servo motor unit which opens or closes the valve to suit various engine speeds. A small microprocessor in the servo motor unit is used to read the engine speed from the CDI unit.

2   On UK versions, the power output legislation governing learner machines of 125 cc and less makes the power valve unnecessary. The modified cylinder barrel is fitted, as is the power valve itself, but the valve is locked in the closed position by a peg in the end cap. The connecting cables and the servo motor are not fitted.

3   With the valve locked in one position, there is no reason to disturb it under normal circumstances. The only possible exception may be if there has been a severe accumulation of carbon in the exhaust port. If normal decarbonisation methods fail to clear the carbon, remove the cylinder barrel as described in Chapter 1, Section 6 and proceed as follows.

4   On the left-hand side of the barrel, remove the single Allen screw which retains the power valve end cap. Lift away the end cap to reveal the end of the valve spool and the head of the long Allen screw which retains the two halves of the valve. Slacken and remove the screw, then withdraw the left-hand half of the valve spool. The two halves of the valve are located by two small pins; take care not to lose them. Note that if the valve is removed with the barrel in position the pins may drop out and fall into the exhaust system or back into the cylinder bore.

5   The right-hand half of the valve spool can be removed after the two Allen screws holding the end cap have been released. Displace the valve together with the end cap and locating sleeve and lift them clear of the barrel. The assembly is retained by the single central retaining bolt, but it should not be necessary to separate the component parts. If the bolt is removed for any reason, mark the position of the valve, sleeve and end cap to ensure that the valve is positioned correctly during assembly.

6   Refit the valve assembly by reversing the above sequence. Lubricate the end cap O-ring with molybdenum disulphide grease. Tighten the end cap Allen bolts to 0.5 kgf m (3.6 lbf ft) and the long Allen-headed screw which secures the two halves of the valve to 0.6 kgf m (4.3 lbf ft). Remember to ensure that the two small dowel pins are in position when joining the valve halves.

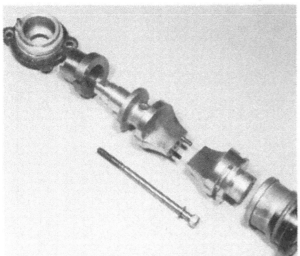

2.2 The YPVS components as fitted to UK models. Note headed sleeve and locating pin (left of picture)

2.4a Left-hand end cap is retained by a single Allen screw

2.4b Long Allen screw secures the two halves of the valve spool

2.4c Lift away the left-hand half of the valve, taking care not to drop pins into exhaust system

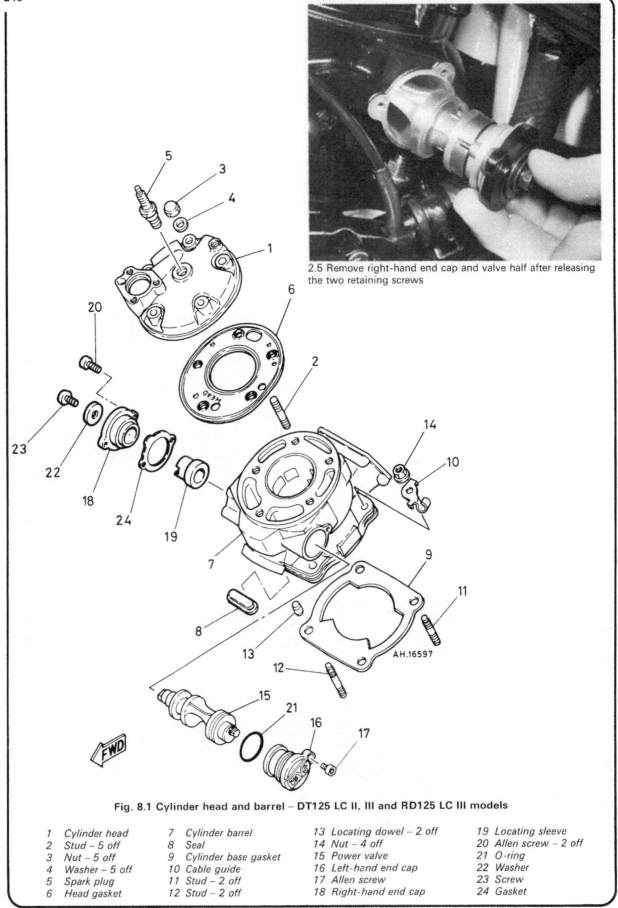

2.5 Remove right-hand end cap and valve half after releasing the two retaining screws

Fig. 8.1 Cylinder head and barrel – DT125 LC II, III and RD125 LC III models

AH.16597

1 Cylinder head	7 Cylinder barrel	13 Locating dowel – 2 off	19 Locating sleeve
2 Stud – 5 off	8 Seal	14 Nut – 4 off	20 Allen screw – 2 off
3 Nut – 5 off	9 Cylinder base gasket	15 Power valve	21 O-ring
4 Washer – 5 off	10 Cable guide	16 Left-hand end cap	22 Washer
5 Spark plug	11 Stud – 2 off	17 Allen screw	23 Screw
6 Head gasket	12 Stud – 2 off	18 Right-hand end cap	24 Gasket

3 Cooling system: modifications – general

The cooling system layout remains basically similar to that of the previous versions described in Chapter 2. In almost every respect the information in Chapter 2 can be applied, but it should be noted that a thermostat has been added to the system on the DT125 LC II and III, and to that of the RD125 LC III. The addition of the thermostat entailed detail alterations to the connecting hoses and radiator, but these do not materially affect the working procedures relating to the cooling system in general. Details of the thermostat and test procedures will be found in Section 4 of this Chapter.

4 Thermostat: general description and testing

1   The thermostat is a temperature-controlled valve used to regulate the flow of coolant around the system. The valve is located in a cast alloy housing between the end of the radiator top hose and the crankcase. When the engine, and thus the coolant, is cold, the thermostat remains closed, effectively stopping the circulation of coolant round the engine. When the engine is started, the lack of coolant flow ensures that the cylinder and cylinder head warm up quickly. This means that the engine reaches an efficient operating temperature much faster, and so engine wear is minimised. At a predetermined temperature the valve opens, allowing coolant to circulate through the system. The valve remains open until the engine is stopped and begins to cool down, when it closes again.
2   Thermostats are generally reliable devices and can be expected to function for several years without attention. It should be noted that this is dependent to some extent on regular maintenance of the cooling system, particularly flushing to remove scale deposits. If neglected, the thermostat may jam open or closed, causing serious over-cooling or overheating problems. A fault of this type should be immediately obvious, but check that any overheating is not due to a simpler problem, like a blocked radiator or trapped hose. In the event of a suspected fault, the thermostat should be removed for testing after the system has been drained as described in Section 2, Chapter 2. Remove the bolts which retain the thermostat housing to the crankcase and lift it away. Remove the thermostat, taking care not to damage the gasket or O-ring.
3   To test the operation of the thermostat you will need an accurate thermometer capable of reading up to 100°C (212°F). Using pieces of wire, suspend the thermostat and the thermometer in the centre of a saucepan or similar. Fill the pan with water so that the thermostat is submerged. Heat the water, stirring it slowly to ensure even heat distribution. Note the operation of the thermostat and at what temperature it opens. At 80.5° – 83.5°C (177° – 182°F) the thermostat should begin to open, and it should be fully open by the time the water temperature reaches 95°C (203°F).
4   If the thermostat proves faulty it should be renewed without delay. In an emergency, it is preferable to run the machine with the thermostat missing rather than to risk it jamming closed and causing serious overheating. Note that this means that the engine will take much longer than normal to reach its full operating temperature, and it should be ridden gently for the first few miles to avoid the risk of accelerated piston and bore wear. When fitting the thermostat check that the O-ring and housing gasket are sound. Fit the thermostat with the small bypass hole facing towards the rear of the machine. Refit the thermostat housing, tightening the retaining bolts to 0.8 kgf m (5.8 lbf ft), then refill the system and check for leaks.

5 Air filter: examination and renovation – DT models

The DT125 LC II and III feature a revised air filter casing housing a conical foam element. To gain access to the element for cleaning, first remove the seat to reveal the cover of the filter casing. Remove the two screws and lift the cover away, then withdraw the element and its plastic support frame. The element should be cleaned, examined and re-oiled as described in Section 13 of Chapter 3. When refitting the element, make sure that it locates correctly in the casing and that the cover seals correctly.

6 Carburettor: modifications – general

1   The carburettor fitted to both the RD and DT models remains very similar to that used on the previous models, the only obvious external change being the adoption of a small horizontal float bowl drain plug in place of the large vertical one used previously. In the case of the DT125 LC II, III and RD125 III the main jet holder (see item 21 in Fig. 3.3) is dispensed with, the main jet screwing directly into the base of the needle jet.

4.2a Remove drain bolt to empty the cooling system ...

4.2b ... then release housing and remove thermostat from head

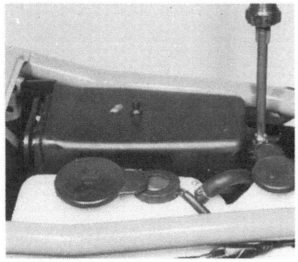

5.1a Air filter cover is retained by two screws

5.1b Element and its support frame can be lifted away for cleaning

2   As might be expected, the carburettor jet sizes and settings have been altered to suit various changes to the engine, exhaust and intake systems, and full details are given in the Specifications at the start of this Chapter. In other respects the relevant Sections of Chapter 3 should be followed.

7   Ignition system: modifications – general

1   There have been a number of changes in the ignition system, largely due to the use of components of different manufacture from those originally specified. These changes can be seen by comparing the accompanying ignition system specifications with those shown at the start of Chapter 4. Whilst this has little effect on the operation of the ignition system, some of the test connections and values are affected. These are summarised below, together with the Section in Chapter 4 to which they relate.
2   The main change in the system is the adoption of a revised flywheel generator on both models. In each case the basic arrangement is similar, but on the DT model the system has been converted to 12 volt operation and external pickup coils have been added. These are triggered from a magnet embedded in the outer face of the rotor, as in the RD arrangement.

7.2 Generator on DT model now features external pickup coil

8   Source coil: testing (Chapter 4, Section 5)

RD models
1   Remove the fuel tank and trace the wiring from the generator to the 3-pin connector (brown, sky blue and black leads) and the two single bullet connectors (red/black and yellow leads). Disconnect them, then perform the following resistance tests on the two source coils.
2   The resistance of source coil 1 (red/black to black leads) should show 650 ohms ± 20% at 20°C (68°F). The resistance figure for source coil 2 (yellow lead to brown lead) is 63 ohms ± 20% at 20°C (68°F). If either is significantly outside these limits it will be necessary to renew the stator as an assembly. Note that source coil 1 provides the charging current for the CDI capacitor, whilst source coil 2 controls ignition advance.

DT models
3   Remove the fuel tank and trace the wiring from the generator to the 2-pin connector (brown and black leads) and the single bullet connector (red/black lead). Measure the resistance of the source coil by connecting the meter probes between the brown lead and the black/red lead. The correct figure is 355 ohms ± 20% at 20°C (68°F). As was the case with the earlier DT models, it is possible to renew individual coils in case of failure.

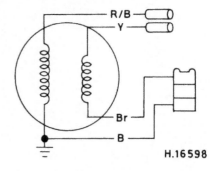

Fig. 8.2 Source coil test circuit diagram

Wire colour key

B	Black	R/B	Red and black
Br	Brown	Y	Yellow

9 Pickup coil: testing (Chapter 4, Section 6)

Remove the fuel tank, then trace the pickup wiring back to the 2-pin connector (white/red and white/green leads). Separate the connector and measure the resistance between the two leads. The specified resistance is 350 ohms $\pm$ 20% at 20°C (68°F). If the resistance falls outside these limits the pickup coil should be renewed.

10 Ignition HT coil: testing (Chapter 4, Section 7)

The HT coil can be tested in the same way as has been described for the earlier models in Chapter 4, by measuring the primary and secondary winding resistances. The test is performed in an identical fashion, but the values obtained should be as follows:

Primary winding
resistance .... 1.6 ohms $\pm$ 10% at 20°C (68°F)
Secondary winding
resistance .... 6.6 k ohms $\pm$ 20% at 20°C (68°F)

10.1 Ignition HT coil location – DT model

11 CDI unit: testing (Chapter 4, Section 8)

As with the earlier models, no detail specifications for the CDI unit are supplied by Yamaha, and this means that a suspect unit can only be checked by substitution. In the case of the RD125 LC, the CDI unit is mounted on the frame, just to the rear of the crankcase, whilst on the DT125 LC it is located below the fuel tank.

12 Ignition timing: checking (Chapter 4, Section 9)

The ignition system fitted to the later models is similar to that described for the earlier RD125 LC models. Again, no adjustment is possible, but the timing advance can be checked using a stroboscopic timing lamp (strobe) in the manner described in the first part of Section 9, Chapter 4. It should be noted that the advance figures and the engine speeds at which the readings should be taken vary between the two models and are shown below:

RD model:
    At 1300 rpm .... 16°
    At 3000 rpm .... 30°
DT model:
    At 1350 rpm .... 8°
    At 4000 rpm .... 30°

11.1 CDI unit location – DT model

13 Suspension: modifications – general

1 As has been mentioned, the models sold in the UK differ most significantly from the earlier models in the suspension area. On both machines the front forks have been changed, but whilst this affects the response of the front suspension it has little material effect on the approach to dismantling, overhaul or reassembly. Removal and refitting of the forks is affected to some extent by the inclusion of a fork brace, and on the RD model, the modified top yoke and handlebar.
2 In the case of the rear suspension, the RD model retains the simple monoshock cantilever suspension of its predecessor, but a completely new rising-rate arrangement is fitted to the DT model, and this is described in Sections 16 to 20.

14 Front forks: removal and refitting

1 The front fork legs can be removed and refitted as described in Chapter 5, Section 2, noting the following points.

RD models
2 Before attempting to remove the front wheel spindle note that a pinch bolt has been added at the bottom of the right-hand lower leg. This should be slackened after the spindle nut has been removed. Note also that a fork brace is fitted between the two lower legs. This is retained to each leg by two bolts and should be removed at the same time as the front mudguard. When preparing to remove the legs from the yokes, slacken the upper and lower pinch bolts, and also the handlebar mounting bolts. During reassembly, note the revised torque settings and capacities shown in the specifications at the beginning of this Chapter.

DT125 LC III

3   The adoption of a disc front brake has a marginal effect on the front wheel removal procedure on the later DT125 LC models. Before the wheel can be detached it is necessary to remove the brake disc cover. This is secured to the left-hand fork leg by two screws. After it has been detached the wheel can be removed, leaving the brake caliper attached to the left-hand lower leg. When installing the wheel, make sure that the cover is refitted. The cover serves to keep dirt away from the disc and caliper, and must not be omitted if the machine is used off-road.

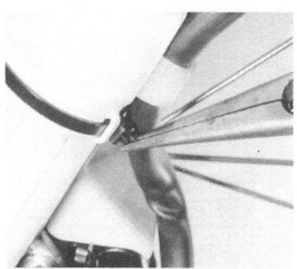

14.3 On later DT125 LC III models, plastic fork covers protect the lower legs. Covers are held in place by electrical cable ties

15 Rear suspension assembly: general – RD models

The RD model retains the same cantilever subframe and single rear suspension unit of its predecessors. For details refer to Sections 9-11 of Chapter 5.

16 Rear suspension assembly: general – DT models

1   The DT models feature a completely revised rear suspension assembly. The previous cantilever subframe is replaced by a box section swinging arm. This is connected to the single De Carbon type suspension unit by way of a relay arm and a short connecting rod. The arrangement provides rising rate suspension, the operation of which is as follows.
2   During the initial stages of rear wheel deflection, the rear suspension unit is compressed only slightly for a given amount of swinging arm movement. As deflection increases, the rate at which the suspension unit is compressed becomes progressively greater. The result is very soft initial movement, becoming progressively stiffer as movement increases. It follows that small irregularities can be absorbed easily, whilst larger bumps are contained by the stiffer final rate of the suspension.

17 Swinging arm and suspension linkage: removal and refitting – DT models

1   Disconnect the final drive chain and remove the rear wheel as described in Chapter 6, Section 15, noting that the security bolts in the ends of the swinging arm should be removed to permit the wheel to be pulled clear. It is preferable to remove the chainguard and the white plastic chain guide from the underside of the swinging arm, though this is not essential.
2   Remove the nut from the swinging arm pivot bolt. Remove the suspension unit lower mounting bolt and the relay arm connecting rod to frame bolt. Support the swinging arm, then displace the pivot bolt and lift the assembly clear of the frame.
3   Installation is a reversal of the above sequence, noting that the shims and end caps must be refitted correctly. If in any doubt as to the accuracy of the shimming, refer to Section 18 below. Note that all of the linkage bolts are tightened to 4.2 kgf m (30 lbf ft) and that the swinging arm pivot bolt is secured to 8.0 kgf m (58 lbf ft).

17.2a Relay arm connecting rod to frame mounting bolt (arrowed)

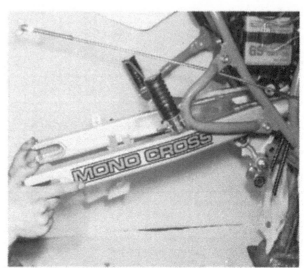

17.2b Remove pivot bolt and lift swinging arm clear of frame

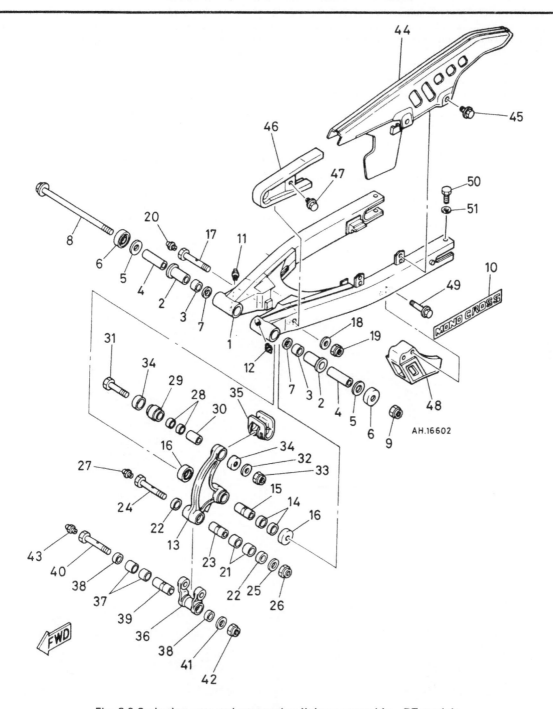

Fig. 8.3 Swinging arm and suspension linkage assembly – DT models

1	Swinging arm	15	Inner sleeve	28	Bush – 2 off	40	Bolt
2	Bush – 2 off	16	End cap – 2 off	29	Bush	41	Washer
3	Bush – 2 off	17	Bolt	30	Inner sleeve	42	Nut
4	Inner sleeve	18	Washer	31	Bolt	43	Grease nipple
5	Shim – as required	19	Nut	32	Washer	44	Chain guard
6	End cap – 2 off	20	Grease nipple	33	Nut	45	Bolt and washer – 2 off
7	Oil seal – 2 off	21	Bush – 2 off	34	End cap – 2 off	46	Chain guide
8	Pivot bolt	22	Oil seal – 2 off	35	Boot	47	Bolt and washer
9	Nut	23	Inner sleeve	36	Relay arm connecting	48	Chain guard
10	Emblem	24	Bolt		rod	49	Bolt – 2 off
11	Grease nipple	25	Washer	37	Bush – 2 off	50	Bolt – 2 off
12	Grease nipple	26	Nut	38	Oil seal – 2 off	51	Washer – 2 off
13	Relay arm	27	Grease nipple	39	Inner sleeve		
14	Bush – 2 off						

18 Swinging arm and suspension linkage: examination and renovation – DT models

1   The swinging arm side clearance is checked as follows. Referring to the accompanying line drawings, measure the dimensions shown using a vernier caliper. The two bushes should be checked for length (A1 and A2). The specified length is 68.75 – 69.05 mm (2.707 – 2.719 in). If the bushes have worn to less than the lower figure, they should be renewed. The old bushes can be driven out as described in Chapter 5, Section 10, taking care to support the swinging arm boss to avoid any risk of distortion. Use a drawbolt arrangement to fit the new bushes. Note that a gap of exactly 4 mm (0.08 in) should be left between the end of the bush and the inner edge of each swinging arm pivot boss to leave room for the seal.

2   To calculate the side clearance, add dimensions A1 and A2 together, and also dimensions B1 and B2. Subtract the lower figure from the higher one to obtain the clearance. If this is outside the range 0.4 – 0.7 mm (0.016 – 0.028 in) the clearance should be adjusted by adding or subtracting shims. The shims are 0.3 mm (0.012 in) in thickness. If only one shim is used, fit it on the left-hand side. If more than one is used, spread them evenly on both sides to keep the swinging arm central in the frame.

3   The remaining bushes, seals and end caps on the relay arm and its connecting rod can be checked for wear or damage, and where necessary, renewed in the same way as described for the swinging arm pivot bushes. Before reassembling the various parts, make sure that all traces of road dirt are removed. Lubricate the bushes using a water-resistant lithium-based wheel bearing grease. If the seals or end caps are loose, damaged or distorted, renew them. Refer to the accompanying line drawing when refitting the relay arm and connecting rod to the swinging arm, noting the relative position of each part, and in particular the 'REAR' marking on the connecting rod. Tighten the various pivot bolts to the torque wrench settings shown. For further general information, refer to Chapter 5, Section 10.

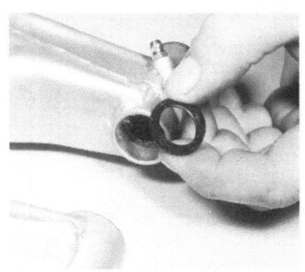

18.1 Note 4 mm gap between end of bush and inner face of boss

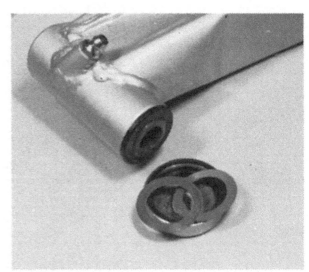

18.2 Spread shims evenly between ends of bosses to keep the swinging arm central

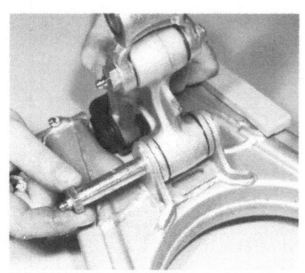

18.3a Remaining pivots can be dismantled and dealt with in the same way as swinging arm

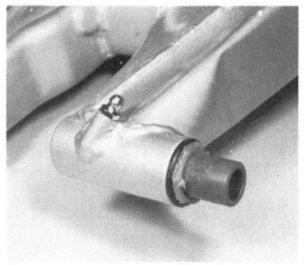

18.3b Grease all bushes with water-resistant lithium based grease

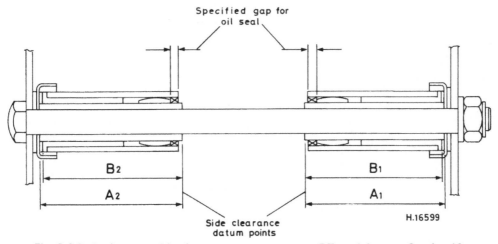

Fig. 8.4 Swinging arm side clearance measurement – DT models – see Section 18

19   Rear suspension unit: removal and refitting – DT models

1   The rear suspension unit can be removed with the remaining rear suspension components in place. Start by removing the seat, fuel tank and the rear wheel, having first supported the machine on a suitable stand. Remove the upper mounting bolt, followed by the lower mounting bolt. The unit can now be manoeuvred clear of the frame via the cutout in the rear mudguard, a rubber flap being provided to permit this.

2   Before refitting the unit, clean off any accumulated dirt. It is advisable to make any spring preload changes prior to refitting, access being much easier. This is carried out as described for the earlier DT models in Chapter 5, Section 11 paragraph 12, noting the following dimensions:

Rear suspension unit spring dimensions (installed):
    Standard .... 225 mm (8.9 in)
    Minimum .... 213 mm (8.4 in)
    Maximum .... 233 mm (9.2 in)

3   As with the earlier models, each turn of the adjuster alters the spring length by 1 mm (0.04 in). Adjustments should be made in increments of 2 mm (0.08 in), or two turns at a time. Once adjusted, tighten the locknut to 5.5 kgf m (40 lbf ft). Refit

the unit by reversing the removal sequence, noting that the torque setting for both top and bottom mounting bolts is 4.2 kgf m (30 lbf ft).

19.1a Remove tank and seat to gain access to suspension unit upper mounting bolt

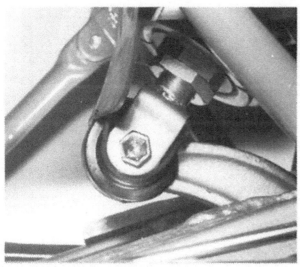

19.1b Remove the lower mounting bolt ...

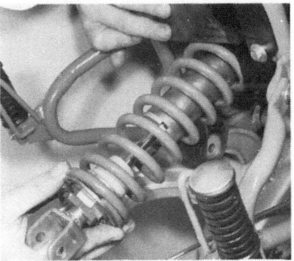

19.1c ... and manoeuvre unit out through access flap

19.3 Note preload adjuster and locknut near lower mounting

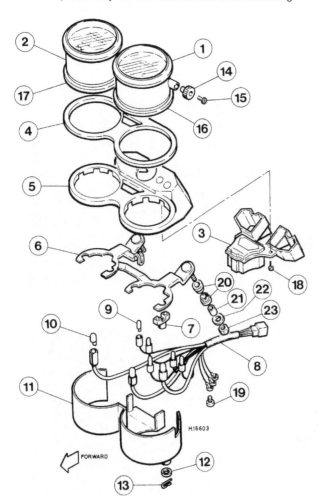

20 Rear suspension unit: examination and renovation – DT models

As on previous models, there is little that can be done to repair a damaged or leaking rear suspension unit. It is of sealed construction, and thus cannot be dismantled or overhauled. The general remarks in Chapter 5, Section 11 can be applied, as can the procedure for releasing nitrogen pressure before discarding a defective unit. Note, however, that the hole should be drilled at a point 25 – 30 mm (1.0 – 1.2 in) from the bottom of the unit.

21 Speedometer and tachometer heads: removal and refitting

1   The instrument heads fitted to the RD model have been revised as part of the general cosmetic update, and as such require a slightly different approach to removal and refitting. In the case of the DT model, the instrument panel is generally similar to that fitted to previous models, and can be removed in the sequence described in Chapter 5, Section 16.

RD models
2   The instrument panel can be removed as described in Chapter 5, Section 16, bearing in mind the following points. The panel assembly is supported on a rubber-bushed double mounting bracket attached to the top yoke. The two retaining nuts can be reached from the underside, rather than from the top as on previous models. To gain access to the underside of the instruments it is necessary to release the lower cover. This is secured by two wire clips and special washers. Once removed, access is gained to the various bulbholders, and the instrument heads can be freed from the mounting bracket.

21.1a Remove single fixing bolt and unclip headlamp nacelle to gain access to instruments – DT model

Fig. 8.5 Speedometer and tachometer heads – RD models

1 Speedometer	7 Damping rubber – 4 off	13 Wire clip – 4 off	19 Screw and washer – 3 off
2 Tachometer	8 Bulbholders and wiring	14 Reset knob	20 Rubber bush – 4 off
3 Water temperature gauge	9 3W bulb – 5 off	15 Screw	21 Spacer – 2 off
4 Plate	10 3.4W bulb – 3 off	16 Damping ring	22 Washer – 2 off
5 Cover	11 Lower cover	17 Damping ring	23 Nut – 2 off
6 Mounting bracket	12 Special washer – 4 off	18 Screw – 4 off	

21.1b Pull off R-pin clip and washers to free instrument heads

21.1c Lift instrument head to gain access to bulbs

22 Front disc brake: general – RD models

1 The general arrangement of the front disc brake is similar to that used on earlier models with the exception of the caliper. On earlier models a single piston type was fitted in which the caliper was permitted to move along pins under pressure from the single piston, thus equalising pressure from each of the two pads. On the new arrangement the caliper body is rigidly positioned on the fork leg, each pad being moved against the disc surface by its own piston. This means that pressure is equalised hydraulically, rather than mechanically.

2 The new caliper design means that there are a number of changes relating to Sections 7 and 11 of Chapter 6, these being described below. In other respects the text in Chapter 6 can be applied.

23 Front disc brake: pad renewal – RD models

1 Pad renewal can be carried out as described in Section 7 of Chapter 6, noting the following points. Start by removing the two caliper mounting bolts and lift the caliper clear of the disc and fork leg. **Do not** slacken or remove the caliper bridge bolts; these pass through the caliper halves from the inner face of the caliper.

2 Remove the two Allen-headed pad retaining bolts to free the pad assembly. Lift out the two pads, noting the position and direction of the anti-squeal shims fitted between each pad and piston. Lift out the pad spring, noting its position and direction of fitting.

3 Clean and examine the caliper and pads as described in Chapter 6, paying particular attention to signs of fluid leakage or pad contamination; problems of this nature must be rectified before proceeding further. If new pads are to be fitted, push each piston back into the caliper bore to permit the thicker friction material to be accommodated. Remember that this will cause the fluid level to rise in the reservoir, so take care that excess fluid is removed without spillage.

4 Fit the pad spring in its original position, then fit the anti-squeal shims to the backs of their respective pads. Note that the cutaway side of the shim must face downwards and to the rear of the machine. Refit the pads and their retaining bolts, tightening the latter to 1.0 kgf m (7.2 lbf ft). Offer up the caliper and fit the retaining bolts loosely. Operate the brake lever a few times to centralise the pads, then tighten the caliper mounting bolts to 3.5 kgf m (25 lbf ft). Finally, check and if necessary adjust, the fluid level in the reservoir.

Fig. 8.6 Front brake caliper – RD models

1   Brake disc
2   Bolt – 6 off
3   Caliper body
4   Allen bolt – 2 off
5   Bleed nipple
6   Cap
7   Bolt – 2 off
8   Piston – 2 off
9   Piston seals – 4 off
10 Anti squeal shims
11 Brake pads
12 Pad spring

AH.16600

24 Front brake disc: general – RD models

In line with the redesigned caliper, a new brake disc has been fitted. In general, the remarks in Chapter 6, Section 9 can be applied, but note that the minimum thickness of the disc is changed to 3.0 mm (0.12 in). The disc is retained by six (rather than four) mounting bolts; they should be tightened to 2.0 kgf m (14.5 lbf ft).

25 Front disc brake: overhauling the caliper unit – RD models

Note: *The twin piston caliper unit is manufactured in two halves, these being joined by two bridge bolts fitted from the inner face of the caliper. On no account slacken these bolts or attempt to separate the caliper halves.*

The procedure for overhauling the twin piston caliper is generally similar to that described in Chapter 6, Section 11 for the earlier single piston type. An obvious exception to this is that two pistons and bores are involved. These will have worn to suit each other, and care should be taken not to interchange components from one bore to the other. Unlike the earlier caliper, it is possible to buy the seals separately from the pistons, and in view of the relatively low cost it is recommended that these be renewed whenever the caliper is dismantled, as a precautionary measure. The remarks concerning the caliper mounting bracket pins used on single piston calipers do not apply to the new version, for obvious reasons.

26 Front disc brake: DT125 LC III model

General description
1   The DT125 LC III is equipped with an hydraulic disc brake assembly in place of the drum brake arrangement used on previous UK models. The master cylinder is of the same type as that fitted to the RD models, and reference should be made to Section 10 of Chapter 5, and to Fig. 6.5 for details.
2   The caliper is of the sliding, single-piston type, and is thus similar to the earlier RD unit described in Chapter 5, Section 11. For most purposes, reference should be made to Chapter 5 when working on the system, bearing in mind the following points.

Pad renewal
3   Clean all accumulated dirt from the caliper, noting that this is especially important where the machine is used off-road. Remove the bolt which secures the lower edge of the caliper to the support bracket. The caliper can now be pivoted upwards and slid off the support pin.
4   Examine the pads for wear or contamination. If contaminated with oil, remember to locate and remedy the cause of the contamination before cleaning and degreasing the caliper and fitting new pads. Check the thickness of the pad friction material. If this has worn to 0.8 mm (0.031 in) or less on either pad, renew them as a set, together with the pad springs. The pads are held on the caliper bracket by the springs, the exact arrangement being shown in the accompanying photographs.
5   Clean off all dirt from the caliper bracket and from inside the caliper opening. Fit the new pads, ensuring that the curved face of each is rearmost. Retain the pads on the bracket using new pad springs. To accommodate the new pad thickness it is usually necessary to push the caliper piston inwards. As this is done, the hydraulic fluid will be displaced along the brake hose to the reservoir; make sure that this does not cause it to overflow. If necessary, remove the reservoir lid and drain off excess fluid.
6   Grease the pivot pin and the caliper retaining bolt. Refit the caliper on the pin and swing it down over the pads. Fit the retaining bolt and tighten it to 1.8 kgf m (13.0 lbf ft). Operate the brake lever several times until the caliper piston adjusts to the new pads and braking action feels normal.

Caliper overhaul
7   The caliper overhaul procedure is generally similar to that described in Chapter 5, the only differences being those relating to the design of the caliper unit. To clarify these points refer to the accompanying line drawing during the overhaul.
8   Note that it is especially important to ensure that the caliper unit is cleaned properly before dismantling commences. Any off-road use will normally result in a build-up of dirt and mud around the caliper, and on no account must any of this find its way into the caliper.

Brake disc
9   Although well proven in off-road applications, the disc brake assembly is in some ways more vulnerable than a drum unit, and it is especially important that the caliper and disc are kept clean. The disc is protected by a cover over its forward edge, and this is secured by two screws to the fork left-hand lower leg. If the cover becomes damaged, fit a new one immediately. **Never** use the machine off-road with the cover missing or rapid wear of the pads and disc surfaces is likely to result.
10   When new, the disc has a thickness of 3.5 mm (0.14in). If it wears to 3.0 mm (0.12 in) or less, or if it becomes badly scored, a new disc must be fitted. Note the pattern of holes in the disc surface before removing it; the new disc must be fitted so that the hole pattern runs in the correct direction.

26.3a DT125 LC III – access to brake pads is gained after lifting the caliper unit clear of the mounting bracket

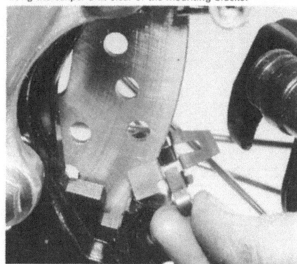

26.3b Pad springs locate at top and bottom of caliper bracket, as shown

26.3c Note how the tangs on the pads are located by the pad springs

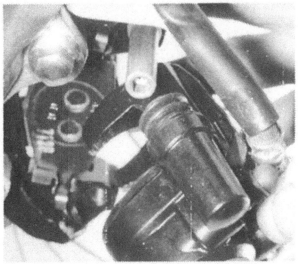

26.6a Grease the caliper pivot pin, then slide the caliper over it

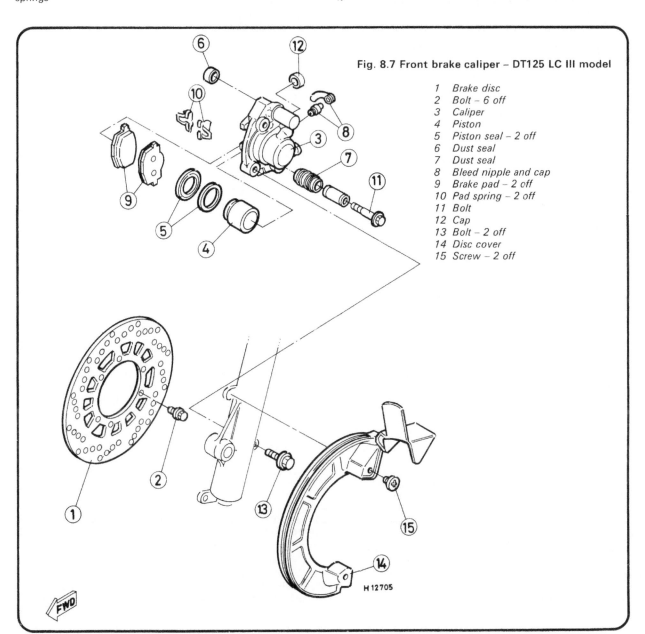

Fig. 8.7 Front brake caliper – DT125 LC III model

1   Brake disc
2   Bolt – 6 off
3   Caliper
4   Piston
5   Piston seal – 2 off
6   Dust seal
7   Dust seal
8   Bleed nipple and cap
9   Brake pad – 2 off
10 Pad spring – 2 off
11 Bolt
12 Cap
13 Bolt – 2 off
14 Disc cover
15 Screw – 2 off

H 12705

FWD

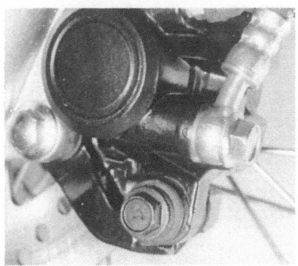

26.6b Swing the caliper down over the pads, then fit and tighten the retaining bolt

27 Rear wheel: removal and refitting – DT models

1   The procedure for rear wheel removal is generally similar to that described in Chapter 6, Section 15. Note, however, that the redesigned end of the swinging arm requires a slightly modified approach. Start by slackening the rear wheel spindle, then turn the snail cam adjusters so that the final drive chain can be lifted clear of the sprocket. It is now necessary to remove the two security bolts which close the end of the swinging arm forks. Once these have been removed the wheel, complete with spindle and spacers, can be withdrawn. Note also that the rear wheel spindle is fitted from the left-hand side of the machine, unlike earlier models. When refitting the rear wheel, refer to the torque settings given at the beginning of this Chapter, and set the final drive chain free play to 35 – 45 mm (1.4 – 1.8 in).
2   In the case of the DT 125 LC III model, note that the dust seal shown as item 5 in Fig. 6.8 has been omitted, and the spacers (items 4 and 21) have been redesigned. Other than this, the rear wheel is similar to that used on the DT 125 LC II.

28 Electrical system: modifications – general

The electrical systems used on the later models are broadly similar to those of their predecessors, the most significant change being the adoption of a full 12 volt system on the DT model. Other changes are largely due to the fitting of components of different manufacture, and in some cases this affects the test figures and procedures described in Chapter 7. In the Sections which follow, these changes are described, the headings giving the Section number of Chapter 7 to which the information relates. Note that where a test requires the engine to be run at various speeds, do not allow it to exceed 6000 rpm for more than one or two seconds.

29 Charging system: checking the output (Chapter 7, Section 4)

RD models

1   Remove the left-hand side panel and disconnect the battery positive (red) lead at the fuse holder. Set the multimeter on the dc amps x 20 range, then connect the positive (+) meter probe to the wiring harness side of the red lead and the negative (–) meter probe to the battery side of the red lead. Start the engine and note the meter readings at 2000 rpm and 8000 rpm. If the readings obtained are significantly different from those shown below, check the generator coil resistances and the condition of the regulator/rectifier unit.

Charging output:
At 2000 rpm .... 5A or more
At 8000 rpm .... 14A or less

DT models – charging system

2   Remove the right-hand side panel and disconnect the battery positive (red) lead at the fuse holder. Set the multimeter on the dc amps x 5 range, then connect the positive (+) meter probe to the wiring harness side of the red lead and the negative (–) meter probe to the battery side of the red lead. Start the engine and note the meter readings at 2000 rpm and 8000 rpm. Now turn the lights on and set the dip switch to the main beam position and repeat the tests. If the readings obtained are significantly different from those shown below, check the generator coil resistances and the condition of the regulator/rectifier unit.

Charging output:
Lights off:
At 2000 rpm .... 0.7A or more
At 8000 rpm .... 1.8A or less
Lights on:
At 2000 rpm .... 0.3A or more
At 8000 rpm .... 1.8 A or less

DT models – lighting system

3   Follow the test sequence described in Chapter 7, Section 4, noting that the meter should be set to the 20V ac scale. The lighting voltage readings obtained should be as follows:

Lighting voltage:
At 2000 rpm .... 12V or more
At 8000 rpm .... 18V or less

30 Generator coils: resistance tests (Chapter 7, Section 5)

RD models

1   Remove the fuel tank and separate the three pin connector from the generator. Set the meter to the ohms x 1 scale and connect one of the probes to any one of the three white output leads. Connect the remaining probe to each of the remaining white leads in turn and note the reading shown. In each case a figure of 0.35 ohms ± 20% at 20°C (68°F) should be obtained. If a coil has failed (open or short circuit) the stator must be renewed.

DT models

2   Remove the side panels and fuel tank, then disconnect the three-pin connector (sky blue, white and yellow/red leads) and the two pin connector (brown and black leads) from the generator. To test the charging coil resistance, set the meter to the ohms x 1 scale and check the resistance between the white lead and the black lead. This should show a resistance of 0.43 ohms ± 20% at 20°C (68°F).
3   The lighting coil resistance is measured in the same way, this time between the yellow/red lead and the black lead. A reading of 0.35 ohms ± 20% at 20°C (68°F) should be obtained.

31 Regulator/rectifier unit: location and testing – RD models (Chapter 7, Section 6)

1   If a test of the charging system has indicated a possible fault in the regulator/rectifier unit, it should be checked as follows. Remove the seat and the left-hand side panel. Trace the wiring back from the unit and separate it at the connector.

Remove the regulator/rectifier from the battery tray and place it on the bench for testing.

2   The rectifier stage contains six diodes, each of which should conduct in one direction only. This can be checked using a multimeter set on the resistance (ohms) scale, following the accompanying test chart and circuit diagram. If any one diode shows either an open or short circuit in both directions, the regulator/rectifier unit must be renewed. Note that no test information on the regulator stage is available; if this is suspect, check by substituting a new unit.

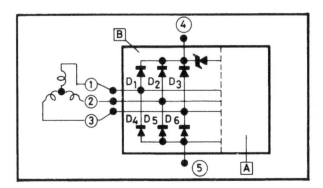

Checking element	Pocket tester connecting point		Good
	(+) (Red)	(−) (Black)	
D1	4	1	O
	1	4	X
D2	4	2	O
	2	4	X
D3	4	3	O
	3	4	X
D4	5	1	X
	1	5	O
D5	5	2	X
	2	5	O
D6	5	3	X
	3	5	O

Continuity (O)   H.16601

No continuity (X)

Fig. 8.8 Rectifier unit testing table – RD models

A   Regulator   3   White wire
B   Rectifier   4   Red wire
1   White wire   5   Earth wire
2   White wire

32 Regulator/rectifier unit: location and testing – DT models (Chapter 7, Section 7)

The regulator/rectifier unit is mounted on the frame below the fuel tank, near the CDI unit. In the absence of test data the only way of assessing the condition of a suspect unit is to substitute a new one, and to this end it is probably best to take the machine to a Yamaha dealer for confirmation.

32.1 Regulator/rectifier location – DT model

33 Battery: general

The information covering battery maintenance and charging procedures shown in Chapter 7, Sections 8 and 9 can be applied to the later models, but note that the DT now employs a 12 volt system and is fitted with a 3Ah battery. This means that when charging the battery, the charger should be set to 12 volts and that the normal charge rate should not exceed 0.3 amps.

34 Water temperature gauge and sender unit: location and testing

1   The procedure for checking the water temperature gauge sender resistance remains the same as described in Chapter 7, Section 22. Note, however, that the resistance figures and temperatures should be as shown below:

Water temperature	Resistance
31 – 49°C (88 – 120°F)	579 ohms
80°C (176°F)	127 ohms
110°C (230°F)	47.9 ohms
109.5 – 120.5°C (229 – 249°F)	41.6 ohms

2   You may have noted that the above test temperatures pose something of a problem in that it is not possible to raise the temperature of water above 100°C (212°F) at normal atmospheric pressure. For all practical purposes, this means that the two highest resistances cannot be checked without a pressurised test rig, something few owners will be able to improvise at home. It is recommended that the first two resistances are checked to give some indication of the condition of the unit. As a last resort, try fitting a new unit and noting whether this resolves the problem.

35 Side stand switch: general – DT125 LC III model

The DT125 LC III models are fitted with a small switch connected to the side stand. The switch is connected to the CDI circuit so that the engine cannot be started until the stand is retracted. This is a fairly common safety feature on many recent machines, and is intended to prevent the machine from being ridden away with the stand down; a very common cause of sometimes fatal accidents. In the event of the switch failing it should be renewed. Note that the sealed construction of the switch rules out any possibility of repair, but try using a silicone-based aerosol such as WD40 before resorting to a new unit.

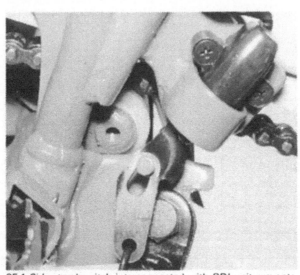

35.1 Side stand switch interconnected with CDI unit prevents the machine being ridden with the side stand down – DT125 LC III

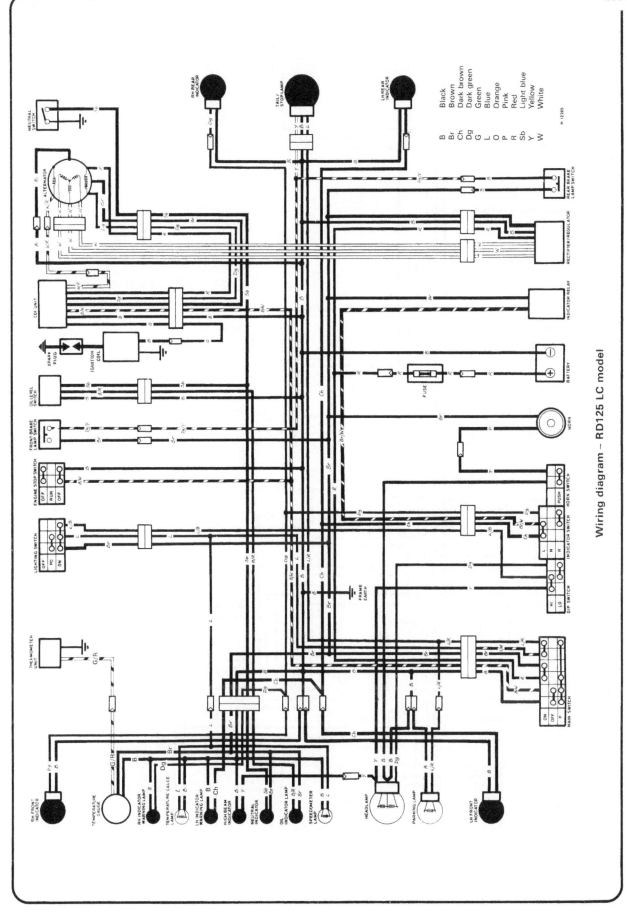

Wiring diagram – RD125 LC model

B — Black
Br — Brown
Ch — Dark brown
Dg — Dark green
G — Green
L — Blue
O — Orange
P — Pink
R — Red
Sb — Light blue
Y — Yellow
W — White

H.12265

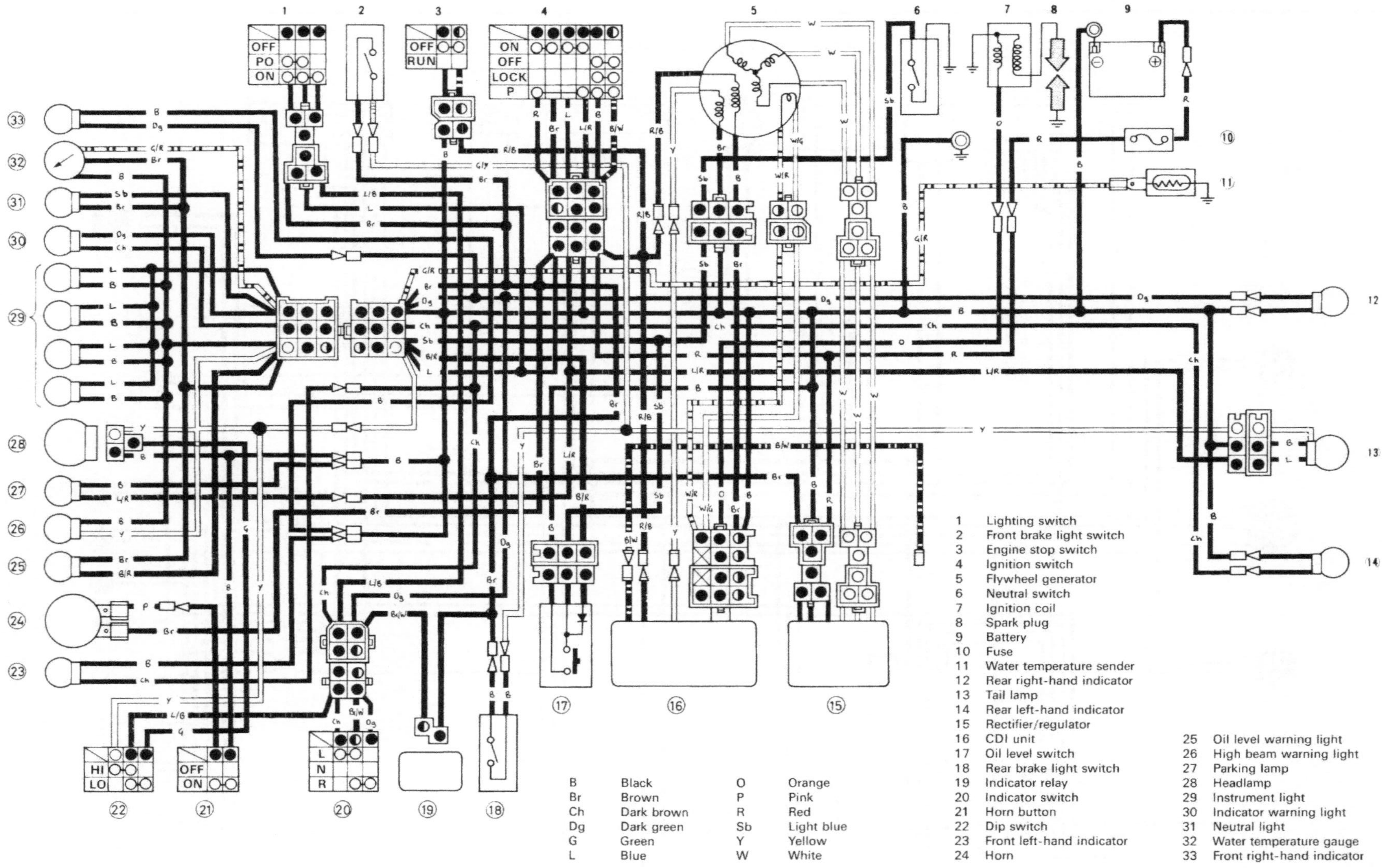

1. Lighting switch
2. Front brake light switch
3. Engine stop switch
4. Ignition switch
5. Flywheel generator
6. Neutral switch
7. Ignition coil
8. Spark plug
9. Battery
10. Fuse
11. Water temperature sender
12. Rear right-hand indicator
13. Tail lamp
14. Rear left-hand indicator
15. Rectifier/regulator
16. CDI unit
17. Oil level switch
18. Rear brake light switch
19. Indicator relay
20. Horn button
21. Dip switch
22. Front left-hand indicator
23. Horn
24. Horn
25. Oil level warning light
26. High beam warning light
27. Parking lamp
28. Headlamp
29. Instrument light
30. Indicator warning light
31. Neutral light
32. Water temperature gauge
33. Front right-hand indicator

B	Black	O	Orange
Br	Brown	P	Pink
Ch	Dark brown	R	Red
Dg	Dark green	Sb	Light blue
G	Green	Y	Yellow
L	Blue	W	White

Wiring diagram – RD125 LC II and III models

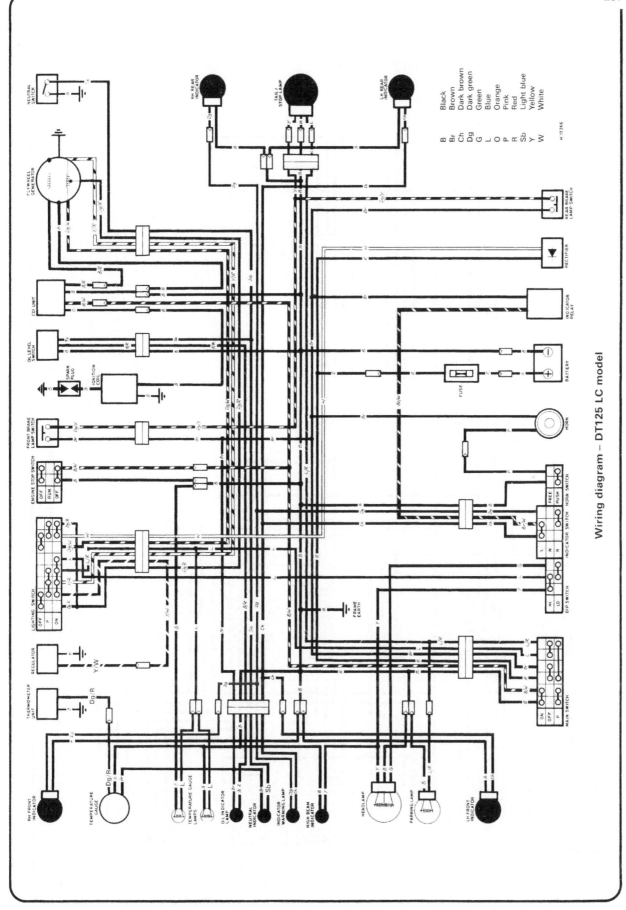

Wiring diagram – DT125 LC model

B	Black
Br	Brown
Ch	Dark brown
Dg	Dark green
G	Green
L	Blue
O	Orange
P	Pink
R	Red
Sb	Light blue
Y	Yellow
W	White

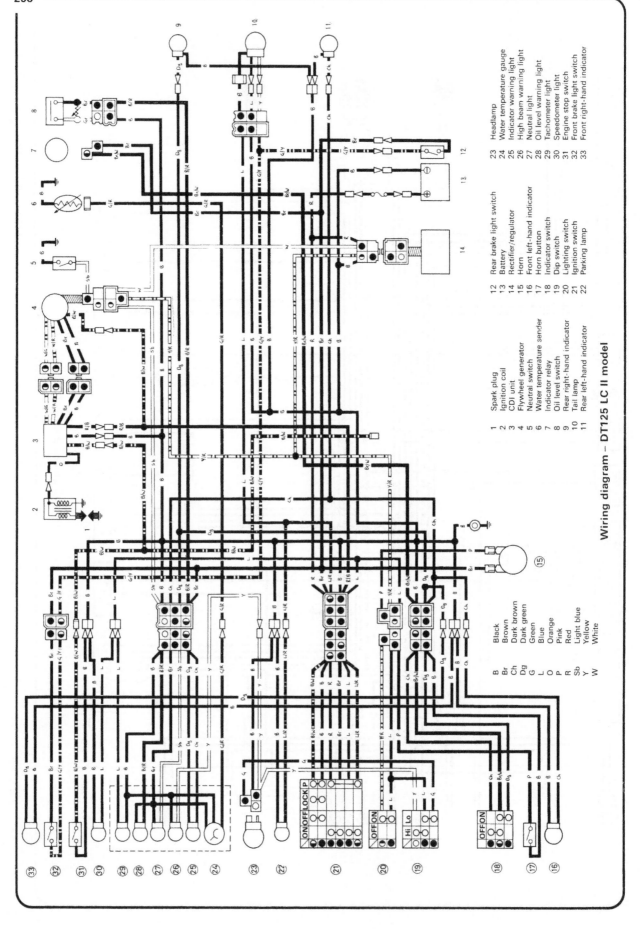

Wiring diagram – DT125 LC II model

1 Spark plug
2 Ignition coil
3 CDI unit
4 Flywheel generator
5 Neutral switch
6 Water temperature sender
7 Indicator relay
8 Oil level switch
9 Rear right-hand indicator
10 Tail lamp
11 Rear left-hand indicator

12 Rear brake light switch
13 Battery
14 Rectifier/regulator
15 Horn
16 Front left-hand indicator
17 Horn button
18 Indicator switch
19 Dip switch
20 Lighting switch
21 Ignition switch
22 Parking lamp

23 Headlamp
24 Water temperature gauge
25 Indicator warning light
26 High beam warning light
27 Neutral light
28 Oil level warning light
29 Tachometer light
30 Speedometer light
31 Engine stop switch
32 Front brake light switch
33 Front right-hand indicator

B   Black
Br   Brown
Ch   Dark brown
Dg   Dark green
G   Green
L   Blue
O   Orange
P   Pink
R   Red
Sb   Light blue
Y   Yellow
W   White

259

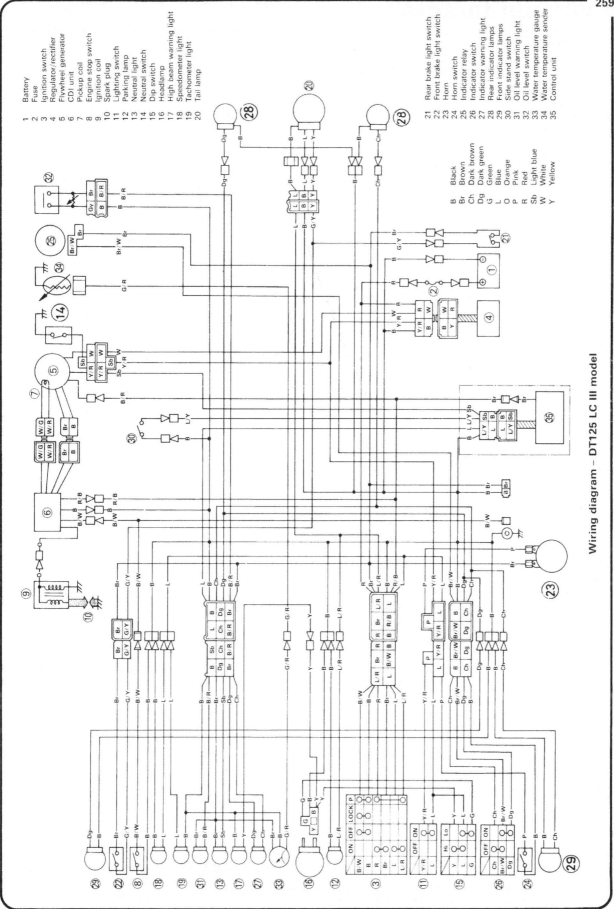

Wiring diagram – DT125 LC III model

1 Battery
2 Fuse
3 Ignition switch
4 Regulator/rectifier
5 Flywheel generator
6 CDI unit
7 Pickup coil
8 Engine stop switch
9 Ignition coil
10 Spark plug
11 Lighting switch
12 Parking lamp
13 Neutral light
14 Neutral switch
15 Dip switch
16 Headlamp
17 High beam warning light
18 Speedometer light
19 Tachometer light
20 Tail lamp

21 Rear brake light switch
22 Front brake light switch
23 Horn
24 Horn switch
25 Indicator relay
26 Indicator switch
27 Indicator warning light
28 Indicator warning lamps
29 Front indicator lamps
30 Rear indicator lamps
31 Side stand switch
32 Oil level switch
33 Oil level warning light
34 Water temperature gauge
35 Water temperature sender
35 Control unit

B Black
Br Brown
Ch Dark brown
Dg Dark green
G Green
L Blue
O Orange
P Pink
R Red
Sb Light blue
W White
Y Yellow

Conversion factors

Length (distance)

Inches (in)	X	25.4	= Millimetres (mm)	X 0.0394	= Inches (in)
Feet (ft)	X	0.305	= Metres (m)	X 3.281	= Feet (ft)
Miles	X	1.609	= Kilometres (km)	X 0.621	= Miles

Volume (capacity)

Cubic inches (cu in; in^3)	X	16.387	= Cubic centimetres (cc; cm^3)	X 0.061	= Cubic inches (cu in; in^3)
Imperial pints (Imp pt)	X	0.568	= Litres (l)	X 1.76	= Imperial pints (Imp pt)
Imperial quarts (Imp qt)	X	1.137	= Litres (l)	X 0.88	= Imperial quarts (Imp qt)
Imperial quarts (Imp qt)	X	1.201	= US quarts (US qt)	X 0.833	= Imperial quarts (Imp qt)
US quarts (US qt)	X	0.946	= Litres (l)	X 1.057	= US quarts (US qt)
Imperial gallons (Imp gal)	X	4.546	= Litres (l)	X 0.22	= Imperial gallons (Imp gal)
Imperial gallons (Imp gal)	X	1.201	= US gallons (US gal)	X 0.833	= Imperial gallons (Imp gal)
US gallons (US gal)	X	3.785	= Litres (l)	X 0.264	= US gallons (US gal)

Mass (weight)

Ounces (oz)	X	28.35	= Grams (g)	X 0.035	= Ounces (oz)
Pounds (lb)	X	0.454	= Kilograms (kg)	X 2.205	= Pounds (lb)

Force

Ounces-force (ozf; oz)	X	0.278	= Newtons (N)	X 3.6	= Ounces-force (ozf; oz)
Pounds-force (lbf; lb)	X	4.448	= Newtons (N)	X 0.225	= Pounds-force (lbf; lb)
Newtons (N)	X	0.1	= Kilograms-force (kgf; kg)	X 9.81	= Newtons (N)

Pressure

Pounds-force per square inch (psi; lbf/in^2; lb/in^2)	X	0.070	= Kilograms-force per square centimetre (kgf/cm^2; kg/cm^2)	X 14.223	= Pounds-force per square inch (psi; lbf/in^2; lb/in^2)
Pounds-force per square inch (psi; lbf/in^2; lb/in^2)	X	0.068	= Atmospheres (atm)	X 14.696	= Pounds-force per square inch (psi; lbf/in^2; lb/in^2)
Pounds-force per square inch (psi; lbf/in^2; lb/in^2)	X	0.069	= Bars	X 14.5	= Pounds-force per square inch (psi; lbf/in^2; lb/in^2)
Pounds-force per square inch (psi; lbf/in^2; lb/in^2)	X	6.895	= Kilopascals (kPa)	X 0.145	= Pounds-force per square inch (psi; lbf/in^2; lb/in^2)
Kilopascals (kPa)	X	0.01	= Kilograms-force per square centimetre (kgf/cm^2; kg/cm^2)	X 98.1	= Kilopascals (kPa)
Millibar (mbar)	X	100	= Pascals (Pa)	X 0.01	= Millibar (mbar)
Millibar (mbar)	X	0.0145	= Pounds-force per square inch (psi; lbf/in^2; lb/in^2)	X 68.947	= Millibar (mbar)
Millibar (mbar)	X	0.75	= Millimetres of mercury (mmHg)	X 1.333	= Millibar (mbar)
Millibar (mbar)	X	0.401	= Inches of water (inH$_2$O)	X 2.491	= Millibar (mbar)
Millimetres of mercury (mmHg)	X	0.535	= Inches of water (inH$_2$O)	X 1.868	= Millimetres of mercury (mmHg)
Inches of water (inH$_2$O)	X	0.036	= Pounds-force per square inch (psi; lbf/in^2; lb/in^2)	X 27.68	= Inches of water (inH$_2$O)

Torque (moment of force)

Pounds-force inches (lbf in; lb in)	X	1.152	= Kilograms-force centimetre (kgf cm; kg cm)	X 0.868	= Pounds-force inches (lbf in; lb in)
Pounds-force inches (lbf in; lb in)	X	0.113	= Newton metres (Nm)	X 8.85	= Pounds-force inches (lbf in; lb in)
Pounds-force inches (lbf in; lb in)	X	0.083	= Pounds-force feet (lbf ft; lb ft)	X 12	= Pounds-force inches (lbf in; lb in)
Pounds-force feet (lbf ft; lb ft)	X	0.138	= Kilograms-force metres (kgf m; kg m)	X 7.233	= Pounds-force feet (lbf ft; lb ft)
Pounds-force feet (lbf ft; lb ft)	X	1.356	= Newton metres (Nm)	X 0.738	= Pounds-force feet (lbf ft; lb ft)
Newton metres (Nm)	X	0.102	= Kilograms-force metres (kgf m; kg m)	X 9.804	= Newton metres (Nm)

Power

Horsepower (hp)	X	745.7	= Watts (W)	X 0.0013	= Horsepower (hp)

Velocity (speed)

Miles per hour (miles/hr; mph)	X	1.609	= Kilometres per hour (km/hr; kph)	X 0.621	= Miles per hour (miles/hr; mph)

Fuel consumption

Miles per gallon, Imperial (mpg)	X	0.354	= Kilometres per litre (km/l)	X 2.825	= Miles per gallon, Imperial (mpg)
Miles per gallon, US (mpg)	X	0.425	= Kilometres per litre (km/l)	X 2.352	= Miles per gallon, US (mpg)

Temperature

Degrees Fahrenheit = (°C x 1.8) + 32

Degrees Celsius (Degrees Centigrade; °C) = (°F – 32) x 0.56

It is common practice to convert from miles per gallon (mpg) to litres/100 kilometres (l/100km), where mpg (Imperial) x l/100 km = 282 and mpg (US) x l/100 km = 235

Index